Family-Oriented Informed Consent

Philosophy and Medicine

VOLUME 121

ASIAN STUDIES IN BIOETHICS AND THE PHILOSOPHY OF MEDICINE 7

For other titles published in this series, go to
http://www.springer.com/series/6414

Ruiping Fan
Editor

Family-Oriented Informed Consent

East Asian and American Perspectives

Springer

Editor
Ruiping Fan
Department of Public Policy
City University of Hong Kong
Hong Kong
Hong Kong SAR

ISSN 0376-7418 ISSN 2215-0080 (electronic)
ISBN 978-3-319-12119-2 ISBN 978-3-319-12120-8 (eBook)
DOI 10.1007/978-3-319-12120-8

Library of Congress Control Number: 2014958883

Springer Cham Heidelberg New York Dordrecht London

Printed on acid-free paper

Springer is part of Springer Science+Business Media (www.springer.com)

Acknowledgments

Most chapters included in this volume were discussed at an international conference titled, "Family-Oriented Informed Consent: East Asian and American Perspectives on a Cardinal Moral Practice," held in Hong Kong, December 13–14, 2012. The conference was sponsored by the Department of Public Policy of the City University of Hong Kong. I am also grateful for funding provided by the Center for the Study of Culture, Ethics, and the Environment in Alaska. I thank Professor Ray Forrest, the Head of the Department of Public Policy at City University of Hong Kong, Professor Brian Partridge, the Director of Center for the Study of Culture, Ethics, and the Environment, and Professor H. Tristram Engelhardt, Jr. for their support of the conference. I am thankful to my research assistant, Ms. Cao Chengcheng, for her assistance in preparing for the conference.

I would like to thank the authors for contributing their essays and for their efforts revising and improving their essays in response to critical feedback after the conference. I appreciate the work of Jennifer Bulcock and Victor Saenz in polishing the English of those chapters contributed by East Asian authors whose mother tongue is not English. I am particularly grateful to Lisa Rasmussen, Associate Editor of the Philosophy and Medicine series, for her arduous efforts made for the publication of this volume.

Ruiping Fan

Contents

Contributors

Jeffery P. Bishop Tenet Endowed Chair in Health Care Ethics, Albert Gnaegi Center for Health Care Ethics, Saint Louis University, St. Louis, MO, USA

Yu Cai Law School, Yunnan University of Finance and Economics, Kunming, Yunnan Province, China

Ho Mun Chan Department of Public Policy, City University of Hong Kong, Kowloon, Hong Kong, China

Mark J. Cherry Department of Philosophy, St. Edward's University, Austin, TX, USA

Kyungsuk Choi School of Law, Ewha Womans University, Seoul, Korea

Chun Kit Chui Department of Public Policy, City University of Hong Kong, Kowloon, Hong Kong, China

Julian Chuk-Ling Lai Department of Applied Social Studies, City University of Hong Kong, Kowloon, Hong Kong, China

Rui Deng College of Humanities and Social Sciences, Shanxi Medical University, Taiyuan, Shanxi, China

M. W. Tse Doris Department of Medicine and Geriatrics/Intensive Care Unit, Caritas Medical Centre, Hospital Authority, Kowloon, Hong Kong, China

Ruiping Fan Department of Public Policy, City University of Hong Kong, Hong Kong, China

Ana S. Iltis Center for Bioethics, Health and Society; Department of Philosophy, Wake Forest University, Winston-Salem, NC, USA

Ilhak Lee (李日學) Department of Medical Law and Ethics, College of Medicine, Asian Institute for Bioethics and Health Law, Yonsei University, Seoul, Seadamun-gu, Korea

Shui Chuen Lee Graduate Institute of Philosophy, National Central University, Chungli, Taiwan

Ryan R. Nash Center for Bioethics, Wexner College of Medicine, The Ohio State University, Columbus, OH, USA

Lisa M. Rasmussen Department of Philosophy, University of North Carolina, Charlotte, NC, USA

Jue Wang Department of Philosophy, Huazhong University of Science and Technology, Wuhan, China

Hon Chung Wong Central Regional Research Ethics Center, China Medical University, Taichung, Taiwan

Kam Hung Wong Department of Clinical Oncology, Queen Elizabeth Hospital, Hospital Authority, Kowloon, Hong Kong, China

Yaning Yang CMI Institute for Hospital Management, Tsinghua University 1/F, Shenzen, China

Kam-Por Yu General Education Centre, The Hong Kong Polytechnic University, Kowloon, Hong Kong, China

Lawrence Y. Y. Yung Department of Public Policy, City University of Hong Kong, Kowloon, Hong Kong, China

Wenqing Zhao Department of Public Policy, City University of Hong Kong, Kowloon, Hong Kong, China

Part I
Introduction

Chapter 1
Informed Consent: Why Family-Oriented?

Ruiping Fan

1.1 Introduction

This volume introduces an East Asian Confucian ethical perspective on the place of the family for informed consent regarding clinical treatment and biomedical research. It offers a unique perspective on an approach to informed consent salient in East Asia but generally unexplored in English-language literature. By drawing on bioethics scholars from Hong Kong, mainland China, South Korea, and Taiwan, this volume offers a cluster of viewpoints that play a major role in law and health-care policy in East Asia, although they are nearly unknown in Western bioethical reflection. Contributions from four American scholars are also offered in order to place the East Asian Confucian perspective in a broader context. Even in the West, there has been a move to address issues of dependency, family care, and shared decision-making that have many similarities with the concerns addressed by East Asian scholars. The themes of the five sections designed for this volume range from theoretical reflections on such concepts as autonomy, individualism, rationality, dependence, independence, and interdependence to the practical explorations of issues like end-of-life decisions, organ donation, research participation, truth telling, and the arrangement of advance directives. The focus of the whole volume is on the role that the family ought to play in the proper exercise of informed consent in biomedical practices.

Regarding the concept of "family," there are roughly two different meanings adopted throughout the chapters of the volume. On the one hand, "family" is used in a loose sense, referring to any well-functioning unit of people connected either by blood or marriage or by alternative bonds, where there is mutual concern among

I wish to thank Jeffrey Bishop, H.T. Engelhardt, Jr., Lisa Rasmussen and an anonymous reviewer for their comments on the previous versions of this essay.

R. Fan (✉)
Department of Public Policy, City University of Hong Kong, Tat Chee Avenue, Kowloon, Hong Kong, China
e-mail: safan@cityu.edu.hk

R. Fan (ed.), *Family-Oriented Informed Consent*, Philosophy and Medicine,
DOI 10.1007/978-3-319-12120-8_1

the various members. This sense of "family" is invoked by some authors in addressing biomedical decision-making issues in contemporary Western countries. On the other hand, most chapters in the volume appeal to the concept of the traditional family—a husband with his wife and their biological children (and the husband's and the wife's parents as well, as they, at least in East Asian regions, are usually involved in medical decision making for the husband and the wife respectively)—to formulate their views and develop their arguments for the appropriate engagement of family members in the process of biomedical decision making in a Confucian cultural context. Indeed, in East Asian regions such as mainland China, a three-generation stem family remains the normal model of the family. This stem family is taken for granted as essential to any informed consent practice for any significant biomedical issues of their family members. This is the case even when the grandparents, in most situations, no longer live under the same roof with their adult child's nuclear family (cf. Deng 2014, pp. 203–218).

This introductory essay does not intend to outline the views and arguments of every chapter one after another. Instead, based on the major concerns of all the chapters, I attempt to defend the merits of the family-oriented model of informed consent in East Asian health care contexts. First, I summarize the major problems faced by individual-directed informed consent. Next I reconstruct the useful ideas of justification conveyed by the authors for the family-oriented model. In the following section I address the critiques and challenges faced by the engagement of family members in the informed consent practice and explore suitable ways to deal with these challenges. In the final section I give some concluding remarks about the importance of the traditional family in contemporary society.

1.2 The Problems of Individual-Directed Informed Consent

The dominant model of informed consent in contemporary Western biomedical practice is individual-directed rather than family-oriented. As H.T. Engelhardt describes this practice: "Those who regard autonomous individualism as the presumptively appropriate relation among persons would require any deviations to be established by explicit statement and agreement. For example, patients would be presumptively treated as autonomous individuals willing and committed to choosing on their own, unless they explicitly demanded to be regarded and treated within a traditional family structure" (Engelhardt 2002, pp. 24–25). Evidently, securing such an autonomous individualist vision has required a progressive political view that recognizes itself as having an adversarial relationship with traditional family structures and forms of familial authority. Indeed, as Mark Cherry points out, liberal individualist advocates often appreciate the family as a major impediment to social implementation of their preferred conceptions of equality and social justice, in which they intend to see the family as no more than a social construct, created through the particular agreements of its participants, with no independent reality of its own (Cherry 2014, pp. 43–62). It is for this reason that the liberal individualist notion of personal autonomy is highlighted as integral to human good and human flourishing in the established American account of bioethics (Beauchamp and Childress 2009).

As a result, the practice of informed consent under this individualistic model is at least family-unfriendly in multiple ways, even if it is not family-hostile in nature. First, the individual is appreciated as possessing sole or exclusive decisional authority in biomedical matters independently of his family. While patients usually do not make truly single or solitary decisions, there is the sense in which the modern Western medical establishment and the social apparatus of decision making expect solitary decisions from patients. They are approached individually in the decision making process as if they were not members of functioning families, even if they wish to be treated as members of functioning families. Family members have no right to participate in the process, unless such participation has been explicitly authorized by the patients through a formal or quasi-formal procedure (Faden and Beauchamp 1986; Wear 1993). In this way, the exclusive authority of the individual in making medical decisions independently of the engagement of the family is secured, implemented and promoted.

Indeed, there is a series of institutional barriers to family involvement in informed consent. As Jeffrey Bishop demonstrates, "the structures and practices of medicine, including informed consent and decisions about removing life support in the critically ill, organize decisions such that the family is structurally marginalized. Individualized informed consent—as well as living wills, durable powers of attorney, privacy laws (such as HIPAA), and even the case-law itself—is set up to support the myth of an isolated individual alone in the world making his own decisions by himself and for himself' (Bishop 2014, pp. 27–42). Such barriers have already been exported to East Asian regions through the design of an individualist informed consent mechanism for biomedical research. As Rui Deng indicates, when families are eager to participate in the decision-making process to protect their family members from the likely risk of biomedical research, they find that they have no role to play in the "standard" implementation of informed consent (Deng 2014, pp. 203–218). Worse yet, contemporary bioethics, healthcare systems, and laws increasingly assume that a family is not in a state of integrity and is not helpful in biomedical practice. As Ryan Nash indicates, this mistrust of family has become a presupposition or at least a practiced assumption in Western society. The family is treated as something to be wary of, a problem to prevent, or even a disease to attempt to cure. The current system of informed consent reinforces the overly simplistic individualist view that decision making is merely between the individual and the medical establishment. Accordingly, as Nash sees it, the current system encourages a practice of familial and cultural shunning (Nash 2014, pp. 219–230).[1]

[1] This is not to deny that many patients in the West regularly include families in their health care decision making. However, this volume is making the point that the policies set up in light of the dominant individualist model form barriers to including families in health care decision making

In addition to its family-unfriendly features, the individual-directed model of informed consent suffers serious conceptual and practical problems. In the first place, the exclusive decisional authority possessed by the patient independently of her family members is not beneficial to the practical reasoning required to make biomedical decisions, as such practical reasoning is inevitably engaged in accord with her moral identity, established character, and/or her long-standing life plan. Under the individual-directed model, informed consent is defined as an individual's *autonomous* authorization of a medical intervention or of participation in research. Since a patient may hold altruistic intentions or otherwise reasonably unselfish values in making her medical decisions, her autonomous choices may not necessarily be in her best medical interest. Thus it need not be the case that an individual's autonomous decision will always be in her best medical interest. Instead, in contemporary bioethics it is generally assumed that "[t]o respect autonomous agents is to acknowledge their right to hold views, to make choices, and to take actions based on their personal values and beliefs" (Beauchamp and Childress 2009, p. 103) This is to say, leaving the best-interest-consideration aside, a patient's "personal values and beliefs" must be taken seriously in order to respect her as an autonomous agent as well as her right to the fostering of autonomous decisions. Now a question arises: is it the case that as long as an agent is psychiatrically judged not incompetent and is independent from any controlling influences by others, then any of her biomedical decisions, whatever they are, should be taken as having been made in accord with her "personal values and beliefs," so as to be definitely autonomous? If the answer is yes, then our notion of autonomy would be overly shallow and trivial. It would be analogous to the notion of "capricious freedom" in the sense that one is free insofar as one has a power of acting without a motive (cf. Sidgwick 1888). I don't think any serious bioethicist should accept this notion of autonomy. If it is discovered that a patient's individual medical decision is egregiously in discord with the essential features of her moral identity, cultivated integrity, or a well-standing life plan that she has long cherished, then it is nonsensical to hold that her decision has been made based on her "personal values and beliefs" and so autonomous.

Indeed, one's "personal values and beliefs" are typically integrated into the features of one's moral identity or character in the exercise of practical reasoning through one's life experience. The genuine way of formulating decisions based on one's "personal values and beliefs" is inevitably an exercise of practical reasoning that one performs in the company of others on whom one depends, and "there is no point then in our development towards and in our exercise of practical reasoning at which we cease altogether to be dependent on particular others" (MacIntyre 1999, p. 97). In the predicament of serious illness, as Jue Wang argues, the essential relevance of dependence on family members to making autonomous choices becomes even more explicit and dominant. To see this point clearly, readers can look

in the West. As Nash points out, "Beauchamp and Childress (2009, pp. 106–107) suggest that the practical way to deal with those requesting family-oriented consent is to ask the individual if this is his preference. This, of course, is the application of an individualistic system and already does violence to any family-oriented consent" (Nash 2014, pp. 219–230).

at an illuminating case that Wang discusses in her chapter (Wang 2014, pp. 65–82). Contrary to the individualist myth that one's moods, beliefs, and intentions are only transparent to oneself and largely opaque to others, the patient's own voice can often sound frustrated, broken, and ambiguous, far from being an indisputable standard for judgment. Instead, the very exercise of practical rationality presupposes a caring network through interdependence and attachment—the paradigm of which is the family—as its foundation that makes practical reasoning in accord with one's convictions and values possible. Accordingly, individualistic informed consent undermines the patient's chance of making autonomous decisions in accord with her values, beliefs and life aspirations, because this model assumes a single decisional authority possessed by the patient alone and regards the effective exertion of the authority as isolated from the patient's significant relations with her family (Wang 2014, pp. 65–82).

Following this line of thought, Lawrence Yung contends, self-interpretation and interpretation of the self by others are interrelated. As Charles Taylor puts it, "we define our identity always in dialogue with, sometimes in struggle against, the things our significant others want to see in us" (Taylor 1994, p. 28). As an individual's self-interpretation is important to his capacity for autonomy, this is all the more important for a patient, since his decisional capacity may be undermined by his internal weaknesses or psychological impediments, such as irrational preferences, false consciousness, a belief in oppressive norms, or deformed desires (Yung 2014, pp. 109–124). From Jue Wang's observation, while the patient himself may be confused due to suffering from illness, the family is in a position to judge whether a certain medical decision is consistent with the patient's life plan. This is because a meaningful medical decision has to be assessed according to a coherent narrative of one's life, and it is through such a narrative that the patient brings unity to his life, by orienting all his decisions and actions towards some core commitments that determine who he is as a person. Accordingly, the patient's personal identity in key part depends on how well he lives out that unity and brings it to completion (Wang 2014, pp. 65–82). As Alasdair MacIntyre powerfully argues, the self has to find its identity in and through its membership in communities such as the family, in the sense that my story is part of the story of other family members just as their story is part of mine (MacIntyre 2007, p. 221). Particularly, in the process of making a medical decision with his family, the patient gains the better understanding of his conditions, his options, and his genuine wishes (Yung 2014, pp. 109–124).

This is to say, a mistaken understanding of autonomy is implicit in the scope of the exclusive individual authority assumed by individual-directed informed consent. As Bishop argues, "despite the reigning myths of the American individual who enacts and rejects treatment at the end of life, in truth the family, who can bring the nuance of spiritual and moral values to the clinical scenarios, is in the best position to make decisions on behalf of an individual rendered dependent by his illness. Dependency necessitates a family to enact a community of care" (Bishop 2014, pp. 27–42). Similarly, Kysungsuk Choi contends that we have to revise the individualistic notion of autonomy so that it may reflect a desirable sense of moral agents as it is embedded in the Eastern way of life, in which the family is just as important as the individual in participating in the process of biomedical decision making. This revision avoids depriving patients of the company of their families—and it is only through and with their families that patients can maintain the integrity of their moral agency in the practice of informed consent (Choi 2014, pp. 83–92).[2]

Another problem of the individual-directed model of informed consent is the mistaken assumption that the patient can sovereignly and independently control his actions and determine his fate in today's high-tech medical contexts, such as the ICU. By drawing on the notion of "technological brinksmanship" coined by Daniel Callahan, Bishop shows why this assumption is illusory. By technological brinksmanship Callahan means the drive to seek more aggressive care, thinking that, at some point in the course of the illness, an individual will be able to know when he has crossed a threshold. In crossing that threshold, he will know it is time to say 'no' to more aggressive care (Callahan 2000, p. 41). In other words, the social apparatus of decision making in contemporary American medicine holds that the self is in a position to control one's body, life, and death, knowing when to say, 'enough is enough.' As Callahan suggests, "[t]he only evil greater than one's personal death is increasingly taken to be the loss of control of that death" by the individual decision-maker (Callahan 2000, p. 37). The social apparatus of the ICU, along with all the documents that an individual ought to have completed before going into the ICU—living will, durable power of attorney, and consent forms, has at its heart a sovereign and independent individual who sets in motion the aggressive technological features of medicine. This apparatus also assumes that the sovereign individual is capable of conquering the power of this technological environment and bringing his life and death issues under his own control (Bishop 2005, 2014, pp. 27–42).

However, because of the sheer power of the apparatus of the ICU, it is not at all clear that the individual will be in a position to know precisely when he has crossed over the brink so as to be able to give an explicit directive to withhold or withdraw aggressive technological treatment. Even if the individual knows that a time will come when the machines may no longer be to her benefit and has completed all the documents—such as a living will or advance directive—those machines might not be rightly turned off in terms of her genuine wishes. As Bishop indicates,

> For the most part, living wills and the consent that they imply, say something similar to this: "in the event that I am unable to participate in decision making, and if my attending physician thinks that aggressive treatment will not result in return of function, I authorize the doctor to stop" various aggressive treatments. However, anyone who has ever practiced medicine knows that the complicated physiological nature of seriously ill patients does not offer bright lines that are baldly apparent to anyone. Rather, there are nuances of clinical judgment and interpretation that are part of the decision-making process, and *having a*

[2] Indeed, in this volume the Western reader finds for the first time in English a presentation of the grounds of East Asian approaches to individual autonomy and informed consent and their differences from what have generally prevailed in Western bioethics. Without such an understanding, East Asian practices would to the Western bioethicists appear puzzling at the very least.

> *family there who can help to make these decisions is so much more important than having a living will, or a document of informed consent that has been individually executed.* The truth is that, when we are at our most frail, for the vast majority of cases it is the family that knows its members best and the family that knows how best to reason with and to inform the doctors of what is best for the sick individual (Bishop 2014, pp. 27–42).

This is to say, though the general ethos of contemporary bioethics promotes the attitude of technological brinksmanship and individual-directed informed consent, the actual circumstances and practices of care at the end of life can never place the patient in a position to be capable sovereignly and independently of controlling her actions and determining her fate. The patient cannot *independently* know when it is the right time to say 'no.' Neither can a living will, set out in the general delimited circumstances of critical care, explicitly tell the doctor what the patient precisely wants. The engagement of family members is not only beneficial but also inevitable.

In addition, individual-directed informed consent does not fit the reality of East Asian cultural circumstances in general and their practices of biomedical decision making in particular. It will even be ethically inimical if this individualist model is forced on East Asian patients. For example, as Kysungsuk Choi reports, most elderly people in Korea feel a moral obligation to their offspring. When they say they do not want any life-sustaining treatments, they are probably concerned with sparing such treatments for saving money that can be used by their offspring for better lives (Choi 2014, pp. 83–92). Similarly, based on their research findings, Ho Mun Chan et al. find that a strong ethos of familism is still manifested in current Hong Kongese society. In this ethos, people are often willing to sacrifice their own interests, such as health, material interests or other self-centered interests, for the sake of promoting the interests of other family members. This means that interests of the self may include the interests of other family members. In other words, the interests of other family members are inseparable from one's own interests, and so one's own interests have to be sacrificed for the sake of promoting the interests of other family members. Accordingly, in the process of medical decision making, the patient should not be conceived as an independent or isolated being; rather, the patient must be appreciated as a person situated in a family network, and the patient's wishes should not be dictated solely by what he wants for himself but should sometimes be overridden by the family's decision for the patient's benefit (Chan et al. 2014, pp. 151–170).

In Taiwan, from Shui Chuen Lee's observations, family relations are still very intimate, and family involvement in all kinds of personal activities remains common practice. Medical decisions are no exception. Family members can provide the most needed help and trust when one is sick and vulnerable. Family members are also usually a reliable source of the patient's values and preferences, which provide the best guidance for treating the patient and ground the medical decisions for the health professionals. The medical law in Taiwan is even written in such a way that diagnoses and results can be disclosed either to the patient or to the family, and in some cases the diagnosis is not made known to the patient at all. In cases of terminal illness, the family's legitimate participation is critical because it is often the case that the wishes of an incompetent or comatose patient are unclear for health professionals trying to determine which alternatives to deploy. The family as a whole is usually competent to make such a decision. This may help health professionals avoid some of the hard moral dilemmas. On the other hand, family participation is also useful for preventing the abuse of weak and lonely patients by health professionals. In short, as Lee sees it, it would be a non-starter for Taiwan to change to the individual-directed model of informed consent (Lee 2014b, pp. 125–136).[3]

Finally, individual-directed informed consent is characteristic of a type of principlism that is not sensitive to the biomedical complexities and nuances that call for context-relative virtuous practices. The most popular contemporary bioethical principlism holds as a starting assumption that "no more basic moral content exists than the collection of rules and general moral judgments that are developed from the four clusters of principles": the principles of respect for autonomy, nonmaleficence, beneficence, and justice (Beauchamp and Childress 2009, p. 387). These four principles, from this principlist view, not only "function as an analytical framework intended to express general norms of the common morality that are a suitable starting point for biomedical ethics," but also "function as general guidelines for the formulation of the more specific rules" (Beauchamp and Childress 2009, p. 12). In contrast, there remains within East Asian culture a quite different Confucian approach to bioethical issues that does not take general principles as the starting point. Instead, this approach is rooted in Confucian understandings of a life of virtue (德) and is embedded in a way of life sustained by rituals or rites (禮) towards virtue. The focus is not on offering an exhaustive set of principles to formulate specific guiding rules for all cases, but on cultivating virtue through ritual exercise which is inevitably context-relative and practice-dependent. In short, Confucian thought appreciates that the moral and bioethical life is not just directed by general principles, but is learned and manifested in concrete manners in which specific ritual practices play a fundamental function in attending to the nuances and complexities of biomedical matters in the virtuous way.[4]

By analyzing a recent Chinese case in which a wife conceals the cancer diagnosis from her husband, Wenqing Zhao argues that the principlism implicit in individual-directed informed consent is bioethically misleading. From Zhao's view, the case of concealing illness from the patient partly reflects the Confucian way of

[3] Here again puzzles regarding East Asian approaches to the bioethics of informed consent can be resolved. These approaches do not discount the individual, but rather appreciate the individual patient within the context of the dependency of illness and the thick social relations that many patients have within their family.

[4] Here one has a light shed from the East on the one-sided and incomplete appreciation of general principles as envisaged within some Western accounts of medical decision making within the practice of informed consent. To act freely is always to act in a thick social context. From a Confucian view, only in the *constitutive* ritual or quasi-ritual practices of the life world (as Confucius disclosed 2500 years ago), can general principles play their secondary *regulative* function (cf. Fan 2012).

making medical decisions, which is primarily not general principle- or rule-guided, but is ritual- and virtue-based and family-oriented. In this case, on the one hand, the wife treated her husband according to her knowledge of what was best for him, which shows a Confucian understanding of caring. Her shouldering of all of the fear, stress, and suffering arising from her husband's worsening condition is, according to Confucian norms, regarded as manifesting her virtue as a wife in treating her ill husband with great courage and love. From the Confucian point of view, a virtuous person need not obey the rule of truth telling regardless of the mental and physical condition of the patient. It is a much more complicated process of weighing the consequences of different decisions, carefully evaluating the patients' mental and physical condition, and most importantly listening to the will of the patient and other family members. In this story, the husband showed a strong will to live, and the wife was afraid that knowing the bad news would diminish her husband's courage to fight the cancer. Thus, one could hardly say that informing the husband of his condition would really be doing what was best for his care. There are no Confucian rules or requirements stating that the family must or must not conceal information from the patient. The right decision is highly contextual. From Zhao's view, taking such a Confucian approach is often much harder than simply following the principles of beneficence, justice, and, most importantly, respect for autonomy. If one always acts by deferring to some general principles or rules, such as that of telling the truth to the patient, one would have shied away from one's *virtuous* responsibilities to significant others in the complex bioethical context (Zhao 2014, pp. 231–244).[5]

1.3 The Justification of Family-Oriented Informed Consent

Family-oriented informed consent denies what the individual-directed model assumes: that the patient has exclusive decisional authority. From an East Asian Confucian perspective, the patient does not have an individual right to include or exclude family decision making. Instead, this perspective stresses the naturalness,

[5] Some Western readers may not find Zhao's example compelling. They may contend that it is one thing to affirm that the wife was well-intended, but quite another to judge whether hiding the truth from the husband was actually in his "best interest": just because the wife thought it would be in his best interest, doesn't mean it was so. However, it should be noted that Zhao's main purpose of addressing this case is to indicate a fundamental cultural difference, which invites Western readers to consider how Chinese people view this situation. As Zhao observes, this case was met with general approval and admiration in mainland China. In addition, it is at least theoretically possible that it was in his best interest—in the same way that the Western physician sometimes invokes the "therapeutic privilege" condition for non-disclosure. Importantly, as Zhao points out, in order to judge best interest, we need to appeal to a particular conception of the good shared in a culture. While she does not offer a full-brown Confucian conception of the good, she emphasizes the Confucian experience that "to be able to rely on one's family members in the final stage of life is the good thing to do, and it makes the patient happy" (Zhao 2014, pp. 231–244).

usefulness and normalness of the engagement of family members in a patient's biomedical decision making, thus acknowledging a shared decisional authority enjoyed by both the patient and his family. In East Asian regions, such a family-oriented model requires the physician to recognize the family as a fundamental unit for making medical decisions for the patient. Can such a family-oriented (rather than individual-directed) model of informed consent be justified? The first consideration of justification would be to attend to relevant sociological facts and social customs existing in society. In this regard, Ana Iltis summarizes two main reasons in favor of family engagement in biomedical decision making even in the American context: family members may have a stake in each other's well-being, and may also be concerned with advancing and protecting each other's interests because they care about each other's well-being. She warns that clinicians and researchers should not assume that most patients and potential research participants exist outside of an intimate family whose participation in health-related decisions is unnecessary or even inappropriate. Rather, these two considerations give us important reasons to take families seriously in the health care setting. As Iltis remarks, these considerations are especially important when patients are faced with severe diagnoses or critical-care situations or individuals are asked to assume health risks primarily or exclusively for the benefit of others (Iltis 2014, pp. 171–186).

The practical benefits of the involvement of family members in medical decision making are significant and multi-dimensional. First, as Lawrence Yung points out in his chapter, a lamentable characteristic of modern medicine is that the physician-patient relationship has become highly formalized, regulated and constrained due to the application of ever-changing medical technologies and ever-increasing health care bureaucratization. Thus a huge gap has been formed between the patient and the physician, and this gap prevents them from forming a personal, communicative and intimate relationship. A family platform may provide the kind of caring and emotional support that a patient needs most during illness. It may also empower the patient when dealing with difficulties from a highly bureaucratized relationship with his physician, so as to mitigate the gap between them (Yung 2014, pp. 109–124). In other words, the involvement of family members is capable of creating and sustaining a space in which both patients and their family members find themselves commonly situated to pull together in face of crisis. Their shared life history, memories and commitments put them in the best position to inform and reason with the physician, so as to get the physician appropriately engaged in the treatment of the patient. In short, the involvement of family members creates a locus that helps bring the physician into a closer, more committed, and more "biographical based" relationship with the patient, against the background of highly specialized and bureaucratized medical practice (Wang 2014, pp. 65–82).

Moreover, as Bishop observes, the family seems to be best suited to take on this role of decision making for at least three specific practical factors. First, the family has a special role in caring for the material needs of its members. Second, the family is in the best position to appreciate and preserve the values of its members due to a common or shared life history. Finally, the family is able to carry out the methods of decision-making and methods of discernment utilized in the biomedical setting.

And these methods are much more likely to render decisions that are consistent with the moral and meaningful life-world of the family-member who has become ill. Not only are the family-members' values enacted, but both the family and the individual family member learn how to make these sorts of decisions in a family-oriented model of informed consent (Bishop 2014, pp. 27–42).

The turn to family-oriented informed consent could also contribute to the solution of certain thorny biomedical issues, such as cadaveric organ donation. In her chapter, Yu Cai proposes that the individual-directed legal model adopted in mainland China for cadaveric organ donations should be converted to a family-oriented model, where the express consent of the individual to cadaveric donation during her lifetime should be a decision jointly made with her close family members, and this requirement should be reflected on a consent document to be signed by both the donor and a representative of her family. Although the currently adopted individualist legal model in China only requires individual consent, it is simply impossible—culturally and socially—that the doctor could remove a donor's organs after his death without having to seek out the extra informed consent of his family members. From Cai's view, only by changing to a family-oriented policy as she proposes, can a Chinese citizen's right to dispose of his corpse and his family's right to the ownership of his body (as it is culturally appreciated) be both respected and fulfilled. In this way doctors could avoid the embarrassment of having to ask for the permission of family members and would not risk harvesting organs without the family's consent (Cai 2014, pp. 187–199).[6] Similarly, for biomedical experiments using human subjects in mainland China, Rui Deng proposes "a family-based binary decision model" to protect Chinese subjects. In this binary decision model, the family has a right to affirm or deny a family member's consent to being a research subject. The family, however, does not have the right to require that a family member participate in a trial. That is, for any trial, if a person decides not to participate, the family cannot require her to participate. However, for any significant experiment, if a person decides to participate, she still needs to get the consent of her family; if the family denies consent, the person should not be allowed to participate in the experiment. As Deng states, this right of the family's participation in the decision making is necessary to consider the risks of scientific research from more angles so as to protect the important interests of individual family members as subjects (Deng 2014, pp. 203–218).

[6] What is the way of resolving disagreement if one family member thinks differently? This is certainly an inevitable question from Western readers. However, Western readers should be informed of a general Confucian cultural ethos in which people have formed a familist mentality and attitude: family members, since they are from one family, should reach agreement and make one decision about any important matters (such as donating organs after death) confronting any of their family members. This "one-decision" strategy for each family is understood as the basic way of manifesting the integrity and solidarity of the family in the process of protecting the morality and interest of each family member, and all family members, in their life experiences, have been cultivated by Confucian virtues (such as harmony (*he*) and relevant rituals (such as family meetings) to practice this "one-decision" strategy. Thus, agreement will usually be achieved through familial discussion. Those families that cannot reach agreement will be despised by others as wanting of necessary familial virtue and integrity (Fan 2011; Fan and Chen 2010).

In addition to being accepted as a normal social custom, the engagement of family members in biomedical decisions in East Asian contexts has also been supported by the Confucian moral virtues prevalent in the region. Indeed, from a Confucian virtue perspective, the family has a moral obligation to protect the interests of its members who are ill or interested in participating in biomedical research. The family cannot undertake this obligation without being actively involved in the decision-making process for its members. For example, as Ilhak Lee indicates, Confucian filial duty (孝) has served as the backbone of social relationships and guided children in the care of their parents in Korean society. The younger generation regards it as their moral duty to care for their elderly parents in return for the grace they received in the early days of their lives. Since this sense of moral responsibility and gratitude to one's parents and grandparents is taken as a given in South Korea, it is morally impossible for adult children not to be included as principal decision makers to protect their parents' and grandparents' interests (Lee 2014a, pp. 137–148).[7] From this Confucian moral perspective, as Yu Cai points out, the family stands as a morally united entity for mutual responsibility, and in this united entity every member's health care or life-sustaining treatment is no longer an individual matter, but is a common matter, for which every member is morally responsible (Cai 2014, pp. 187–199). Echoing this view, Rui Deng states that everyone's moral identity is formed and confirmed in the family. To make morally important and good decisions in the course of one's life, such as matters of education, employment, marriage, health, hospitalization, and medical treatment, one's family members are morally obliged to join in the deliberating and decision-making process (Deng 2014, pp. 203–218).

Finally, family participation in biomedical decision making gains support from the ontological-metaphysical account of the family that is still prevalent in East Asian contexts. In this account, as H.T. Engelhardt presents it in a recent essay (2013), the family is appreciated as a deep, normative social entity, and this appreciation has traditionally been endorsed by such religions as Confucianism and Christianity. Differing from the libertarian/liberal construal of the family that has been adopted by fashionable contemporary bioethical strategies in the West, the ontological-metaphysical account of the family recognizes the family as a normative reality that should be realized by particular families as far as possible. This account brings with it pre-existing roles for husbands and wives, fathers, mothers, and children. The family, in this account, tends to be multi-generational, looking back with respect and support to previous generations, while looking forward with love and concern to succeeding generations. From Engelhardt's view, because the social entity of the family embodies itself in a normative socio-biological unity, it supports the autonomy and integrity of the family, inevitably endorsing family-oriented biomedical decision making and policies of confidentiality in preference to individual-directed ones (Engelhardt 2010, 2013).

[7] The East Asian appreciation of the virtue of filial piety allows one to take what otherwise would be a narrow, rule-based imposition of autonomy in informed consent and see it more amply in terms of a virtue ethics.

Based on this ontological-metaphysical understanding of the family, authors for this volume, such as Jeff Bishop and Jue Wang, further explore the rationale of family engagement in the informed consent practice for the possibility of human flourishing. As Bishop states it, since the individual is typically dependent on (rather than independent from) the family, the family is the first and fundamental unit of care when the contingencies of injury, disease, and other afflictions threaten one of the family-members. It is the family, not the individual nor the *polis*, that does the vast majority of care when injury or disease strikes the individual. In this regard, families are actually the sites where the virtues of acknowledged individual dependence are first and most fundamentally nurtured, and family members shape the specific forms of giving and receiving within families as they strive to meet and to attenuate the contingencies of injury, disease, and other afflictions upon any of the family members. It is for this very reason, Bishop contends, that the family is best situated to take on the role of participating in the decision-making process of its members (Bishop 2014, pp. 27–42). Children, spouses, parents, and grandparents are the context within which bodily dependency is attenuated, intellectual needs are met, and moral values are formed. It is within the context of the family that, as bodily and intellectual needs are met, the moral and existential meanings and purposes of the lives of its members are inculcated (Bishop 2014, pp. 27–42). Similarly, along these lines of consideration, Jue Wang adds that the embodied, vulnerable and dependent condition of the human individual prescribes a common way of human flourishing, that is, to be included and cared for in the family—a network of unconditional taking and giving that constitutes the common good for every individual. Accordingly, if due attention must be paid to the frailties and dependencies of human life in our practical reasoning in biomedical contexts, we will inevitably bear witness to the ontological necessity of the family as well as the moral necessity of family participation in the decision-making process, since human flourishing first and foremost occurs within the family (Wang 2014, pp. 65–82).

1.4 Challenges to Family-Oriented Informed Consent

There are numerous misunderstandings of and challenges to family-oriented informed consent in contemporary bioethics. This section will tackle a few of them that appear prominent. First, it is mistakenly thought that in urging family engagement in informed consent, one is pushed to take family interests over the interests of individual family members. This charge sounds all the more plausible when the word of "familism" is used in addressing biomedical issues as I frequently do in my Confucian bioethical work. As Kam Por Yu suspects in discussing the two examples constructed in his chapter, if one can provide better care to his family as a whole by letting his father die, then from the perspective of "familism," letting his father die may be the right thing to do. Similarly, if one's mother is regarded as a more important member in the family than one's 3-year-old son, it can be argued that killing the son to save the mother is to the greater interest of the family (Yu 2014, pp. 93–106). Does family-oriented informed consent really have such repugnant moral implications? If not, how should the probable tension and conflict between individual and family interests be resolved?

It seems to me that the real issue at stake is what standard should be adopted to guide biomedical decisions, regardless of whether they are made in family-oriented or individual-directed manners. As is well-known, in traditional Western medical ethics as characterized by the Hippocratic Oath, the best interest of the patient was invoked as the standard for the physician to make medical decisions. In the modern West, in contrast, the patient's own wishes, no matter whether they are in his best interest, take pride of place as long as they do not violate the physician's professional integrity. In Confucian culture, a proper medical decision is taken to follow the way of heaven (天道) as manifested in the requirements of the virtues, such as *ren* (loving humans), *yi* (being loyal to relatives), *xiao* (being filial to parents), and *he* (seeking harmony). Hence, although the Confucian ethical tradition has been robustly familistic rather than individualistic, it has never held a utilitarian principle of interest calculation or utility maximization. When the best interests of the patient and the family come to conflict with each other, the Confucian approach to medical decision making does not provide a guiding formula that requires either family interest or individual interest to trump the other. Rather, the Confucian standard is always following virtue by performing the rituals (禮) that have been affirmed in the tradition, such as mutual concern, family meeting, and deliberation. The result, accordingly, has to be contextual. The ideal is to integrate patient interest and family interest into a possibly harmonious system that is structured by the guidance of the virtues (Fan 2011).

It could be charged, on the other hand, that family-oriented informed consent advocates a kind of family-paternalism—family members may, in the name of pursuing the best interest of the patient, be allowed to impose their conception of the good upon the patient who actually holds very different values and preferences. Such medical family-paternalism is morally objectionable not only because it deprives the patient of her decisional authority, but also because it allows family members to substitute their judgments for those of the patient as if they knew better what was conducive to the interest and wellbeing of the patient (Yung 2014, pp. 109–124). This charge, which may be made against any form of family-oriented informed consent, is especially serious to the robust form of family-oriented informed consent embedded in the so-called Confucian bioethics exercised in East Asian regions. I don't have enough space here to address this challenge in detail, but it suffices to point out one crucial point to demonstrate that Confucian bioethics is not committed to family-paternalism in the general sense. While the robust form of Confucian family-oriented informed consent stresses "family sovereignty" or "family autonomy" rather than "individual sovereignty" or "individual autonomy," it by no means excludes the patient from sharing the authority of the family. In other words, though Confucian bioethics does not accept that the patient has a sole decisional authority for his biomedical matters, neither does it hold that his family members possess an exclusive power. What is actually stipulated is that both the patient and his family members share the "family" authority to make proper medical decisions for the patient. Of course a key issue is what should be done when the patient and his family members disagree. Confucian bioethics does not want to establish any "universal" principle to demand that either the patient's preferences or his family members' wishes be always followed. Rather, in such situations the physician is usually called for to exercise medical professionalism by favoring one side in order to protect the best medical interest of the patient according to his professional judgment. At the end of the day, unless the situation is urgent, the patient must be persuaded, rather than coerced, to accept a medical intervention. In short, because the family members are not taken to have unique authority to "force" the patient to accept medical treatment for the patient's benefit, this type of family-oriented informed consent cannot reasonably be criticized as family-paternalism (Fan and Chen 2010).

When the patient becomes incompetent, how should surrogate decisions be made for the patient? Wouldn't family-paternalism become prominent in such cases? Surely there are different standards that have been suggested in the literature regarding surrogate decision making. The best interest standard holds that a surrogate decision should be made by deciding what would be best for the patient. On the other hand, the proxy judgment standard holds that a proxy (who should be appointed by the patient before becoming incapacitated) should make surrogate decisions according to the patient's expressed wishes or other knowledge of the patient's values. Evidently, the proxy judgment standard has nowadays become dominant in Western bioethical accounts. In contrast, in East Asian regions, the best interest standard is still prevalent: family members are making decisions for their incompetent patients according to their judgment of the patient's best interests. Sometimes they do appeal to their own judgment of the patient's best interests to trump the patient's previously expressed wishes, and this indeed is a kind of family-paternalism. In his chapter, Lawrence Yung attempts to defend this kind of family-paternalism by arguing that the patient could have been persuaded to accept the family members' decision if he were competent and possessed the relevant information of his new condition (Yung 2014, pp. 109–124). Indeed, some American bioethicists have discovered that even most terminally ill patients in the United States prefer to have surrogate decisions be made by balancing the patient's own wishes, the proxy's judgments about the patient's best interests, and the input of the physician, rather than demand that their wishes be absolutely followed (Sulmasy et al. 2007).

Of course, medical family-paternalism could be abused in practice, not to be executed to the genuine benefit of the patient. Let us consider East Asian Confucian cultural contexts, in which adults are morally obliged by filial duty to take care of their elderly parents, including making medical decisions to maintain or withdraw their life-sustaining treatment. In this connection Ilhak Lee uses the South Korean situation to address relevant issues. Generally children seem to feel that it is their duty to provide every possible means of curative care for a parent, regardless of the likelihood of recovery. Even in cases where parents explicitly refuse aggressive medical interventions, children still feel (or would be expected to feel) guilty if they cease to request every possible medical treatment for their parents. They will sacrifice their savings, jobs and houses to pay the hospital expenses allowing their parents to remain in the hospital longer. Indeed, filial duty requires a child to think from the perspective of the welfare and happiness of one's parent in order to make decisions. But Confucian ethics and bioethics have yet to provide updated, specific or "quantitative" criterion of filial duty that can be applied by adult children to make appropriate end-of-life decisions for their parents in contemporary high-tech medical contexts. Children do not want to risk being blamed or scrutinized for "killing" or "abandoning" their parents. Therefore, their decisions are usually made in a very cautious manner: e.g., the parent remains connected to a ventilator for as long as possible (Lee 2014a, pp. 137–148).

The abuse of elderly care can be generated by various factors and improper incentives. I heard some Chinese cases in which the patients in persistent vegetative state were kept ventilated for years in order for their relatives continuously to enjoy certain material privileges assigned to the patients by government policy as long as the patients were "alive." Such decisions were not made in the genuine interests of the patients. What we should be concerned with here is whether such abuse is necessarily caused by family-oriented informed consent and will be ruled out by changing to individual-directed informed consent. From my understanding, similar abuse can take place under the individualist model as well, because the patient may individually require keeping his life no matter what. Thus, it may not be that the family-oriented informed consent or the Confucian virtue of filial piety is necessarily to blame. Rather, as Ilhak Lee recognizes, people do not need to abandon their ethical tradition to be ethical. In the Korean case, what needs to be done, from his view, is for adult children to understand that aggressive life-sustaining treatment is not always useful but can be harmful to their parents; accordingly, suitable palliative care is not abandoning their parents and should not be seen as a failure of filial duty. At the same time, elderly patients should be encouraged to participate in the decision-making process to exchange their views expressively with their children and medical professionals so that appropriately shared medical decisions can be made for their care. In the Confucian tradition, this is a more active way of seeking the parent's wishes and letting them conform to the way of heaven, so that family-oriented informed consent can be maintained and perfected in South Korea in contemporary times (Lee 2014a, pp. 137–148).

Another challenge to family-oriented informed consent concerns the issue of truth-telling to the patient. The critic would contend that at least in East Asian medical practices, family-oriented informed consent has the effect of encouraging the family members (and subsequently the physician) to hide the truth from the patient so as to violate his right to know his own health condition and choose his own treatment. To meet this challenge, the case discussion and theoretical exploration offered by Wenqing Zhao in her chapter are illuminating (Zhao 2014, pp. 231–244). Briefly put, Confucian ethics and bioethics do not accept a robust liberal individualist conception of human rights. Instead, from a Confucian perspective, while it is helpful in contemporary society to set down Confucian rights as legal requirements to protect Confucian individuals' basic legitimate self-interests, Confucian rights as entitlements may be necessary only as a fallback apparatus. This is to say, rights are needed only when the virtues fail to obtain or people's personal relationships break down

(Chan 1999; Fan 2010, p. 58). Accordingly, in the Confucian-cultural context, the patient's right to medical truth serves only as a fallback right: when the doctor finds evidence that family members do not care for the best interest of the patient and/or that their decisions violate the best medical interest of the patient according to the doctor's professional judgment, the doctor is morally obliged to take the initiative to intervene and tell the truth to the patient (Fan and Li 2004, p. 179).

What if, after the patient becomes incompetent, family members cannot arrive at a consensus among themselves? They may have significant disagreements about what is best for the patient. Here, from Shui Chuen Lee's observation of Taiwanese experience, medical professionals can assist in conducting and moderating family meetings. As he remarks, medical advice is important for preventing family members' misunderstandings and unfounded worries about the patient's medical possibilities. Meanwhile, through open discussion with family members, medical professionals can learn about the differences expressed among family members, the family's structure, as well as the values and preferences of the patient and the family as a whole. With such understanding, medical professionals would be better able to protect the best interest of the patient in cases of deadlock among family members. In rare cases in which the patient is maliciously manipulated by his family members, medical professionals can serve as powerful regulators and protectors of the vulnerable patient (Lee 2014b, pp. 125–136).

Should an ethics committee be set up to check the family's decision and make the final decision for the incompetent patient in the case of, say, DNRs? Shui Chuen Lee, using Taiwan's Hospice Palliative Care Act as an example, answers the question in the negative. In a previous version of the Act, there was the requirement that an ethics committee examine and approve the family's decision before it could be implemented. This was required to prevent the possible abuse of family members and physicians to the detriment of the patient. However, it was subsequently discovered that the ethics committees refused reasonable decisions arrived at by the head physicians and the major family members for DNRs simply because the committee members worried about accusations by other opposing family members. Unlike the patient's major family members who are actually sharing in the suffering of the patient or the patient's medical professionals who are committed to the care of the patient, such committee members are detached and removed from the patient so that they tend to be conservative in their decisions in order to protect themselves from possible troubles, at least in the Taiwan context. In this way, the ethics committee became a mechanism that furthered the pain and suffering of both the family members and the health professionals. Thus, this provision in the Act was deleted from its most recent version in 2012, so that decisions made by major family members supported by the medical professionals have become final (Lee 2014b, pp. 125–136).

Finally, how should the family-oriented model of informed consent arrange for the patient to prepare an advance directive before she becomes incompetent to assess her future treatment? In addressing this issue, both Hon Chong Wong and Yaning Yang in their respective chapters agree that such documents are helpful for families to know better about their patient's wishes and preferences so as to facilitate their decisions after the patients become incompetent. From Wong's view, encouraging the patient to work out an advance directive individually and appoint one of her family members to be her formal proxy does not in any sense undermine the value of the family or the intimate relationships she has with her family members (Wong 2014, pp. 245–256). On the other hand, Yang argues that the process of appointing the durable power of one family member as one's attorney by the patient must have the involvement of all of the patient's major family members. The surrogate decision maker should be recognized not only by the patient herself but also by her other major family members. In fact, Yang thinks that all family members should participate in determining who should serve as the formal proxy on behalf of the entire family, and the living will document or the durable power of attorney form should be signed by both the patient and all major family members to empower its function. From Yang's view, this Confucian family-oriented model of advance directives has two advantages. First, it can encourage patients, especially elderly patients, to discuss their end-of-life care and decision making with their family members so that their family members can better understand their desires and wishes. Moreover, it can also help the patient and her family members engage in harmonious cooperative interactions to arrive at important medical decisions, so as to prevent possible conflicts between the surrogate decision maker and other family members after the patient becomes incompetent (Yang 2014, pp. 109–124).

1.5 Concluding Remarks

The force of the reflections in this volume is to bring bioethicists in East Asia to appreciate better the family as the central ground of their approach to consent for medical treatment. It also makes available in English a basis for appreciating less one-sidedly and incompletely informed consent within Western bioethics. The importance of this task is underscored by the American contributors to this volume. These reflections on the East Asian emphasis of family-located informed consent interestingly reflect the recent discovery that many Americans (especially native Americans) wish to have consent for their treatment embedded within a family context. It is hoped that, having made this discussion available in English, a rich and critical response will develop. Indeed, part of the response has been made by Lisa Rasmussen in her thoughtful Epilogue to this volume.

The traditional family has largely been broken in the West. In his chapter, Mark Cherry cites a series of statistics and figures to show the severity of this crisis. For example, in the United States, unmarried birth rates in 2011 tracking the race of the mother were as follows: Black—72.3 %; Hispanic—53.3 %; American Indian or Alaska Native—66.2 %; White—29.1 %; Asian or Pacific Islander—17.2 % (Cherry 2014, pp. 43–62). Such data ought to raise significant concern because children reared outside of the traditional family environment face real disadvantages (Cherry 2010). On the other hand, out-of-wedlock birth rates and single parent families in East Asian regions are fortunately still low: for example, in Hong

Kong, there were 81,705 single parents in 2011 (among a population of a bit over 7 million) and the average number of dependent children for single parents was 1.3 (Census and Statistics Department of HKSAR 2012); in South Korea, there were approximately 16,000 unwed mothers raising children in 2010 (D'Itri 2010); in Taiwan, the proportion of single-parent households has grown by 50% over the last decade, numbering 560,000 and accounting for 7.6% of the totally 7.41 million households of Taiwan in 2012 (Taiwan Insights 2012). Accordingly, while East Asian bioethicists still have a great deal to learn from the ethics and practice of the modern West, they should not find everything worthy to learn. Instead, they should be cautions and critical about the changes of modern Western families. Indeed, they should be warned of the lamentable damaging effects of dominant and fashionable modern ideologies on the traditional family as Michael Oakeshott remarks: "First, we do our best to destroy parental authority (because of its alleged abuse), then we sentimentally deplore the scarcity of 'good homes,' and we end by creating substitutes which complete the work of destruction" (Oakeshott 1991, p. 41).

Indeed, as Bishop observes, "the excesses of the West have been to spend too much time thinking about the *independence* of practical reasoners, that is to say too much time on the autonomous agent as an individual on the one hand, and too much time thinking about the political nature of human thriving on the other, and all to the neglect of the family" (Bishop 2014, pp. 27–42). By focusing on the debate between individual-directed vs. family-oriented informed consent in the biomedical sphere, the chapters covered in this volume bring to the fore the fundamental place and value of the family embodied in East Asian Confucian biomedical commitments and practices as well as the fresh American bioethical reflections on the role of the family for long-term human interests and flourishing. Taken as a whole, this volume argues that the family-oriented model of informed consent should be maintained and developed in East Asian contexts. This familist model should not be forced to change to the individualist model. Of course, numerous specific issues around the legislation and exercise of informed consent in relation to the family cannot be settled once and for all. But East Asian readers should be reminded that whether individual-directed or family-oriented solutions will be accepted and promoted in their biomedical contexts bears on quite different futures for their societies as well as for their bioethics.

References

Beauchamp, T. L., and J. F. Childress. 2009. *Principles of biomedical ethics*. 6th ed. New York: Oxford University Press.

Bishop, J. 2005. Autonomy and informed consent: A much misunderstood relationship. *The Journal of Value Inquiry* 38:383–391.

Bishop, J. 2014. Dependency, decisions, and a family of care. In *Family-oriented informed consent: East Asian & American perspectives*, ed. R. Fan, 27–42. Dordrecht: Springer.

Cai, Y. 2014. On family informed consent in the legislation of organ donation in China. In *Family-oriented informed consent: East Asian & American perspectives*, ed. R. Fan, 187–199. Dordrecht: Springer.

Callahan, D. 2000. *The troubled dream of life: In search of a peaceful death*. Washington, D.C.: Georgetown University Press.

Census and Statistics Department of HKSAR. 2012. Thematic report: Single parents. *Census and Statistics Department*. http://www.census2011.gov.hk/pdf/single-parents.pdf. Accessed 6 Dec 2013.

Chan, J. 1999. A Confucian perspective on human rights on contemporary China. In *East Asian challenge for human rights*, ed. J. R. Bauer and D. Bell, 212–237. Cambridge: Cambridge University Press.

Chan, H. M., D. M. W. Tse, K. H. Wong, J. C. L. Lai, and C. K. Chui. 2014. End-of-life decision making in Hong Kong: The appeal of the shared decision making model. In *Family-oriented informed consent: East Asian & American perspectives*, ed. R. Fan, 151–170. Dordrecht: Springer.

Cherry, M. J. 2010. Parental authority and pediatric decision making. *The Journal of Medicine and Philosophy* 35 (5): 553–572.

Cherry, M. J. 2014. Individually directed informed consent and the decline of the family in the West. In *Family-oriented informed consent: East Asian & American perspectives*, ed. R. Fan, 43–62. Dordrecht: Springer.

Choi, K. 2014. The ideal of autonomy and its misuse. In *Family-oriented informed consent: East Asian & American perspectives*, ed. R. Fan, 83–92. Dordrecht: Springer.

Deng, R. 2014. The informed consent of human medical research in mainland China: A family-based binary decision model. In *Family-oriented informed consent: East Asian & American perspectives*, ed. R. Fan, 203–218. Dordrecht: Springer.

D'Itri, J. 2010. South Korea's unwed mothers organize. *Foreign Policy Blogs*. http://foreignpolicyblogs.com/2010/02/25/south-koreas-unwed-mothers-organize/. Accessed 6 Dec 2013.

Engelhardt, H. T. 2002. Morality, universality, and particularity: Rethinking the role of community in the foundations of bioethics. In *Cross-cultural perspectives on the (im) possibility of global bioethics*, ed. J. Tao Lai Po-wah, 19–38. Dordrecht: Kluwer Academic Publishers.

Engelhardt, H. T. 2010. Beyond the best interests of children: Four views of the family and of foundational disagreements regarding pediatric decision-making. *The Journal of Medicine and Philosophy* 35 (5): 499–517.

Engelhardt, H. T. 2013. The family: Crucial to and divisive in bioethics. *International Journal of Chinese & Comparative Philosophy of Medicine* 11 (2): 113–127.

Faden, R., and T. L. Beauchamp. 1986. *A history and theory of informed consent*. New York: Oxford University Press.

Fan, R. 2010. *Reconstructionist Confucianism: Rethinking morality after the West*. Dordrecht: Springer.

Fan, R. 2011. Family-oriented surrogate decision making and Confucian bioethics. *Bioethical Law & Ethics* 5 (2): 1–20.

Fan, R. 2012. Confucian reflective equilibrium: Why principlism is misleading for Chinese bioethical decision-making? *Asian Bioethics Review* 4 (1): 4–13.

Fan, R., and B. Li. 2004. Truth telling in medicine: The Confucian view. *Journal of Medicine and Philosophy* 29 (2): 179–193.

Fan, R., and X. Chen. 2010. The family and harmonious medical decision making: Cherishing appropriate Confucian moral balance. *Journal of Medicine and Philosophy* 35 (5): 573–586.

Iltis, A. 2014. Families and medical decisions to assume risks for the benefit of others. In *Family-oriented informed consent: East Asian & American perspectives*, ed. R. Fan, 171–186. Dordrecht: Springer.

Lee, I. 2014a. Filial duty as the moral foundation of caring for the elderly: Its possibility and limitations. In *Family-oriented informed consent: East Asian & American perspectives*, ed. R. Fan, 137–148. Dordrecht: Springer.

Lee, S. C. 2014b. Family consent in medical decision-making in Taiwan: The implications of the new revisions of the Hospice Palliative Care Act. In *Family-oriented informed consent: East Asian & American perspectives*, ed. R. Fan, 125–136. Dordrecht: Springer.

MacIntyre, A. 1999. *Dependent rational animals: Why human beings need the virtues*. Chicago: Open Court Press.

MacIntyre, A. 2007. *After virtue*. 2nd ed. Notre Dame: University of Notre Dame Press.

Nash, R. 2014. Toward a shared decision: Against the fiction of the autonomous individual. In *Family-oriented informed consent: East Asian & American perspectives,* ed. R. Fan, 219–230. Dordrecht: Springer.

Oakeshott, M. 1991. *Rationalism in politics and other essays*. Indianapolis: Liberty Fund.

Sidgwick, H. 1888. The Kantian conception of free will. *Mind* 13 (5): 405–412.

Sulmasy, D. P., M. T. Hughes, R. E. Thompson, A. B. Astrow, P. B. Terry, J. Kub, and M. T. Nolan. 2007. How would terminally ill patients have others make decisions for them in the event of decisional incapacity? A longitudinal study. *Journal of the American Geriatrics Society* 55:1981–1988.

Taiwan Insights. 2012. Singles and single-parent families on the rise. *Taiwan Insights*. http://www.taiwaninsights.com/2012/12/17/singles-and-single-parent-families-on-the-rise. Accessed 6 Dec 2013.

Taylor, C. 1994. The politics of recognition. In *Multiculturalism: Examining the politics of recognition,* ed. A. Guttman, 25–74. Princeton: Princeton University Press.

Wang, J. 2014. Family and autonomy: Towards share medical decision-making in light of Confucianism. In *Family-oriented informed consent: East Asian & American perspectives,* ed. R. Fan, 65–82. Dordrecht: Springer.

Wear, S. 1993. *Informed consent: Patient autonomy and physician beneficence within clinical medicine*. Dordrecht: Kluwer Academic Publishers.

Wong, H. C. 2014. Towards a good practice of family-oriented consent: Reflections on medical practice in Taiwan. In *Family-oriented informed consent: East Asian & American perspectives,* ed. R. Fan, 245–256. Dordrecht: Springer.

Yang, Y. 2014. A family-oriented Confucian approach to advance directives in end-of-life decision making for incompetent elderly patients. In *Family-oriented informed consent: East Asian & American perspectives,* ed. R. Fan, 109–124. Dordrecht: Springer.

Yu, K. P. 2014. The Confucian alternative to the individual-oriented model of informed consent: Family and beyond. In *Family-oriented informed consent: East Asian & American perspectives,* ed. R. Fan, 93–106. Dordrecht: Springer.

Yung, L. 2014. The East Asian family-oriented principle and the concept of autonomy. In *Family-oriented informed consent: East Asian & American perspectives,* ed. R. Fan, 109–124. Dordrecht: Springer.

Zhao, W. 2014. A Confucian worldview and family-based informed consent: A case of concealing illness from the patient in China. In *Family-oriented informed consent: East Asian & American perspectives,* ed. R. Fan, 231–244. Dordrecht: Springer.

Part II
Dependency, Autonomy, and the Role of the Family

Chapter 2
Dependency, Decisions, and a Family of Care

Jeffery P. Bishop

2.1 Introduction

In American popular culture, the reigning myth is one in which the individual is a cowboy, alone in the wild and capable of sustaining his own life without assistance from others. One thinks of Western movie classics, such as "Shane," or "The Man with No Name" as quintessentially American. In fact, the good life is the solitary life, one lived alone on the frontier without interference from others. This solitary life as the paradigmatic life can be found even in contemporary American popular culture, for example in movies like "Spiderman" or the Batman series. The hero is solitary, alone in the world, and he alone can protect the society from those who would destroy it. These quintessential American heroes cannot have a family. They can neither love others, nor can they be loved by others; for to be loved and to love leaves them vulnerable to their enemies, who might exploit them and weaken them from the task of creating and protecting the *polis*.

This myth of the individual moves from the American frontier and into the medical arena, where the individual reigns supreme over his or her body. This mythical understanding of the individual creates a strange sort of relationship of oneself to oneself. I have elsewhere referred to self-made person as the sovereign subject, the one who is both the sovereign and the subject of that sovereignty (Bishop 2011). In other words, the patient, alone and solitary makes his own decisions for himself. Certainly this is why many studies related to both clinical and research informed consent, chastise medical practitioners for not adequately making information clear enough for patients. While patients do not make truly solitary decisions, there is the sense in which the medical establishment expects solitary decisions from patients, even when he or she turns to engage with their families. Of course, strong families—even in the United States—participate in the practices of informed consent,

J. P. Bishop (✉)
Tenet Endowed Chair in Health Care Ethics, Albert Gnaegi Center for Health Care Ethics, Saint Louis University, 3545 Lafayette Ave, Room 527, St. Louis, MO 63104, USA
e-mail: jbisho12@slu.edu

R. Fan (ed.), *Family-Oriented Informed Consent*, Philosophy and Medicine,
DOI 10.1007/978-3-319-12120-8_2

especially at the end of life; however, the structures and practices of medicine, including informed consent and decisions about removing life support in the critically ill, organize decisions such that the family is structurally marginalized. Individualized informed consent—as well as living wills, durable powers of attorney, privacy laws (such as the Health Insurance Portability and Accountability Act of 1996 (HIPPA)), and even the case-law itself—is set up to support the myth of an isolated individual alone in the world making his own decisions by himself and for himself.

In the first part of this essay, I shall unpack and expand Daniel Callahan's critique of end of life decision-making (Callahan 2000). Callahan notes that decision-making on the individualist model creates a game of brinksmanship with oneself, in which the person making the decision knows when he must strive to stay alive, and precisely when he must decide to allow the disease to take his life. Such a model is flawed. It is highly unlikely that even a healthy and clearly thinking person can ever know when he has crossed over the brink, even when the decision-making capacity is not altered. Moreover, dependency abounds and that dependency is best attenuated through familial support. In the second part of this essay, I shall draw on (and offer a slight critique of) the work of Alasdair MacIntyre (1999) in *Dependent Rational Animals*. MacIntyre's point is that our biological frailty requires a community of care committed to the virtues especially in our development, as we move from dependent animals in our youth to independent practical reasoners. However, in illness we move from independent rational thinkers into a state of dependency, even if intellectual capacities are preserved. It is during these times of illness and dependency when community is necessary, and family is the closest form of community. However, MacIntyre does not understand that the primordial community is the family.

Finally, I shall conclude with a reflection on why family-oriented decision-making ought to be the norm. After all, our human frailties are clearest at the beginning and ending of human life. Those frailties are mitigated by family-oriented care at the beginning of life, and they ought also to be mitigated by the family at the end of life. After all, it is family that creates and sustains the values of its members, and the family is in the best position to offer the right kind of care to its members in their hour of dependency and need. In other words, what is actually the norm in most families, even American families, is a family of care—the family as the foundational context for living and even for dying, despite the institutionalized structures and mythologies of American culture. Thus, all decision-making, even within the structures of informed consent, ought to emerge from the context of the family, where the frailties and dependencies of human life have always been the site of care, as well as its origin. Informed consent ought therefore to emerge from the family of care and not from the corrosive myth of the individual.

2.2 Definition of a Family

Before getting into the thick of the argument, I must first define how I understand "family." Any time one appeals to a social concept, like the family, especially in the West, one is immediately asked, "what does 'family' mean?" Any definition of a well-functioning family, where there is mutual respect among the various members, where there is concern for body, spirit, and emotion—however one wishes to define the elements of one's humanity—is immediately attacked from multiple and all angles. "Just where does such a family exist?" one is asked. It is just this sort of family that I mean: a well-functioning unit, where there is mutual respect among the various members and where there is concern for bodily, spiritual, and emotional well-being. Of course, as a community, any particular family may have a different definition of body, soul, spirit, and emotion, on what well-being may mean, and on what normal functioning may look like.

Certainly, robust notions of family are dissolving in the West, not least because so many of the institutions of the West seem designed to corrode any traditional or normative communal understandings in the West. Given the demographic data in the West, it is therefore difficult to give an easily circumscribed definition of the family. However, for the purposes of this paper, I will define a family as those who are bound to each other in a way that their material, spiritual, and emotional frailties and dependencies are met by one another and are ordered to some normative notion of well-being.

The family is typically the site where the adult members have emotional attachment, and in the case of parents, both emotional and sexual relationships. It is also typically the social site where children are conceived, born, and raised and are ordered toward some notion of the meaning and purpose of life as understood by the family. Yet, when I use the term "family" in this paper, I do not necessarily mean a nuclear family of mother, father, and children. I also mean traditional and extended families as well, including those second tier relations such as grandparents, siblings, aunts, and uncles, who take responsibility to meet the bodily, spiritual, and emotional needs of each other. Typically, the relational members of such families are fathers, mothers, children, siblings, grandparents, aunts, uncles, and cousins.

In addition, the members of a family need not be, but usually are, genetically-related or blood-related to one another. I take the family to be a relational unit that creates and sustains the material, spiritual, and emotional existence of those who are bound together mutually to meet the most basic demands of bodily existence, but at the same time the most basic emotional, moral, intellectual, and spiritual needs are being sustained, nurtured, and created. Thus, I take as family the well-functioning unit that supports the bodily, spiritual, and emotional; but I also understand the family to be the site where, when meeting each other's needs, much more is going on. Because when bodily, spiritual, and emotional needs are being met, the family is at the same time creating, nurturing, and sustaining moral, intellectual, and spiritual worlds of meaning and purpose for its members. And in meeting these needs and creating these life-worlds, the ties that make the family unit a whole are also being created, nurtured and sustained.

This working definition of a family should not be seen as existing in some ideal form. Every family has its failings. Certainly, one or more members of a family may fail in the family's understanding of what it means to be a good member of that family, for example a good father or a good mother. Because families tend to have normative weight, the various members are able to recognize when, for example a father or a mother fails. While any particular member/relationship may be missing or absent from a family, typically others are capable of filling in for the lost role or at least work to attenuate the deficiencies which that member fails to achieve. The fact that any one relationship might be abusive or fail in achieving some virtuous norm does not mean that we cannot speak normatively about the family. For families always carry understandings of the normativity of the various roles within them. Most families carry as part of their local and particular definition what it means for a father or mother to live up to his or her obligations. Put differently, in the work of meeting the material needs and of attenuating the bodily frailties of its members, a certain normative dimension is enacted by the family. And so we are capable of speaking of the virtues of its various members, and the virtue of this functional unit ordered toward the attenuation of bodily, spiritual, and emotional well-being, which we call the family.

2.3 Anecdotes of Individualism

Having given a working definition—however inadequate it may be—of family, let me describe two cases from my own medical practice where I think individualism has adversely affected patient care, and thus where families might have been of tremendous importance: the first demonstrates the way that individualism structures the policies and procedures of medical culture, and the second demonstrates the way that individualism affects the way that patients think of themselves. While these instances are anecdotal, there will be many doctors who will recognize these sorts of cases. Each example elevates individualism over against the importance of family.

As a resident physician in training, I had a Vietnamese patient in her late 1960s, whom I shall call Thuy. She had come to the clinic with her daughter, Phuong. Phuong had been the translator for her mother, who spoke no English. Thuy had fevers, night sweats, malaise, and fatigue for the past eight months. She ignored her symptoms for the first five months because she was preparing to travel to the United States in order to visit Phuong and her family. For two months after her arrival, Thuy's fevers, sweats and fatigue had worsened and she had lost another eight pounds. Chest x-ray showed bilateral hilar adenopathy. A CT scan of her chest and abdomen showed diffuse adenopathy in the hilum and in the periaortic region of the abdomen. Her spleen was also slightly enlarged. The diagnosis appeared to be consistent with a lymphoma. When I suggested this to Phuong, she requested that if Thuy's disease turned out to be cancer or some other deadly disease, she did not want me to tell her mother. She stated that discussions about the disease, prognosis, and therapy should only be directed to her, as she is Thuy's eldest child. Phuong stated that this is the "way it is done in our country."

When I went to my attending physician and presented the case to him, he could not believe what he was hearing. He did not like that Phuong had requested this of us and immediately assumed that Phuong was up to nefarious purposes. When I suggested that I did not think so, he asked me if I had gotten an independent translator. I told him that I had not, and he berated me for not having done so. He suggested that the patient had a right to know what was going on, and ignored my suggestion that perhaps this is the way decisions are made in Thuy's rural Vietnamese village. After all, he said, the individualism of American culture is what made Western medicine great, and that if Thuy wanted to benefit from American medicine, then she would have to participate in our rules about individual decision-making. He insisted that I get a translator and that I ask Thuy if she wanted her daughter to make all decisions, including decisions about treatment options for diagnoses of lymphoma. I was able to finesse this situation without disclosing to Thuy that she very likely had a cancer; however, Thuy was very puzzled by all the fuss over what her decisions would be, as it was clear to her that her daughter should be helping the medical team make all of the decisions.

Or consider a second case. Bill (not his real name) was in his early 1970s and his wife, whom I shall call Mildred, was in her later 1960s. Bill had metastatic transitional cell carcinoma of the bladder. He had multiple complications including frequent episodes of sepsis syndrome, some renal failure, pneumonia that was resistant to treatment, and had to make repeated visits in the Intensive Care Unit (ICU). He was in and out of the ICU for weeks. I personally had several conversations with him about his overall poor prognosis. He understood that he was in the final weeks to months of his life, but he continued to pursue aggressive care, "because my wife cannot handle my death." Bill understood that he was dying, and he actually did not want all of the aggressive care. Yet, he pursued aggressive treatment that would ultimately prove futile for the sake of his wife. In the meantime, I had conversations with Mildred. She understood that Bill was dying, and she did not want to see him struggling to live in pain and suffering, if he was going to die soon anyway. However, each refused to allow me to talk with the other person about their individual thoughts and feelings about Bill's illness and impending death. I tried on several occasions to bring up the topic with them both in the room, and inevitably they would agree that aggressive treatment was the best option, even when, if we could have reasoned together, we likely would have come to a different conclusion. The myth of the individual had become so much a part of their individual identities, that Bill would not authorize a shift to hospice care because his wife could not cope with this decision, and Mildred would not challenge Bill, as it was his decision.

Each of these experiences from my early career in the practice of medicine illustrates the individualism of American culture and American medicine. In the first instance, the American legal system and the hospital policies were such that the attending physician could not imagine medicine being practiced in any other way than to disclose to Thuy that she had lymphoma, or at least to get Thuy's authorization for her daughter's request to share diagnoses, prognoses, nor treatment plans only with her daughter. What is also revealing is that the attending physician assumed that Phuong was up to something nefarious, perhaps seeking to inherit all of her mother's earthly possessions. The attending physician could not imagine a world in which the family was the primary unit of decision-making, and that consent for diagnostic and treatment procedures ought to emerge from the family, and not from the individual.

In the second case, Bill and Mildred were products of American culture with its individualistic values. Each had so taken in the American culture that Bill could not imagine that Mildred would be able to assist him in making his decisions; and Mildred could not imagine that her input on Bill's decisions would be wanted or heeded. Moreover, the legal structures of American society, along with the policies of an American hospital would not allow me to speak directly about Bill's care to Mildred, even though Mildred was able to surmise just how sick Bill was. Bill and Mildred were like individual ships passing in the night each not knowing the intimate details of the other, even while they were married. This case not only illustrates the influence of individualism on decision-making in health care, but the corrosive effects of individualism on family structure in the West. Rawls's desire to see the family undermined seems to have come into being[1] (Rawls 1999, p. 64, 265, and 448).

2.4 Individualism and Technological Brinksmanship

In part, I would argue that it is this notion of individualism that leads to what Daniel Callahan has referred to as "technological brinksmanship." Callahan means by technological brinksmanship the drive to seek more aggressive care thinking that, at some point in the course of the illness, an individual will be able to know when he has crossed a threshold. In crossing that threshold, he will know it is time to say no more aggressive care (Callahan 2000). Callahan defines brinksmanship as the effort to get "as close to that line as possible before the cessation or abatement of treatment" is called for (2000, p. 41). Thus, the myth of the individual, combined with aggressive medical technology leads us to believe that each individual will know when enough is enough, and will be able to end the aggressive care in order to stop life-extending treatment. Callahan concludes that this mode of aggressive care initiated by the individual "assumes an ability to manage technology and its consequences with a delicacy and precision that medicine simply does not possess" (2000, p. 41). Callahan's assessment has a certain kind of autonomous agency in mind. He envisions an individual that first embraces an all-out technological intervention upon the dying body in order to save it, and an equally strong individual will that can push back against the aggressive care on the dying body. Autonomous individual agency is only partly able to counter the aggressive treatment of the ICU,

[1] It should be noted that John Rawls, in his *Theory of Justice*, thought that families created and sustained inequality of opportunity (Rawls 1999). He falls short of calling for the full-scale removal of families from the *polis*; his hard-nosed realism seemed to keep him from calling for the dissolution of the family as seen in the city in speech (Plato 1991). Whereas Plato seems to be speaking about impossibilities in the city in speech and seems know that because of realities of families there is no real possibility for perfect justice, I read Rawls as realizing that one cannot simply call for the dissolution of all intermediary institutions. In other words, I read Rawls as being more cunning, suggesting but not calling for the dissolution of the family, and thereby initiating the dissolution of the family by stealth through subtle state action. This essay is not the place to argue for this point.

because the ICU is governed, not by individual human agency, but by the agency of a social apparatus of the ICU that has as part of its structure the myth of the individual. I have argued elsewhere that the power of the social apparatus of the ICU is too strong and too seductive for any one person to easily control it (Bishop 2011, pp. 110–118).

In other words, the social apparatus of decision-making in contemporary American medicine holds that the self is in a position to know when to say, enough is enough. Callahan suggests that, "[t]he only evil greater than one's personal death is increasingly taken to be the loss of control of that death," (2000, p. 37) by the individual decision-maker. The social apparatus of the ICU, along with all the documents—living will, durable power of attorney, and consent forms—that an individual ought to have completed before going into the ICU, have at their heart a sovereign individual who sets in motion the aggressive technological features of medicine, that once deployed cannot be easily brought back under his own control. Even while the individual who has completed all the documents, and who has consented to the aggressive treatment in the ICU knows that a time will come when the machines may no longer be to his benefit, the documents and procedures of the ICU give the semblance that at any time those machines may be turned off. Yet, because of the sheer power of the apparatus of the ICU, it is not at all clear that the individual will be in a position to know when he has crossed over the brink, nor is it clear that any individual will be able to overcome the power of the ICU apparatus that sustains his life.

Finally, despite the structures and practices of care at the end of life that promote the attitude of individualism and technological brinksmanship, most medical personnel would rather have family-members with whom they can reason rather than a living will that tells the doctor what the patient wants in the general delimited circumstances set out in a living will. For the most part, living wills and the consent that they imply, say something similar to this: "in the event that I am unable to participate in decision-making, and if my attending physician thinks that aggressive treatment will not result in return of function, I authorize the doctor to stop" various aggressive treatments. However, anyone who has ever practiced medicine knows that the complicated physiological nature of seriously ill patients does not offer bright lines that are baldly apparent to anyone. Rather, there are nuances of clinical judgment and interpretation that are part of the decision-making process, and having a family there who can help to make these decisions is so much more important than having a living will, or a document of informed consent that has been individually executed. The truth is that, when we are at our most frail, for the vast majority of cases it is the family that knows its members best and the family that knows how best to reason with and to inform the doctors of what is best for the sick individual. In other words, despite the reigning mythos of the American individual who enacts and rejects treatment at the end of life, in truth the family, who can bring the nuance of spiritual and moral values to the clinical scenarios, is in the best position to make decisions on behalf of an individual rendered dependent by his illness. Dependency necessitates a family to enact a community of care.

2.5 Dependency and the Community of Care

Alasdair MacIntyre in his book, *Dependent Rational Animals* makes some very important points about dependency and the necessity of a community of care (MacIntyre 1999). Aristotle's ethics informs MacIntyre's approach. Like Aristotle in the *Politics*, MacIntyre takes seriously the idea that humans are social animals (Aristotle 1984, pp. 1253a3–4). Not only that, but our sociality is part of our natural state for Aristotle and for MacIntyre. MacIntyre claims that it is by virtue of our dependency that various forms of social enterprise, language not being the least, have evolved. Exploring scientific literature on dolphin behavior, he describes the ways that humans have ascribed language to dolphins. MacIntyre goes to great lengths, and through careful analyses to be precise about what can be said about the human animal with language. He notes that we humans are in large part dependent upon "modes of belief and activity that we share" with other animals (MacIntyre 1999, p. 41). Moreover, he shows that even if they do not have language, the prelinguistic forms of animal communication still assist in the thriving of the community of dolphins (MacIntyre 1999, p. 51). Thus, MacIntyre argues that some species of animals are practical reasoners and able to solve problems of survival. Thus, we can speak of dolphins as having complex and goal directed activities directed toward something that is good for them qua dolphins, and we can refer to that which is good for them qua dolphins as flourishing. That is to say, dolphins can be said to have species dependent goods that allow them to over-come the various vicissitudes that threaten dolphin flourishing.

What can be said of dolphins, MacIntyre claims, can be said of human beings: humans are practical reasoners about goods. Like dolphins, the social dimension for flourishing is absolutely essential to the success of humans in thriving; and also like dolphins, there are goods for the human species as such. Certainly humans are different from dolphins in that we humans also have role-dependent goods due to the sophistication of our social capacities. Quite different from dolphins, human beings can argue about what the goods are and what the goods ought to be; and we can even argue for the possibility of *the* good for humans. And, MacIntyre claims, since we can give reasons for our activities, and since we can ascribe reason-giving as a good, reason itself participates in the good for human beings.

Tracing the development of children, it is certainly true that children are unable to give reasons for their actions. Over time, however, a child slowly moves from dependency on others for his or her flourishing and for his or her learning, to independence from others. A child achieves, in time, the ability to give good reasons for his or her goal-directed activity. These goal-directed activities are directed toward something that is, at least relatively good for him or her. Ultimately, the hope is that the child will be engaged in activities that are aimed at his or her own flourishing as an embodied creature, flourishing as persons occupying a role, and flourishing as someone in pursuit of various goods, and even directed at the good for humans as such. MacIntyre names this creature that flourishes and is able to give good reasons for his or her goal-directed activities to be an independent practical reasoner.

Yet, MacIntyre is quick to note that the social dimension of independence remains a necessary condition for the possibility of independent reasoners; and independent reasoners learn the virtues aimed at the common goods of the community.

> Independent practical reasoners contribute to the formation and sustaining of their social relationships, as infants do not, and to learn how to become an independent practical reasoner is to learn how to cooperate with others in forming and sustaining those same relationships that make possible the achievement of common goods by independent practical reasoners. (MacIntyre 1999, p. 74)

In other words, the possibility for any member of a community to become an independent reasoner requires the community that supports those who are not yet independent reasoners. In a symbiotic relationship, the independent reasoner also learns the kinds of goods necessary to create communities of care that create the conditions for the possibility of independent reasoners. Thus there arise a series of moral and intellectual virtues that are aimed both at the flourishing of the community and at the flourishing of independent practical reasoners.

Independent reasoners are therefore dependent on the community, which in turn is dependent upon independent reasoners to sustain the community. Put differently, there are several virtues of acknowledged dependence (MacIntyre 1999, pp. 119–128). Put differently again, there are a whole host of giving and receiving relationships and practices that are involved in acknowledged dependence. Acknowledged dependence is the condition for the possibility of independent reasoners. "The care that we ourselves need from others and the care that they need from us require a commitment and a regard that is not conditional upon the contingencies of injury, disease, and other afflictions" (MacIntyre 1999, p. 128). MacIntyre spends about two pages on the family in a chapter entitled "The political and social structures of the common good," but does not develop in detail the foundational nature of the family. The family is first and foremost the fundamental unit of care when the contingencies of injury, disease, and other afflictions threaten one of the family-members.[2] MacIntyre almost immediately moves away from the family to the political dimension and does not develop a thesis as to why families are foundation (1999, pp. 133–135).

Certainly, MacIntyre is aware that families can thrive even in the face of severe social and political unrest, as has been depicted by Chinese film-maker Yimou Zhang in his 1994 film *To Live* (Zhang 1994). However, MacIntyre follows the typical approach of Western political philosophy, and avoids any sustained philosophical reflection on the family. In fact, he moves almost immediately from the importance of family to the community of care and the social relations that are necessary for flourishing. For MacIntyre, the family can only be sustained by the larger communal, social, and political contexts, even while he acknowledges that families also act to constitute the larger communal, social, and political contexts. In other words, MacIntyre does not spend enough time on how and why families are actually the sites where the virtues of acknowledged dependence are first and

[2] I have elsewhere reflected on the lacunae of the family in Western philosophical reflection (Bishop 2012).

most fundamentally nurtured, nor does he reflect on the specific forms of giving and receiving within families as they strive to meet and to attenuate the contingencies of injury, disease, and other afflictions.

I shall argue that it is for this very reason that the family is best situated to act on behalf of its members. In fact, the excesses of the West have been to spend too much time thinking about the *independence* of practical reasoners, that is to say, too much time on the autonomous agent as an individual, on the one hand, and too much time thinking about the political nature of human thriving, on the other, and all to the neglect of the family. For it is the family that acts as the single-most important unit that creates the conditions for the possibility of independent practical reasoners, but also the family that must be strong for communities to be strong. And for our purposes, it is the family, not the individual, nor the *polis*, that in fact does the vast majority of care when injury or disease strikes the individual.

2.6 Dependency and the Family of Care

Not only does Western culture, particularly in its American stripe, elevate the individual over the community and the family, it tends to emphasize youth and children over the family and the elderly. American culture worships youth, and thus fears anything that reminds it of death (McGill 1987). In addition, the state typically expects the family to make decisions in the best interests of any individual, especially a child, even if the decision might undermine the familial structures. The best interest standard is what is supposed to guide physicians and the health care team, such that if the family does not make decisions in the best interest of an individual child, the child can be removed from the family. The doctrine of *parens patriae* allows the state to intervene in the best interests of the child, even if the family is making a calculation that the intervention will disrupt familial cohesion and the survival of the familial structure.[3] Yet in every other decision, most families make calculations—say for example in what part of a city the family should dwell, or where children should be sent to school—about costs to the family so as to preserve the familial structures. After all, as we have seen, the community in general and the family in particular are conditions for the possibility of flourishing for any one individual member of a family; yet the family is a communal unit that must weigh up the relative costs and benefits on matters of material existence.

[3] The Doctrine of *parens patriae* gives the state authority to act as the parent of any member of the state who is unable to make decisions in the best interests of the individual. This doctrine permits the state to remove children from the family and allows the state to act as the child's parent. There is a fascinating, and oddly inconsistent aspect to the application of this doctrine. When organ transplant teams decide that a child in need of a transplant does not have the familial and social structure in place to justify the use of the scarce resource, the doctrine of *parens patriae* is not deployed. In other words, organ transplant teams, acting as decisional authorities can make life and death decisions based on familial contexts, but families cannot.

A child is born into a family and requires the family spend innumerable hours caring for and nurturing the child. Of course, a larger communal context is necessary to support the family, but, for the most part, the child's parents, grandparents, aunts and uncles, and older siblings are responsible for the direct care of a child. Without the family to offer the necessities of feeding and cleanliness, no child would ever survive. And while the child can learn to feed himself and bathe himself by year 4, the habits for eating a healthy diet or maintaining cleanliness are usually not in place until well into the child's adolescence. Whereas other mammals do not require long and intensive periods of care for development to full adult activity, the human animal exhibits extreme bodily dependency for at least a decade and a half.

Much more care is needed for the development of a robust intellectual life for the child. While in most contemporary industrialized nations education is usually funded by the state, educators routinely point out that familial involvement in the child's education is one of the most important factors in the child's academic success. However, in the early years children do not yet have sufficient cognitive development to understand the value of education. In fact they must develop habits of education, such as reading and writing, in order to maintain those skills into adulthood. In fact, it is not until the child has reached his or her teenage years that higher thinking and evaluative skills are acquired. It is only after the intellectual habits are acquired that the adolescent can begin to choose to engage in scholarly activities knowing that the benefits outweigh their desires for immediate gratification sometimes associated with the neglect of future success (Casey et al. 2005, 2008, 2011; Galvan et al. 2007; Reyna et al. 2005; Reyna and Farley 2006). Moreover studies have shown repeatedly, the full brain development may continue well into the child's 1920s, and children studying at University remain dependent on the family for financial resources related to education (Sallie Mae 2012; Discover 2012).

In short, the child learns that his activities, both those directed at the care of the body and those directed at the care of the intellectual capacities are not merely directed to some immediate material goal or even some merely intellectual goal, but are directed at other higher goals and goods, and possibly even to *the good* for humans qua human as MacIntyre has noted. The child learns more than how to care for his body and his mind, when her body and mind are being cared-for by the family. Indeed, much more about the goods as understood by a particular family are being communicated than the goods of mere bodily and intellectual life. It is this ability to weigh different goods across many different domains of life that requires development, from child to adolescent to independent reasoner, which does not appear to be consistently in place by the mid-1920s (Casey et al. 2005, 2008). In other words, both bodily and intellectual development requires over two decades before a child can fully begin to demonstrate independently the values that have been inculcated by the family. It is only after the child has undergone these bodily, intellectually, and morally formative practices that he or she develops the ability to discern moral and existential meaning.

Therefore, the labor of the family directed at meeting material needs enables the meeting of the intellectual needs of the child, but also the moral valence of care. And it is through the activity of meeting the bodily and the intellectual needs that the moral and existential formation of a child and adolescent is possible. Contrary to Hannah Arendt's (1958, pp. 136–174) claim that labor within the family is directed solely at meeting material existence, at the same time that the family is providing for the bodily needs, the meanings of the body are also being inculcated in the child. Thus, much more than meeting the demands of mere material existence is being done for the child. The family, in meeting the bodily needs of the child, is communicating the spiritual and moral significance of the body, as understood by that particular family. In fact, the social, moral and existential life-worlds are being shaped even as his bodily and intellectual needs are being met (Bishop 2012, pp. 518–522).

While it is certainly clear that rearing children is an important function of the family, the bias of the West tends to be directed at the independence of the child. However, that independence is drastically undermined as an individual family-member ages or when illness strikes. In fact, there is also evidence that the sick role taken on by an individual permits the previously independent reasoner to rely on his family members for tasks he would normally be expected to carry out for himself (Parsons 1975). While we can quibble with Parsons on many points (cf. Frank 2013), it seems clear that different familial structures and different members of a family have different expectations for their behaviors. Yet, when illness strikes, it is not uncommon to see the teenage child "mothering" his or her mother—mother, having been given permission not to act as mother. It is within the family that the child learns to give care because he or she has received such care. I am here using the notion of "sick role" in a way very different from Parsons's original usage.

The scope of this paper does not require a detailed sociological analysis of the "sick role;" however, it seems obvious that when illness strikes a teenage child takes on different roles, sometimes including the "mothering role," and the mother takes on a more dependent role, just as an example. The child learns to give care from the care he or she has received. He or she will have learned those "mothering" behaviors from the mother's offering of care. Of course, I realize that the offering of care is not something that all or only mothers do. The point is that the giving and receiving of care occurs within the family, and from within this structure certain normative practices originate. In other words, we can speak of the virtues of giving and receiving care, virtues inculcated in the child by virtue of being a member of the family. It does not matter whether or not the child offers particular kinds of care in the way that he or she has been reared. The point is that the practices of receiving and giving care—at least care as it has been understood by this particular family—will have been learned within the context of the family. The values and meanings of care, even care offered in the face of illness or injury, are learned in the context of the family. Thus, when illness or injury strike an individual, the family is still best suited to have access to the values and meanings of care being offered, and how that care fits within the family's overall understanding of the meaning of the body and the purpose of that life so stricken.

The family seems to be best suited to take on this role of decision-making, since the family has a special role in caring for the material needs of its members. The family is in the best position to carry out not only the values of its members, but also the methods of decision-making and methods of discernment utilized by the family.

And these methods are much more likely to render decisions that are consistent with the moral and meaningful life-world of the family-member who has become ill. And it is precisely in the practices of families coming to the assistance of another family-member—even if that illness is not life threatening—that prepares the family to make these large and important decisions for one another. In other words, when an individual family-member has a minor illness, and when he takes on the sick role—whether taken on willingly or unwillingly—the family comes in to assist with the care that the sick patient could very likely carry out on his own. And while in the sick role and giving up some aspects of decision-making, lessons are being learned by both the person with the illness and also by the rest of the family. Not only are the family-members' values enacted, but the both the family and the individual family member learns how to make these sorts of decisions. The family of care is itself a practice.

Thus, when a grave illness reduces a previously independent person to a state of dependency, it is the family that has had the most practice at offering care to that person. The family is in the best situation because the family has been practiced at meeting the demands of material dependency. Children, spouses, parents, grandparents, siblings, aunts and uncles, and cousins are the context within which bodily dependency is attenuated, intellectual needs are met, and moral values are formed. It is within the context of the family that, as bodily and intellectual needs are met, the moral and existential meanings and purposes of the lives of its members are inculcated. They have had practice at giving and receiving care—at least as understood by a particular family. When grave illness comes along, the family is best situated to enact the wishes and values of the individual-family member because they have had practice in doing so.

2.7 Conclusions

Thus, it seems to me that American practices of informed consent—as well as living wills and powers of attorney—that are directed at the individual, act to corrode family practices of decision-making. The fact that Bill and Mildred did not know how to communicate and to give and receive care as a family is the result of years of emphasis on the individual as the sovereign decision-maker, as well as institutionalized policies that emphasize individual decision-making. The fact that the medical establishment believed that I should undermine the familial decision-making structures of Thuy's family also suggests that these practices are corrosive to traditional families. The various institutions within Western political systems that have emphasized the individual, remove the family as part of the decision making process, resulting in families that do not learn how to give and receive care according to the moorings of that particular family. That means that the members of a family do not have the occasion to learn to reason together as a family about the morally salient features of medical decision-making.

I have argued that in fact families are the primary unit that creates and sustains the practices of procuring practical bodily needs from the time that a child is brought into the world. And in robust families, the child learns to become independent to some degree. Whereas MacIntyre has shown that independence is sustained by a larger community of care, I have argued that because the family supports the bodily and intellectual needs of the child, and because within the context of the family, the child learns how to make evaluative decisions of moral and meaningful value. In other words, we should be speaking of the family of care as much, if not more so than a community of care. Moreover, a particular family member suffering illness is best suited to make decisions within the context of the family. Thus, families ought to be present whenever consent is obtained.

So families ought to be included in all procedures of informed consent, the articulation of living wills, and any other procedures when value-laden decisions are being made. Even for someone with full capacity to make decisions for him or herself, there are already practices in place such that the sick person learns to receive care from those trusted family-members. During these times of illness, the various members of the family are learning to give and receive care within the context of a family that has for decades provided bodily care within the meaningful context created by the family. It is because of these practices of giving and receiving care that, when severe illness strikes, families are capable of making informed decisions for other family members. When a particular family member is unable to make decisions for him or herself in consultation with his or her family, then the family ought to be the natural place to turn for those decisions.

Rather than presuming individualized decision-making, we should presume familial decision-making and informed consent. The current policies of decision-making found in American hospitals, with their emphasis on individual decision, corrode the practice of familial decision-making. I have argued that the practices of Western individualized medicine, with their individualized decision-making, have resulted in the brinksmanship kind of medicine critiqued by Daniel Callahan. American families are out of practice in giving and receiving care, and thus they have forgotten how to offer care and to make decisions for particular family-members. One way to remedy this situation is for American hospitals to begin to think in terms of family-centered informed consent. After all, families are the primary community for meeting the bodily needs of its members, but also the family is the font from which the values and practices of evaluating moral decisions spring forth. The family of care then is best situated to give informed consent, because the family of care is the source and the strength of the family's values.

References

Arendt, H. 1958. *The human condition*. 2nd ed. Chicago: University of Chicago Press.

Aristotle, B. 1984. *The complete works of Aristotle*, ed. J. Barnes. Princeton: Princeton University Press.

Bishop, J. P. 2011. *The anticipatory corpse: Medicine, power and the care of the dying*. South Bend: University of Notre Dame Press.

Bishop, J. P. 2012. Of life-worlds and savings accounts: Toward a familial philosophy of health care financing. *Medicine and Philosophy* 33:10–4.

Callahan, D. 2000. *The troubled dream of life: In search of a peaceful death* Washington, DC: Georgetown University Press.

Casey, B. J., R. M. Jones, and L. H. Sommerville. 2011. Braking and accelerating the adolescent brain. *Journal of Research on Adolescence* 21:21–33.

Casey, B. J., A. Galvan, and T. A. Hare. 2005. Changes in cerebral functional organization during cognitive development. *Current Opinion in Neurobiology* 15:239–244.

Casey, B. J., S. Getz, and A. Galvan. 2008. The adolescent brain. *Developmental Review* 28:62–77.

Discover. 2012. Investor relations. *Discover*. http://investorrelations.discoverfinancial.com/phoenix.zhtml?c=204177&p=RssLanding&cat=news&id=1717007. Accessed 25 Nov 2014.

Frank, A. W. 2013. From sick role to practices of health and illness. *Medical Education* 47 (1): 18–25.

Galvan, A., T. Hare, H. Voss, G. Glover, and B. J. Casey. 2007. Risk-taking and the adolescent brain: Who is at risk? *Developmental Science* 10:F8–14.

MacIntyre, A. 1999. *Dependent rational animals: Why human beings need the virtues*. Chicago: Open Court.

McGill, A. C. 1987. *Death and life*. Philadelphia: Fortress Press.

Parsons, T. 1975. The sick role and the role of the physician reconsidered. *The Milbank Quarterly* 53 (3): 257–278.

Plato. 1991. *The republic of plato*. 2nd ed. Trans: A. Bloom. New York: Basic Books.

Rawls, J. 1999. *A theory of justice*. Rev. ed. Cambridge: Belknap Press of Harvard University Press.

Reyna, V. F., and F. Farley. 2006. Risk and rationality in adolescent decision making: Implications for theory, practice, and public policy. *Psychological Science in the Public Interest* 7:1–44.

Reyna, V. F., M. B. Adam, K. M. Poirier, C. W. LeCroy, and C. J. Brainerd. 2005. Risky decision-making in childhood and adolescence: A fuzzy-trace theory approach. *The development of judgment and decision making in children and adolescents*, ed. J. Jacobs and P. Klaczynski, 77–106. Mahwah: Erlbaum.

Sallie Mae. 2012. How America pays for college. A national study by Sallie Mae and Ipsos. *Salie Mae*. https://www1.salliemae.com/about/news_info/research/how_america_pays_2012/. Accessed 25 Nov 2014.

Zhang, Y. 1994. *Huozhe* [To live]. China: Shanghai Film Studio and Era International.

Chapter 3
Individually Directed Informed Consent and the Decline of the Family in the West

Mark J. Cherry

3.1 Introduction

The traditional family—a husband and wife, together with their biological children—provides young children, adolescents, and adults with well-documented social, economic and adaptive advantages. Yet, in the West, this form of the family is in decline. A growing percentage of men and women choose not to be bound by the moral and social expectations of marriage and traditional family life and an ever more significant number of children are being born to single mothers. More than 40% of all births in the United States in 2011 were to unmarried women (Hamilton et al. 2012, Table 1). Such demographic shifts are associated with important changes in underlying taken-for-granted social and sexual mores. They also reflect public policies that instantiate a hermeneutic of suspicion against the traditional family. For example, the individualistic character of the social-democratic egalitarian ideology that underlies current dominant approaches to health care policy and medical decision-making in Western Europe and North America is associated with a decline in family stability (Akerlof and Yellin 1996, p. 21). Individually directed informed consent, for example, accents an unqualified affirmation of persons as the source of authority over themselves. This practice of informed consent tends to present persons outside of any social context in general and outside of their families in particular. The burden of proof is placed on the family to demonstrate that it acts with legitimate authority and in the best interests of individual members.

This paper critically engages moral and political pressures that have been brought to bear on the family, through such practices as individual-oriented informed consent. Whereas individual-oriented approaches to medical decision-making accent an ethos of personal autonomy, the family-oriented procedures typical of Hong Kong and China acknowledge the central social and moral reality of the family.

M. J. Cherry (✉)
Department of Philosophy, St. Edward's University, 3001 S. Congress Avenue, Box 844, Austin, TX 78704, USA
e-mail: markc@stedwards.edu

R. Fan (ed.), *Family-Oriented Informed Consent*, Philosophy and Medicine,
DOI 10.1007/978-3-319-12120-8_3

Here, the family ought to be appreciated as more than simply a network of personal relations. It possesses a being that is social and moral, such that it realizes particular elements of the good and preserves necessary conditions for core areas of human flourishing. Moreover, the family and its core relationships are necessary for adequately appreciating the ways in which men and women come together to reproduce and the ways in which they successfully raise children. As Ruiping Fan, Xiaoyang Chen and Yongfu Cao rightly note: "the family is the cardinal intermediate institution between the individual and the state. It is a biologically based community that produces children, nurtures and educates children, and tends to amass capital for its own purposes and goals" (2012, p. 509; see also Fan 2010). Family-oriented approaches to informed consent for medical treatment acknowledge this foundational social and moral reality. The family is appreciated as properly possessing more-or-less significant authority over their members, as well as being the appropriate locus for determining their best interests. Consequently, from the perspective of family members, those who would interfere in the conduct of the family are judged as bearing the burden of proof, with a potentially very significant standard of proof, to impede family decisions. Throughout the analysis, I argue for the centrality of the family for human flourishing and, consequentially, for the importance of sustaining (or re-establishing) family-oriented understandings of informed consent to medical treatment.

3.2 The Family and Human Flourishing

The family has traditionally been regarded as a normative form of social being, a morally regulative category of social life over and above a simple biological fact of the matter.[1] This core social reality, however, is evermore called into question. Even to refer to the traditional biological family as properly normative has become politically controversial. Yet, the family exists not simply as the creation of individuals or the state, but rather possesses a reality *sui generis* as a category of social existence. The dominant ways in which men and women create long-term social units for companionship, procreation and raising children, preserving and allocating resources, while also preferentially caring for and expending resources on their own biologically related children, elderly parents, and other close blood relatives, routinely document that humans are organized into families. The family is a key aspect of the fabric of social reality. The family's being discloses facts of the matter and normative commitments, including social and moral obligations, while

[1] That the family is normative and central to human flourishing does not imply that family-based duties do not need to be carefully explicated. As Ilhak Lee notes: "filial duty has been (mis)understood as an unconditional, unlimited commitment and sacrifice, meaning children should do what, in the opinion of 'others,' is best for their parents, not what the parents prefer. It seems in this case children have little chance of demonstrating obedience, which is the proper understanding of filial duty. They also seem to have little chance for a discussion with their parents about the treatment they prefer, or what the parent would want" (2014).

sustaining the necessary conditions for central areas of human flourishing. Through the family one encounters and can know a domain of goods and virtues, as well as experiences of human flourishing, that are only understandable through its living reality. Appreciating the family in this fashion provides a foundational condition for the possibility of experiencing, conceptualizing, and appreciating a major domain of human experience and human good.

For example, the sociobiological empirical data demonstrates that traditional family structures and familial relationships are central to securing core areas of human flourishing. In part, family life permits persons as individuals better to achieve longer, more fulfilling lives. The high quality social relationships typical of traditional forms of family life, for example, are associated in significant ways with positive mental health, as well as with decreases in morbidity and early mortality.[2] Strong familial relations also increase social integration, social regulation and support, even reduce the risk of suicide (Kposowa 2009; see also Kunce and Anderson 2002; Neumayer 2003; Breault 1986). In part also, within the family, persons discover themselves as already sustained by and within a web of pre-existing responsibilities and moral obligations (see, e.g., Lee 2014a, b; Deng 2014; Yu 2014; Wong 2014). Parents and children are not isolated individuals. They are encountered as a social unity and must be appreciated in terms of their differing roles, duties and obligations within the family. As H. Tristram Engelhardt, Jr. concludes: "The obligations that connect parents and children are such to which they may never have committed themselves and to which they need never have consented in order for the obligations to have moral force" (Engelhardt 2010, p. 508).[3] In the absence of the intact traditional family central possibilities for mutual acknowledgment, social life, and human flourishing go unrealized (Engelhardt 2010, p. 508; Cherry, 2010).

Significant empirical evidence has accumulated, for example, that children who are raised in families with both biological parents present, have social, emotional, psychological and financial advantages over children raised in other types of environments. For example, children reared in single-parent households are statistically more likely to be impoverished, to engage in delinquency as adolescents and criminality as adults, to drop out of school, to get pregnant as a teenager, and to have poorer emotional and psychological health with concomitant difficulties in later

---

[2] Julianne Holt-Lunstad, Timothy Smith and J. Bradley Layton, in a meta-analytic review across 148 studies with some 308,849 participants documented that strong traditional social relationships, such as the family, indicate a 50 % or greater increased likelihood of survival across a wide range of causes of death. "Cumulative empirical evidence across 148 independent studies indicates that individuals' experiences within social relationships significantly predict mortality. The overall effect size corresponds with a 50 % increase in odds of survival as a function of social relationships. Multidimensional assessments of social integration yielded an even stronger association: a 91 % increase in odds of survival…Results also remained consistent across a number of factors, including age, sex, initial health status, follow-up period, and cause of death, suggesting that the association between social relationships and mortality may be generalized" (Holt-Lunstad et al. 2010, p. 9. See also House et al. 1988; Norval et al. 2009).

[3] As Ana Iltis notes: "Families have particular interests both because they are stakeholders in family members' well-being and because they ordinarily want to protect the interests of individuals in the family (and of the family over all)" (2014). On this point see also Bishop (2014).

life.[4] Children from divorced families are more likely to end their own marriages in divorce, to attempt suicide as teenagers (Weisfeld et al. 1987), to misuse alcohol and narcotics (Norval et al. 2002; Weitoft et al. 2003; Defoe, 2003). Boys raised without a father are more likely to commit crimes or to be delinquent (Norva et al. 2002, p. 42). Girls raised in single parent families are twice as likely to give birth while a teenager and to drop out of high school (Rhoads 2004, p. 80).

The statistics on child abuse and neglect are significantly worse for children being raised in single parent homes or homes with stepparents or a non-spousal partner, such a boyfriend or girlfriend, over households in which children are raised by two biological parents. Consider, for example, the results of an empirical study conducted by Martin Daly and Margo Wilson, both professors of evolutionary psychology and former presidents of the Human Behavior and Evolution Society. When looking at child abuse data in the United States, Daly and Wilson were initially surprised by the over-representation of stepfamilies. Concerned that their data reflected an artifact, such as the under-reporting of child abuse by biological parents, they narrowed search criteria to the most unmistakable cases of abuse, such as those with fatal outcomes. They were forced to conclude:

> But as we made our abuse criteria increasingly stringent and narrowed the sample down to the most unmistakable cases, the over-representation of stepfamilies did not diminish. Quite the contrary, in fact, by the time we had reduced the cases under consideration from the full file of 87,789 validated maltreatment reports to the 279 fatal child-abuse cases, the estimated rates in step-parent-plus-genetic-parent households had grown to approximately *one hundred times* greater than in two-genetic-parent households (Daly and Wilson 1999, p. 28).

Even when there is no abuse or neglect, stepparents statistically spend less time with their stepchildren than do biological parents who raise their own children.

Single mothers, who have never been married, produce statistically worse outcomes for children when compared to those reared by their married biological mother and father (Aronson and Huston 2004; Fomby and Cherlin 2007). As Charles Murray summarizes:

> No matter what the outcome being examined—the quality of the mother-infant relationship, externalizing behavior in childhood (aggression, delinquency, and hyperactivity), delinquency in adolescence, criminality as adults, illness and injury in childhood, early mortality, sexual decision making in adolescence, school problems and dropping out, emotional health, or any other measure of how well or poorly children do in life—the family structure that produces the best outcomes for children, on average, are two biological parents who remain married. Divorced parents produce the next-best outcomes…Never-married women produce the worst outcomes (Murray 2012, p. 158).

Such empirical outcomes remain even after controlling for other family characteristics, such as parents' race, income, and socioeconomic status (Gallagher and Waite 2000, p. 125; see also Fagan and Rector 2000; Parcel and Dufur 2001; Rountree and

[4] "Children raised in single-parent households are, on average, more likely to be poor, to have health problems and psychological disorders, to commit crimes and exhibit other conduct disorders, have somewhat poorer relationships with both family and peers, and as adults eventually get fewer years of education and enjoy less stable marriages and lower occupational statuses than children whose parents got and stayed married" (Gallagher and Waite 2000, p. 125).

Warner 1999; Cookston 1999; Osgood and Chambers 2000; Flanagan et al. 1999). More generally, children in single parent families: "... have negative life outcomes at two to three times the rate of children in married, two parent families" (Wilcox and Marquardt 2011, p. 87; see also Wilcox et al. 2011; Parke 2003).

Children raised outside of the biological family are more vulnerable to a wide range of social, psychological, and economic challenges. The transition from adolescence to adulthood, for example, can in the best of circumstances be traumatic.[5] Families typically nurture their teenage and young adult members, providing ongoing lifestyle and career guidance, financial support, and even the possibility of a temporary move back home with mom and dad if necessary. Children raised outside of the family environment are less likely to experience such advantages. Youths raised in non-family environments, such as state care and foster homes, typically must leave these settings once they become too old to be eligible for these types of social welfare services, even though the need for emotional and economic support continues.

Learning to function in the world as a responsible and effective adult can be a slow and arduous process. Children rely on their parents often well into their twenties for financial resources, a place to live, employment and educational assistance, emotional support and personal guidance. Unless children fully separate themselves from their families, this transitional period is most accurately described as taking place over a spectrum of semi-autonomy, during which the now adult child achieves additional autonomy, with significant assistance and guidance from parents and other family members.[6] Indeed, in functioning extended families, full independent autonomy may never be realized or sought. Access to the financial, intellectual, emotional and psychological resources of the family promotes positive outcomes while softening the consequences of the less than optimal judgments typical of early adulthood (Settersten and Ray 2010, p. 33). The roles that the male and female biological parents together play in the successful raising of their children cannot be adequately reproduced by third-parties, social institutions, such as group care settings, foster care, or governmental agencies. In short, the family helps to guide and smooth the transition from childhood to adulthood.[7]

[5] "How an adolescent fares during the transition to adulthood has long-term repercussions. Earning a college degree leads to a higher-paying and more prestigious job, while early parenthood, unsuccessful marriage at a young age, and involvement in crime or problematic substance use all foretell difficulties in finances, family relationships, and beyond" (Osgood et al. 2010, p. 210).

[6] Richard Settersten and Barbara Ray note, for example, that "both in the United States and in many European countries, the process of becoming an adult is more gradual and varied today than it was half a century ago. Social timetables that were widely observed in that era no longer seem relevant, and young people are taking longer to achieve economic and psychological autonomy than their counterparts did then" (Settersten and Ray 2010, p. 20).

[7] "Even if the transition to adulthood had not become so demanding, members of these vulnerable groups [children raised in foster care or in a group care setting] would face exceptional challenges finding employment, attending college, and marrying and starting a family. Many struggle with emotional or behavioral problems; many have histories of problems in school and the community. Often their families are unable or unwilling to provide the support that most families provide to

Despite such advantages, demographics indicate a shift away from traditional family life. Individuals are becoming increasingly isolated from the rich and intimate social connections of the family.[8] Taken-for-granted background social mores and moral expectations have altered. Adults, for example, have become much less likely to marry than in past decades. This does not mean that they are choosing to remain chaste outside of the marriage of husband and wife. Rather, adults are much more likely to live unmarried with a sexual partner, or simply to live as sexually active singles (Wilcox and Marquardt 2009, pp. 69–70). In the United States, between 1960 and 2010, unmarried cohabitation, couples who live together as unmarried sexual partners, increased by more than a factor of seventeen (Wilcox and Marquardt 2011, p. 75). Some studies estimate that approximately 25 % of unmarried women between the ages of 25 and 29 live with a sexual partner, and "an additional quarter have lived with a partner at some time in the past" (Wilcox and Marquardt 2011, p. 75). Moreover, greater than "… 60 % of first marriages are now preceded by living together" (Wilcox and Marquardt 2011, p. 75) and over 40 % of U.S. households with a co-habitating unmarried couple contain children (Wilcox and Marquardt 2011, p. 76).

Among the more predictable consequences of such a shift in sexual morality is that a growing percentage of children are born outside of marriage. In 2011, unmarried birth rates tracking the race of the mother in the United States were as follows: Black—72.3 %; Hispanic—53.3 %; American Indian or Alaska Native—66.2 %; White—29.1 %; Asian or Pacific Islander—17.2 % (Hamilton et al. 2011, Table 1). Such data ought to raise significant concern, since, as noted, children reared outside of the traditional family environment face real disadvantages. A number of studies have found that such disadvantages appear even when the biological father and biological mother cohabitate without getting married.[9] Despite well documented

their children during this transition—funding for college, child care that permits work or schooling for young parents, a place to live when times are hard" (Osgood et al. 2010, p. 211).

[8] "Current evidence also indicates that the quantity and/or quality of social relationships in industrialized societies are decreasing. For instance, trends reveal reduced intergenerational living, greater social mobility, delayed marriage, dual-career families, increased single-residence households, and increased age-related disabilities. More specifically, over the last two decades there has been a three-fold increase in the number of Americans who report having no confidant—now the modal response. Such findings suggest that despite increases in technology and globalization that would presumably foster social connections, people are becoming increasingly more socially isolated" (Holt-Lunstad et al. 2010, p. 2; see also McPherson and Smith-Lovin 2006; Putnam 2000).

[9] "The differences begin in infancy, when most of the cohabiting couples are still living together and the child has a two-parent family. Stacey Aronson and Aletha Huston used data from a study of early child care conducted by the National Institute of Child Health and Human Development to assess the mother-infant relationship and the home environment for children at ages 6 months and 15 months. On both measures and at both ages, the children of married couples did significantly better than the children of cohabiting parents, who in turn had scores that were only fractionally higher than the children of single mothers. … The disadvantages of being born to cohabiting parents extend into childhood and adolescence, even when the cohabiting couple still consists of the two biological parents. Susan Brown used the 1999 cohort for the National Survey of America's Families to examine behavioral and emotional problems and school engagement among six- to eleven year-olds and twelve to seventeen-year-olds. Same story: Having two unmarried biological

social, economic and adaptive advantages, however, an evermore significant number of couples find there to be little justification to be bound by the traditional expectations of marriage and so act in the light of their own particular agreements to live together, engage in sexual activity, and perhaps reproduce.[10]

3.3 Western Bioethics and the Undermining of the Family

By its very nature, the empirical data surveyed is statistical. As a result, not all particular cases will fit the statistical descriptions. Exceptions complicate but do not undermine statistical observations. The suggestion is neither that all traditional biological two-parent families are perfect, nor that all single mothers are poor parents. The character of the empirical data, however, strongly recommends taking seriously the cardinal role and reality of the traditional biological family and its life-world, for sustaining important elements of human flourishing. Such traditional family structures have a demonstrably positive impact on the successful raising of children. Demographic data routinely indicate that the negative impact on children of other types of living arrangements is far from negligible. Moreover, such family structures protect against poverty and provide a safety net for children, adults and the elderly.[11]

Consequently, one should be concerned when the character of public policy contributes to a decline in family integrity or to social shifts away from traditional family life. Here, a significant challenge is that liberal advocates often appreciate

parents was associated with worse outcomes than having two married biological parents, and the outcomes were rarely better than those for children living with a single parent or in a 'cohabiting stepparent family'" (Murray 2012, pp. 164–165, citing Aronson and Huston 2004; Brown 2004).

[10] The prevalence of sexually transmitted disease has also increased. Roughly 16 % of Americans between the ages of 14 and 49, for example, are infected with genital herpes, one of the most common sexually transmitted diseases. The infection rates are worst for African-American women (about 48 %) and African-Americans generally (about 39 %); for women (about 21 %), than for men (about 11.5 %). According to the Centers for Disease Control, treating sexually transmitted diseases costs the United States healthcare system some $ 16 billion annually (see Allen 2010). The World Health Organization issued an alert in June 2012 expressing their concerns regarding new forms of antibiotic resistant gonorrhea (Shepherd 2012).

[11] Perhaps, as Wenqing Zhao (2014), argues, it would be beneficial to turn to the family as a whole to help assure proper medical decision-making. Moreover, as Yaning Yang (2014), argues, family-based accounts of advanced directives would likely benefit the elderly. For insights into the situation in Taiwan see Lee (2014b, pp. 125–136), who notes that the goal in family-based decision making at the end of life in Taiwan is family consensus: "in actual practice, family consensus is the target. It is usually upheld by medical professionals. In many cases, the presentation of all the relevant family members, for example, in the decision for parent's medical treatment, the decisions of all brothers and sisters are usually requested. Even the spouse and elders were present, medical professionals would insist that other known family members be present before the final decision is made. In cases of not a matter of urgency, medical professionals usually ask the family members to hold meetings to arrive at a consensus decision before taking any further action."

the family as a major impediment to social implementation of their preferred conceptions of equality and social justice. To speak of the family as founded on the monogamous, heterosexual union of husband and wife, together with their biological children, clashes with the increasingly dominant political view of the family as fashioned around the equal partnership of free and equal men and women. For the progressive liberal, the idea of the family has become increasingly nominalistic; the family is seen as no more than a social construct, created through the particular agreements of its participants, with no independent reality of its own. Family members, of whichever sex, are presumed to be of equal authority and, as far as possible, as having interchangeable intra-familial social roles. Many feminists, homosexual activists, and other defenders of post-traditional social structures support just such a social-constructivist account of the family. The goal is to emancipate the family from what are judged to be the inappropriate and illiberal confines of traditional cultural, social and religious norms.

This progressive political vision recognizes itself as having an adversarial relationship with traditional family structures and forms of familial authority. The family as a foundational social institution is placed fully within a hermeneutic of suspicion. Elements of the feminist movement, for example, critically judge the family to be an institution of unequal power relationships and female subordination.[12] Susan Moller Okin, for example, disparagingly characterized "the sentimental family":

> …the family had become characterized as entirely distinct from the outside world. Allegedly united in its affections and interests, this special sphere of life was held to depend for its health on the total dedication of women, suited for these special tasks on account of the very qualities that made them unsuited for the harsh world of commerce, learning, and power. Thus anyone who wished to register an objection to the subordinate position of women had now to take considerable care not to be branded as an enemy of that newly hallowed institution—the sentimental family (Okin 1982, p. 88; see also Okin 1994).

As Okin rightly perceives, the traditional family embodies particular understandings of proper family structures, including appropriate, albeit different, roles for men and for women. Such perceptions are among her reasons for concluding that traditional religious groups should not be permitted to nurture and educate their children within the religion itself. Okin decries such pedagogy as indoctrination (Okin 2002, pp. 218, 226). Securing her particular progressive vision requires decoupling morality and personal choice from cultural and religious viewpoints that recognize traditional family structures as presumptively authoritative.

The established American account of bioethics, similarly systematically seeks to limit the authority of families in medical decision making. As Beauchamp and Childress summarize their viewpoint: "…the authority of the family is not final or ultimate.…Health care professionals should seek to disqualify any decision makers

[12] "It was the contribution of the women's movement to attempt such a synthesis by placing the family in the center of social analysis. Feminists identified the family as a crucial institution in the reproduction of social relationships generally, and decisive for women's subordination. Hence, in theory and practice, the women's movement adopted a critical stance toward family life" (Breines et al. 1978, p. 43).

who … have a conflict of interest. Serious conflicts of interest in the family may be more common than either physicians or the courts have generally appreciated" (2009, p. 188; see also 2012). The goal is a secular ethic that begins with the privileged presumption of the sovereignty of the individual. The dominant approaches to bioethics and health care law tend to support an unqualified affirmation of persons as the source of authority over themselves. Medical decision making, it is claimed, rightly rests with the individual patient (Beauchamp and Childress, 2009, p. 106). As a result, persons tend to appear as atomic individuals endowed with a right to determine their own futures (Applebaum et al. 1987; President's Commission for the Study of Ethical Problems in Medicine and Biomedical and Behavioral Research 1982). Individual liberty, conceptualized as autonomous self-determination, is assigned cardinal moral value. As rational moral beings, persons are to choose to be autonomous and self-determining individuals, who shape their moral values and perceptions of the good for themselves. Such an individualistic conception of personal autonomy is judged to be integral to human good and human flourishing.

Health care law, in turn, is framed to support individualistic consent, reflecting the centrality of the individual and ensuring adequate opportunity for persons to free themselves from traditional familial relationships. This individual-oriented practice of informed consent approaches patients as if they are not members of intact functioning families, unless there is good evidence to the contrary. Patients who wish to be treated as members of families, such as a parent who wishes to have his children make medical decisions on his behalf, are usually required explicitly to authorize the involvement of family members in the decision making process (Faden and Beauchamp 1986; Wear 1993). As Engelhardt marks this practice:

> Those who regard autonomous individualism as the presumptively appropriate relation among persons would require any deviations to be established by explicit statement and agreement. For example, patients would be presumptively treated as autonomous individuals willing and committed to choosing on their own, unless they explicitly demanded to be regarded and treated within a traditional family structure (Engelhardt 2002, pp. 24–25).[13]

It is the individual, rather than the family, who is appreciated as possessing decisional authority.[14]

---

[13] Engelhardt continues: "… consider the contrast between those who favor autonomous individualism and those who would give moral priority to family life. Those who regard autonomous individualism as the presumptively appropriate relation among persons would require any deviations to be established by explicit statement and agreement. For example, patients would be presumptively treated as autonomous individuals willing and committed to choosing on their own, unless they explicitly demanded to be regarded and treated within a traditional family structure… On the other hand, if one considered life within a traditional family structure as the presumptively appropriate relation among persons, the burden of proof shifts. Persons are approached as nested within the thick expectations of traditional family structures, unless they explicitly state that they wish to be regarded and treated as isolated individuals" (Engelhardt 2002, pp. 24–25).

[14] For a detailed account of autonomy and family-based autonomy in the Korean medical context see Kyungsuk Choi who argues that "The individual (the self) and the family (a community) should be balanced. The family can be considered a community in a basic sense. From a traditional Eastern perspective, the family, rather than the individual, has been the basic unit for society and the state. However, it was not long before Korean society began to recognize an individual as

This progressive ethos similarly assumes individualistic approaches to patient confidentiality and access to medical information. For example, rather than being understood as united in marriage, spouses are treated as fully separate and separable individuals, with privacy rights vis-à-vis each other, and limited or no authority over the other. This legal circumstance has led to the limitation of spousal access to medical, financial, and educational records as well as limitations on the ability to grant permission on behalf of one's spouse for medical treatment or financial decisions. The United States Health Insurance Portability and Accountability Act (HIPPA), for example, restricts the sharing of medical information with spouses and children, unless the patient provides specific authorization. Such restrictions tend to be harmful to patients and destructive for family relationships:

> Not only does HIPAA impose extravagant costs for exiguous benefits, HIPAA's sour assumptions about human nature work positive harm. For instance, HIPAA assumes people (1) want to keep information from their families … HIPAA's rules are structured to serve patients who fit those assumptions. HIPPA's assumptions are wrong. Most people want their families involved in their medical care. … Instead of having the few patients who fit HIPAA's assumptions opt in to restrict privacy rules, HHS requires the huge majority of patients who don't fit the assumptions to opt out of them. This burdens patients. Worse, most patients won't realize they need to act, and few will get around to it (Schneider 2006, p. 11).

Whereas consent in traditional cultures and religions is typically paternalistic and family oriented, Western bioethics seeks legal requirements for patient-oriented confidentiality and individual autonomous decision making, shielding personal information from spouses, children, parents, and other relatives.

With regard to children, emphasis is placed on protecting the child's "best interests"; such interests, in turn, are usually appreciated in terms of the child's liberty and equality interests. Children are to be nurtured towards equal liberty and personal autonomy as soon as possible.[15] Parental authority, in turn, is conceptualized as flowing from the state, which authorizes its proper structure and appropriate limits:

---

having autonomy. We cannot disvalue modernity. In this regard, the value of autonomy should continue to be emphasized, but it must be balanced with other traditional values" (2014).

[15] See, e.g., Rawls 1999. Rawls urges that parents should not be appreciated as possessing moral status or moral authority in themselves, but only insofar as parents and the family function as the preferred social institution for raising children in a well-ordered and just society. "I shall assume that the basic structure of a well-ordered society includes the family in some form, and therefore that children are at first subject to the legitimate authority of their parents. Of course, in a broader inquiry the institution of the family might be questioned, and other arrangements might indeed prove to be preferable" (1999, p. 405). Left to itself, Rawls argues, the family makes it impossible "…in practice to secure equal chances of achievement and culture for those similarly endowed," which implies that for reasons of justice, "… we may want to adopt a principle which recognizes this fact and also mitigates the arbitrary effects of the natural lottery itself" (1999, p. 64). Insofar as families are the most appropriate social institution to integrate children, as they gradually acquire the proper sense of liberty, equality, and social justice, into a well-order society, so much the better; if not, then either the family ought to be regulated and reorganized or other more preferable arrangements found to raise children to become free and equal members of society.

> Parents *are given* the ethical and legal responsibility to make decisions for children provided that they do so in the best interest of the child…Children are rights owners, even if they are not able to express their rights. Everyone dealing with such rights has the duty to promote them, to give voice to them and to become a true child advocate (Maria De Lourdes Levy, Victor Larcher, Ronald Kurz and the members of the Ethics Working Group of the CESP, 2003, p. 630).

Rather than recognizing the normative nature of the family, parental authority has been reconceptualized to give priority to the child's own self-determination, individual equality, and actual or potential autonomy. Emphasis is placed on the importance of developing children into self-possessed moral agents, who undertake their own moral decision-making as soon as possible and as far as feasible. As a result, Western bioethics frequently turns to children themselves to function as independent autonomous decision-makers (Cherry 2010).

For example, there is a growing commitment to augmenting the participation rights of children in medical decision-making, especially insofar as the minor child demonstrates intellectual understanding, affective grasp of the situation and reasonable maturity.[16] Here pediatric decision-making is sometimes appreciated in terms of a sliding scale, where the child's participation rights are given greater standing as the child demonstrates more individual maturity (see Ziner 1995; Kuther 2003; Dickens and Cook 2005). Children who physicians judge to be "mature minors" are appreciated as having rights to confidentiality, and to healthcare treatment in accordance with their own wishes, rather than in terms of what their parents' choose (Hickey 2007; Downie and Randall 1997; Zawistowski and Frader 2003). The Ethics Working Group of the Confederation of European Specialists in Paediatrics argued, for example, that:

> Younger children may, according to laws of individual states, be able to consent to treatment especially if they have enough maturity and ability to understand the benefits and risks of the proposed treatment and its alternatives. The concept of a "mature minor" has been introduced by some authorities to include groups of children whose age ranges in most EU countries from 14–18 years and who are often regarded as being mature enough to give their own consent to treatment. In some countries the age at which children are considered to be potentially competent is even lower….doctors should always question themselves if the child is mature enough to give consent or assent (Lourdes Levy, Larcher, Kurz, and the members of the Ethics Working Group of the CESP 2003, p. 631).

In England, the "Gillick test" establishes a set of criteria for determining when children under 16 years-of-age have the capacity to consent.

> In clinical practice, this means assessing whether the young person can understand and appraise the nature, purpose and implications of treatment; any risks there might be, any alternative courses of action and their consequences. In making the assessment, it is

[16] For example, Unguru et al. (2010) argue that more should be done to educate children so that they can give meaningful assent to participate in oncology research. "Tools to assist investigators ascertain that children understand what they are agreeing to when they assent to research and to determine their preferences for inclusion in research may help make assent more meaningful" (e876; see also Unguru et al. 2010). Sinclair argues that adolescents should be permitted a significant role in deciding whether to undergo a life saving heart transplant, perhaps even permitting them to refuse a life saving transplant (Sinclair 2009).

> necessary to consider emotional maturity, intellectual capacity and psychological state (Wright et al. 2009, p. 239).

The Canadian Pediatric Society supports a similar stance:

> Once it is determined that children have the capacity to make decisions, which entails full understanding of a situation, and fully grasping the main purpose of interventions, the consequences of consent and the overall extent of what could occur, they should be the primary decision makers (Whitty-Rogers et al. 2009, p. 748; see also Canadian Pediatric Society 2004).

Many bioethicists urge that physicians should seek the "informed permission" of adolescents regarding the direction of their medical treatment (Zawistowski and Frader 2003; Zinner 1995); others conclude that adolescents should be able independently to consent to clinical trials (Ondrusek et al. 1998). Some advocates argue that even end-of-life decisions, such as refusing cancer treatment or a life-sustaining transplant, should be left up to the judgment of "sufficiently mature" minors (Lemmens 2009; Hickey 2007). Parents and physicians should work more-or-less as equals with the mature minor, who is appreciated as possessing the moral authority to make final treatment decisions.[17]

Children are often identified as independent of their parents even prior to having necessarily achieved maturity, especially regarding areas of life secular bioethics deems private. For example, parental consent to access birth control or sex education is not appreciated as necessary to protect the best interests of children (Cook et al. 2007). Controversial procedures, such as abortion, are routinely left up to the will of the child, often without even the requirement to inform her parents. For example, in Seattle, Washington, the health center at Ballard High School reportedly facilitated a 15-year-old girl's abortion during school hours. The young girl was evidently given a school pass, put into a waiting taxi and sent off to have an abortion, free of charge, without notifying her family. T.J. Cosgrove of the King County Health Department was quoted as stating: "At any age in the state of Washington, an individual can consent to a termination of pregnancy" (KOMO Staff 2010).[18]

---

[17] As the Ethics Working Group of the Confederation of European Specialists in Paediatrics argued: "Firstly, seeking a person's consent/assent respects their basic right to self-determination (autonomy). Individuals are best placed to determine what is in their best interests and the only justification for infringing this right is to prevent harm to others. Secondly, obtaining consent/assent involves treating others in a way in which we would expect to be treated ourselves. The universal need to obtain consent/assent also involves treating people justly. Thirdly, obtaining consent/assent protects patients from the physical and psychological harms which may occur as a result of illness or its treatment. Fourthly, obtaining consent/assent confers benefit by encouraging active participation of individuals in investigation and treatments which are intended to restore their health" (Levy et al. 2003, p. 630).

[18] Similarly, current California law requires neither parental consent nor simple parental notification for a child to obtain an abortion. See Planned Parenthood Parental Consent and Notification Laws. Available: www.plannedparenthood.org/teen-talk/teen-pregnancy/parental-consent-notification-laws-25268.htm. See also California Health and Safety Code, Sect. 123420–123450. "Do I have to get my parent's permission to get an abortion? No. You do not need anyone's permission, and the law protects your privacy. No one else has the right to know or do anything about it—not your parents, your boyfriend or partner, or your husband. Even if you are married or under 18, the

"Sufficiently mature" minors, it is urged, should be permitted to decide on their own behalf regarding contraception, abortion and sexual practices with consenting others, and treatment for sexually transmitted diseases. Age thresholds have usually functioned as an established, if somewhat conventional, criterion for the capacity to make competent, mature, and responsible decisions. However, the focus has shifted away from any minimal age threshold to a particular medical professional's personal judgment regarding the child's "sufficient maturity."[19] In short, rather than being appreciated as within the authority of their parents, or other family members, minor children are anticipated as moving as quickly as feasible from giving assent to medical treatment, to giving independent consent and as having moral and legal standing independent of their parents.[20] Children are to exercise their own personal autonomy, to define their own conceptions of proper moral choice and life-style preferences, over against the authority of their parents and other family members, as soon as practicable.

As a result, familial authority is routinely assumed to reach no further than that of trustee of the individual member's best interests and, moreover, as always properly subject to significant state oversight and governmental intervention. As a result, core family relationships are evermore undermined and subject to legal restrictions designed to deflate the significance of, as well as to marginalize the family. Whereas spouses have usually been appreciated as possessing particular authority vis-a-vis each other, and parents have usually been regarded as the best judges for balancing costs and benefits for the family, for determining appropriate life-style choices for themselves, their children, and the family as a whole,[21] these positions have been radically brought into question. As demonstrated, pressure has been brought to bear on parents and families through law and institutional policy, focusing among other concerns on separating children from the sphere of parental and familial authority, as well as spouses from each other's authority. The examples of individually-oriented informed consent, personal privacy rights, and pediatric decision making, illustrate the ways in which the focus of much contemporary moral and political

---

decision is up to you." ACLU, "Your Health; Your Rights" [On-line.] Available: www.teensource.org.

[19] Dickens and Cook argue that: "There is usually no chronological 'age of consent' for medical care, but a condition of consent, meaning capacity for understanding" (Dickens and Cook 2005, p. 179).

[20] "Debates surrounding the rights of adolescents to receive confidential and private reproductive health services have centered around the potentiality conflicting interests of parents and their children. The desire of parents to guide and direct their children's health and development and make health-care decisions for their children is easily understandable. However, the health threat faced by adolescents exposes the tension between public or societal interests in maintaining a healthy population and private or parental interests in maintaining control over their children" (Ringheim 2007, p. 245).

[21] For example, in his detailed Roman Catholic casuistry of parental decision making on behalf of their children, Edwin Healy (1956), argues that parents should and must make decisions regarding ordinary vs. extraordinary care on behalf of their children, including even determining when treatment is too expensive to be obligatory (Healy 1956, pp. 81–89, see especially page 82).

analysis has been to sunder the authority of the family over its members, and to undo the primary loyalty of family members to each other.[22]

3.4 Conclusion

Western accounts of bioethics routinely fail to appreciate the central role and reality of the family in human flourishing. Regardless of its advantages, defenders of post-traditional social structures seek to set aside traditional understandings of the social and moral obligations of family life, to dismiss or ignore the significant benefits of the traditional family, so as to liberate the family from the confines of traditional cultural or religious norms. Such disputes are often embedded in highly ideological frameworks, which give moral priority to secularity and individualism, to personal liberty as a positive entitlement to realize one's own understanding of the good. Such idealizations, however, ignore the actual choices of actual persons, trivializing the real connections between parents and children, among spouses and other family members, so as politically to achieve what is judged to be an ideally liberal, egalitarian, and progressive society. The liberal social-constructivist account of the family simply replaces traditional cultural or religious content with its own full-fledged political ideology. Consequently, such an anti-traditional ethos also contributes to the undermining of the family's nexus of key social relationships and, thereby, to the breakdown of the family.

A core challenge, however, is that this particular liberal ethos routinely conflicts with what would appear to be the long-term good of children as well as of the adults whom children and adolescents are destined to become. There is a considerable body of data, for example, demonstrating the positive impact of the more authorita-

[22] Consider, for example, Susan Moller Okin who argues, "The liberal state … should not only not give special rights or exemptions to cultural and religious groups that discriminate against or oppress women. It should also enforce individual rights against such groups when the opportunity arises and encourage all groups within its borders to cease such practices" (Okin 2002, pp. 229–230). Moreover, education policies, it is argued, should carefully restrict religiously-based education that might encourage children towards traditional sex roles. "If parents are permitted to educate their children in sheltered settings in which they are taught, by example, doctrine, and the content of their curriculum, that it is the will of an omnipotent and punitive God that women's proper role in life is to be an obedient wife and a full-time mother, how can the girls be said to be 'aware of … alternatives' in any meaningful way, to be able to 'assess these alternatives' (or even to think it desirable to do so), or to be able to 'participate effectively' in other roles or ways of life?" (Okin 2002, p. 226) Children, it is asserted, should be educated in the public virtues of justice, equality and tolerance.

Amy Gutmann opines similarly: "Some kinds of social diversity … are anathema to political liberalism. Civic education should educate all children to appreciate the public value of toleration" (1995, p. 559). Gutmann continues: "The basic principles of liberalism, those necessary to protect every person's basic liberties and opportunities, place substantial limits on social diversity. … The limits on racial and gender discrimination, for example, enable many people to pursue ways of life that would otherwise be closed to them by discriminatory practices at the same time as they undermine or at least impede some traditional ways of life" (Gutmann 1995, p. 559).

tive parenting styles and boundary settings typical of traditional family structures on the development of effective, autonomous, decision making. The data support the conclusion that adolescents who grow up with parents who are authoritative, setting limits on the adolescent's behavior and choices, are more likely to become effective adult decision makers. Adolescents raised with permissive parents, in contrast, in which the child himself is treated as the authoritative decision maker, develop significantly poorer effective decision-making skills. Traditional authoritative parenting styles in general support, rather than undermine, the ability of the child to mature into a competent adult decision maker. Authoritative parenting is related to a wide range of positive cognitive and emotional outcomes, including better academic achievement (Dornbush et al. 1987; Weiss and Schwarz 1996; Wintre and Ben-Knaz 2000; Wintre and Yaffe 2000), less psychological distress, fewer adjustment and problem behaviors (Brown et al. 1993; Fuligni and Eccles 1993; Slicker 1998), and higher quality relationships with their peers, as well as greater levels of competence, self-esteem, personal reliance, and even individual autonomy (Baumrind 1991a, b; Buri et al. 1998). Authoritative parenting styles help provide adolescents with the ability to resist peer pressure, to avoid substance abuse and other potentially harmful circumstances (Weiss and Schwarz 1996; Adalbjarnardottir and Hafsteinsson 2001; Huver et al. 2007). As even the United States Supreme Court has come to recognize, the available scientific evidence supports the conclusion that adolescents are qualitatively different types of agents than adults (Cherry 2013).[23] Parents, by setting limits and giving direction, letting adolescents deliberate and choose within limited circumstances, while also withholding for the parents the right to veto adolescent decisions, protect children from the long-term consequences of poor choices. Moreover, as documented, the failure to sustain cardinal concerns focused on the family has lead to increased pregnancy outside of marriage and an ever greater percentage of children raised outside of traditional family life, which is significantly statistically correlated with a wide range of social, psychological and economic disadvantages. Such traditional family structures augment positive characteristics (cognitive, emotional, and adaptive) associated with becoming an effective adult.

In short, traditional forms of the family free people to live in ways that many (both men and women) judge to be central to human flourishing.[24] The family usually functions as the central locus of moral guidance for its members, but especially for the family's children. The empirical data surveyed supports the conclusion that

---

[23] The Supreme Court has reasoned that, on balance, the available scientific evidence, including neuroimaging studies of the relative immaturity of the adolescent brain, does not support the conclusion that adolescents possess adult capacities for personal agency and rational mature choice. See *Roper v. Simmons* p. 541 US 551 (2005); *Graham v. Florida* p. 560 US (2010); and *Miller v. Alabama* p. 567 US (2012).

[24] "'Less marriage means less income and more poverty" (Sawhill 2011, p. 42), reckons Isabel Sawhill, a senior fellow at the Brookings Institution. She and other researchers have linked as much as half of the income inequality in America to changes in family composition: single-parent families (mostly those with a high-school degree or less) are getting poorer while married couples (with educations and dual incomes) are increasingly well-off.

traditional understandings of parental authority benefit children, augment the quality of childhood decision making, helping them to avoid unnecessary risks and to develop into mature autonomous decision-makers. More generally, the rich social structures of traditional family life convey social stability and social integration and support, reducing morbidity and mortality, and offering protection from stressful life events. Given the centrality of the family for sustaining such key areas of human flourishing there are good grounds for sustaining (or re-establishing) more family-oriented understandings of informed consent to medical treatment and for rethinking rules regarding patient confidentiality that restrict access to information from spouses and other family members.

References

Adalbjarnardottir, S., and L. G. Hafsteinsson. 2001. Adolescents' perceived parenting styles and their substance use: Concurrent longitudinal analyses. *Journal of Research on Adolescence* 11:401–423.

Akerlof, G. A., and J. L. Yellin. 1996. New mothers, not married: Technology shock, the demise of shotgun marriage, and the increase in out-of-wedlock births. *The Brookings Review* 14 (4): 18–21.

Allen, J. 2010. U.S. herpes rates remain high—CDC. Reuters. www.reuters.com/article/idUSN0923528620100309 http://www.reuters.com/article/idUSN0923528620100309. Accessed 23 Nov 2013.

American Civil Liberties Union (ACLU). 2013. Your health; Your rights. www.teensource.org http://www.teensource.org/. Accessed 26 Nov 2013.

Applebaum, P., C. Lidz, and A. Meisel. 1987. *Informed consent: Legal theory and clinical practice*. New York: Oxford University Press.

Aronson, S. R., and A. C. Huston. 2004. The mother-infant relationship in single, cohabiting, and married families: A case for marriage? *Journal of Family Psychology* 18 (1): 5–18.

Baumrind, D. 1991a. Parenting styles and adolescent development. In *The encyclopedia of adolescence,* ed. J. Brooks-Gunn, R. Lerner, and A. C. Petersen, 746–58. New York: Garland.

Baumrind, D. 1991b. The influence of parenting style on adolescent competence and substance use. *Journal of Early Adolescence* 11:56–95.

Beauchamp, T. L., and J. F. Childress. 2009. *Principles of biomedical ethics*. 6th ed. New York: Oxford University Press.

Beauchamp, T. L., and J. F. Childress. 2012. *Principles of biomedical ethics*. 7th ed. New York: Oxford University Press.

Bishop, J. 2014. Dependency, decisions and a family of care. In *Family-oriented informed consent: East Asian and American perspectives,* ed. R. Fan, 27–42. Dordrecht: Springer.

Breault, K. D. 1986. Suicide in America: A test of Durkheim's theory of religious and family integration, 1933–1980. *American Journal of Sociology* 92:628–656.

Breines, W., M. Cerullo, and J. Stacey. 1978. Social biology, family studies, and antifeminist backlash. *Feminist Studies* 4 (1): 43–67.

Brown, B. B., N. Mounts, S. D. Lamborn, and L. Steinberg. 1993. Parenting practices and peer group affiliation in adolescence. *Child Development* 64:467–482.

Brown, S. L. 2004. Family structure and child well-being: the significance of parental cohabitation. *Journal of Marriage and the Family* 66:351–367.

Buri, J. R., P. A. Louiselle, T. M. Misukanis, and R. A. Mueller. 1998. Effects of parental authoritarianism and authoritativeness on self-esteem. *Personality and Social Psychology Bulletin* 14:271–82.

Canadian Pediatric Society. 2004. Treatment decisions regarding infants, children and adolescents. *Paediatric Child Health* 9(2): 99–103.

Cherry, M. J. 2010. Parental authority and pediatric decision making. *The Journal of Medicine and Philosophy* 35 (5): 553–572.

Cherry, M. J. 2013. Ignoring the data and endangering children: Why the mature minor standard for medical decision making must be abandoned. *The Journal of Medicine and Philosophy* 38 (3): 315–331.

Choi, K. 2014. The ideal of autonomy and its misimplementation. In *Family-oriented informed consent: East Asian and American perspectives,* ed. R. Fan, 83–92. Dordrecht: Springer.

Cook, R. J., J. N. Erdman, and B. M. Dickens. 2007. Respecting adolescents' confidentiality and reproductive and sexual choices. *International Journal of Gynecology and Obstetrics* 98 (2): 182–187.

Cookston, J. T. 1999. Parental supervision and family structure. *Journal of Divorce and Remarriage* 31 (1/2): 107–127.

Daly, M., and M. Wilson. 1999. *The truth about Cinderella: A Darwinian view of parental love*. New Haven: Yale University Press.

Defoe, B. 2003. *Why there are no good men left*. New York: Broadway Books.

Deng, R. 2014. The informed consent of human medical research in mainland China: A family oriented multi-decision model. In *Family-oriented informed consent: East Asian and American perspectives,* ed. R. Fan, 83–92. Dordrecht: Springer.

Dickens, B. M., and R. J. Cook. 2005. Adolescents and consent to treatment. *International Journal of Gynaecology and Obstetrics* 89 (2): 179–184.

Dornbush, S. M., P. L. Ritter, P. H. Leiderman, D. F. Roberts, and M. J. Fraleigh. 1987. The relation of parenting style to adolescent school performance. *Child Development* 58:1244–1257.

Downie, R. S., and F. Randall. 1997. Parenting and the best interests of minors. *The Journal of Medicine and Philosophy* 22 (3): 219–231.

Engelhardt, Jr., H. T. 2002. Morality, universality, and particularity: Rethinking the role of community in the foundations of bioethics. In *Cross-cultural perspectives on the (Im) possibility of global bioethics,* ed. J. T. L. Po-wah, 19–38. Dordrecht: Kluwer Academic Publishers.

Engelhardt, Jr., H. T. 2010. Beyond the best interests of children: Four views of the family and of foundational disagreements regarding pediatric decision-making. *The Journal of Medicine and Philosophy* 35 (5): 499–517.

Fan, R. 2010. *Reconstructionist Confucianism: Rethinking morality after the west.* Dordrecht: Springer.

Fan, R., X. Chen, and Y. Cao. 2012. Family-oriented health savings accounts: Facing the challenges of health care allocation. *The Journal of Medicine and Philosophy* 37 (6): 507–512.

Faden, R., and T. L. Beauchamp. 1986. *A history and theory of informed consent*. New York: Oxford University Press.

Fagan, P. F., and R. Rector. 2000. The effects of divorce on America (2000). Heritage Foundation Backgrounder. www.heritage.org/Research/Family/BG1373.cfm http://www.heritage.org/Research/Family/BG1373.cfm. Accessed 23 Nov 2013.

Flanagan, C., D. S. Shaw, and E. B. Winslow. 1999. A prospective study of the effects of marital status and family relations on American families. *Child Development* 70:742–755.

Fomby, P., and A. J. Cherlin. 2007. Family instability and child well-being. *American Sociological Review* 72 (April): 181–204.

Fuligni, A. J., and J. S. Eccles. 1993. Perceived parent-child relationships and early adolescents' orientation toward peers. *Developmental Psychology* 29:622–632.

Gallagher, M., and L. Waite. 2000. *The case for marriage*. New York: Random House.

Gutmann, A. 1995. Civic education and social diversity. *Ethics* 105:557–579.

Hamilton, B. E., J. A. Martin, and S. J. Ventura. Centers for Disease Control and Prevention, Division of Vital Statistics. 2012. Births: Preliminary data for 2011. *National Vital Statistics Reports* 61(5).

Healy, E. 1956. *Medical ethics*. Chicago: Loyola University Press.

Hickey, K. 2007. Minors' rights in medical decision making. *Journal of Nursing Administration: Healthcare Law, Ethics, and Regulation* 9 (3): 100–104.

Holt-Lunstad, J., T. B. Smith, and J. B. Layton. 2010. Social relationships and mortality risk: A meta-analytic review. *PLOS Medicine* 7 (7): 1–20. doi:10.1371/journal.pmed.1000316.e1000316

House, J. S., K. R. Landis, and D. Umberson. 1988. Social relationships and health. *Science* 241:540–545.

Huver, R. M. E., R. C. M. E. Engels, G. van Breukelen, and H. de Vries. 2007. Parenting style and adolescent smoking cognitions and behavior. *Psychology and Health* 22:575–93.

Iltis, A. 2014. Families and medical decisions to assume risks for the benefit of others. In *Family-oriented informed consent: East Asian and American perspectives,* ed. R. Fan, 171–186. Dordrecht: Springer.

Kposowa, A. J. 2009. Psychiatrist availability, social disintegration, and suicide deaths in U.S. counties, 1990–1995. *Journal of Community Psychology* 37 (1): 73–87.

Staff, KOMO. 2010. Mother furious after in-school clinic sets up teen's abortion. KOMO News. www.komonews.com/news/local/88971742.html. Accessed 23 Nov 2013.

Kunce, M., and A. L. Anderson. 2002. The impact of socioeconomic factors on state suicide rates: A methodological note. *Urban Studies* 39:155–162.

Kuther, T. L. 2003. Medical decision-making and minors: Issues of consent and assent. *Adolescence* 38 (150): 343–358.

Lee, I. 2014a. Filial duty in Korea and its implication for end-of-life care decision-making. In *Family-oriented informed consent: East Asian and American perspectives,* ed. R. Fan, 137–148. Dordrecht: Springer.

Lee, S. C. 2014b. Family consent in medical decision-making in Taiwan: The implications of the new revision of the law of hospice and related issues. In *Family-oriented informed consent: East Asian and American perspectives,* ed. R. Fan, 125–136. Dordrecht: Springer.

Lemmens, C. 2009. End-of-life decisions and minors: Do minors have the right to refuse life preserving medical treatment? A comparative study. *Medicine and Law* 28 (3): 479–497.

Lourdes, Levy, M. de, V. Larcher, R. Kurz, and The members of the Ethics Working Group of the CESP. 2003. Informed consent/assent in children. Statement of the ethics working group of the Confederation of European Specialists in Paediatrics (CESP). *European Journal of Pediatrics* 162: 629–633

Munoz-Dardé, V. 1998. Rawls, justice in the family and justice of the family. *The Philosophical Quarterly* 48 (192): 335–352.

Murray, C. 2012. *Coming apart: The state of White America, 1960–2010.* New York: Crown Forum.

McPherson, M., and L. Smith-Lovin. 2006. Social isolation in America: Changes in core discussion networks over two decades. *American Sociological Review* 71: 353–375.

Neumayer, E. 2003. Socioeconomic factors and suicide rates at large-unit aggregate levels: A comment. *Urban Studies* 40:2769–2776.

Norval, G., S. Nock, and L. J. Waite. 2002. Why marriage matters: Twenty-one conclusions from the social sciences. *American Experiment Quarterly* 5 (1): 34–44.

Okin, S. M. 1982. Women and the making of the sentimental family. *Philosophy and Public Affairs* 11 (1): 65–88.

Okin, S. M. 1994. Gender inequality and cultural differences. *Political theory* 22 (1): 5–24.

Okin, S. M. 2002. Mistresses of their own destiny: Group rights, gender, and realistic rights of exit. *Ethics* 112:205–230.

Ondrusek, N., R. Abramovitch, P. Pencharz, and G. Koren. 1998. Empirical examination of the ability of children to consent to clinical research. *Journal of Medical Ethics* 24 (3): 158–165.

Osgood, D. W., and J. M. Chambers. 2000. Social disorganization outside the metropolis: An analysis of rural youth violence. *Criminology* 38 (1): 81–115.

Osgood, D. W., E. M. Foster, and M. E. Courtney. 2010. Vulnerable populations and the transition to adulthood. *The Future of Children* 20 (1): 209–229.

Parcel, T. L., and M. J. Dufur. 2001. Capital at home and at school: Effects on student achievement. *Social Forces* 79 (3): 881.

Parke, M. 2003. *Are married parents really better for children?* Washington D.C.: Center for Law and Social Policy.

President's Commission for the Study of Ethical Problems in Medicine and Biomedical and Behavioral Research. 1982. *Making health care decisions*. Washington, DC: U.S. Government Printing Office.

Putnam, R. D. 2000. *Bowling alone: The collapse and revival of American community*. New York: Simon and Schuster.

Rountree, P. W., and B. D. Warner. 1999. Social ties and crime: Is the relationship gendered? *Criminology* 37 (4): 789–813.

Rawls, J. 1999. *A theory of justice*. Cambridge: Harvard University Press.

Rhoads, S. 2004. *Taking sex differences seriously*. San Francisco: Encounter Books.

Ringheim, K. 2007. Ethical and human rights perspectives on providers' obligation to ensure adolescents' rights to privacy. *Studies in Family Planning* 38 (4): 245–252.

Sawhill, I. 2011. The decline of marriage: For richer, for smarter. *The Economist*. http://www.economist.com/node/18867552. Accessed 23 Nov 2013.

Schneider, C. E. 2006. HIPAA-crazy. *Hastings Center Report*. January–February: 10–11.

Settersten, Jr., R. A., and B. Ray. 2010. What's going on with young people today? The long and twisting path to adulthood. *The Future of Children* 20(1): 19–41.

Shepherd, R. 2012. New untreatable gonorrhea could cause epidemic of sexually transmitted infections. Medical News Today. www.medicalnewstoday.com/articles/246268.php http://www.medicalnewstoday.com/articles/246268.php. Accessed 25 Nov 2013.

Sinclair, S. J. 2009. Involvement of adolescents in decision making for heart transplants. *MCN The American Journal of Maternal Child Nursing* 34 (5): 276–281.

Slicker, E. K. 1998. Relationship of parental style to behavioral adjustment in graduating high school seniors. *Journal of Youth and Adolescence* 27:345–372.

Unguru, Y., N. J. Coppes, and N. Kamani. 2008. Rethinking pediatric assent: From requirement to ideal. *Pediatric Clinics of North America* 55 (1): 211–222.

Unguru, Y., A. M. Sill, and N. Kamani. 2010. The experiences of children enrolled in pediatric oncology research: Implications for assent. *Pediatrics* 125 (4): e876–e883.

Wear, S. 1993. *Informed consent: Patient autonomy and physician beneficence within clinical medicine*. Dordrecht: Kluwer Academic Publishers.

Weiss, L. H., and J. C. Schwarz. 1996. The relationship between parenting types and older adolescents' personality, academic achievement, adjustment, and substance use. *Child Development* 67:2101–2114.

Weisfeld, G. E., D. M. Muczenski, C. C. Weisfeld, and D. R. Omark. 1987. Stability of boys' social success among peers over an eleven-year period. In *Interpersonal relations: Family, peers, friends*, ed. J. A. Meacham, 58–80. New York: Karger.

Weitoft, G. R., A. Hjern, B. Haglund, and M. Rosen. 2003. Mortality, severe mortality, and injury in children living with single parents in Sweden: A population based study. *Lancet* 361:289–295.

Whitty-Rogers, J., M. Alex, C. MacDonald, D. P. Gallant, and W. Austin. 2009. Working with children in end-of-life decision making. *Nursing Ethics* 16 (6): 743–758.

Wilcox, W. B., and Marquardt, E., eds. 2009. *The state of our unions 2009: Marriage in America*. Charlottesville: The National Marriage Project.

Wilcox, W. B., and Marquardt, E., eds. 2011. *The state of our unions 2011: Marriage in America*. Charlottesville: The National Marriage Project.

Wilcox, W. B., et al. 2011. *Why marriage matters: Thirty conclusions from the social sciences*. New York: Institute for American Values.

Wintre, M. G., and M. Yaffe. 2000. First-year students' adjustment to university life as a function of relationships with parents. *Journal of Adolescent Research* 15:9–37.

Wintre, M. G., and R. Ben-Knaz. 2000. It's not academic, you're in the army now: Adjustment to the army as a comparative context for adjustment to the university. *Journal of Adolescent Research* 15:145–172.

Wong, H. C. 2014. Towards a good practice of family-oriented consent: Reflections on medical practice in Taiwan. In *Family-oriented informed consent: East Asian and American perspectives*, ed. R. Fan, 245–256. Dordrecht: Springer.

Wright, B., J. Aldridge, K. Wurr, T. Sloper, H. Tomlinson, and M. Miller. 2009. Clinical dilemmas in children with life-limiting illnesses: Decision making and the law. *Palliative Medicine* 23:238–247.

Yang, Y. 2014. A Confucian family-oriented approach to advance directives in end-of-life decision making for incompetent elderly patients. In *Family-oriented informed consent: East Asian and American perspectives,* ed. R. Fan, 257–270. Dordrecht: Springer.

Yu. K. 2014. The individual-oriented model of informed consent and the Confucian alternative: Family and beyond. In *Family-oriented informed consent: East Asian and American perspectives,* ed. R. Fan, 93–106. Dordrecht: Springer.

Zawistowski, C. A., and J. E. Frader. 2003. Ethical problems in pediatric critical care: Consent. *Critical Care Medicine* 31 (5 supplement): s407–s410.

Zinner, S. E. 1995. The elusive goal of informed consent by adolescents. *Theoretical Medicine* 16 (4): 3233–3331.

Zhao, W. 2014. A Confucian worldview and family-based informed consent: A case of concealing illness from the patient in China. In *Family-oriented informed consent: East Asian and American perspectives,* ed. R. Fan, 231–244. Dordrecht: Springer.

Chapter 4
Family and Autonomy: Towards Shared Medical Decision-Making in Light of Confucianism

Jue Wang

4.1 The Patient as Agent: A Desirable Ideal?

One remarkable feature of contemporary bioethics is the triumph of individual-directed autonomy, which has become the characteristic point of departure for considering the rights of patients, the duty of physicians, and the shaping of clinical practice for terminal care. At the same time, this triumph is also a highly controversial one: it is criticized as only temporary (Veatch 1984, p. 38), of very limited implications for medical practice, or even an obsession (Callahan 1984, p. 496).

The doubts concentrate on the individualist orientation of the concept of autonomy: for critics, it is too narrowly "atomic" to pay due attention to the social nature of personal identity as well as the impact of those social relationships on individual autonomy. Beauchamp and Childress (2009, p. 102) in the most recent edition of their *Principles* respond to these criticisms by stressing that "no fundamental inconsistency exists between autonomy and authority if individuals exercise their autonomy in choosing to accept an institution, tradition, or community that they view as a legitimate source of direction." For example, if a patient entrusts her family to handle such matters for her, she could yield decision-making power as long as she gives explicit autonomous delegation. However, this qualifying statement makes the individualistic foundation of their account of autonomy—which also represents the dominant model of autonomy in the West—even more conspicuous: individual choice has been celebrated as the ultimate source of authority. A person' s autonomy is tantamount to his or her independence, primarily expressed in his or her self-contained ability for deciding, analogous to the self-rule of an independent state (Beauchamp and Childress 2009, p. 103). Specifically, to worship the ultimate value of self-determination, two conditions are viewed as necessary:

Supported by the National Social Science Fund of China (Grant No. 14CZX045) and the Social Science Fund of Huazhong University of Science and Technology (Grant No. 2010013).

J. Wang (✉)
Department of Philosophy, Huazhong University of Science and Technology, Wuhan, China
e-mail: wangjue0330@126.com

R. Fan (ed.), *Family-Oriented Informed Consent*, Philosophy and Medicine,
DOI 10.1007/978-3-319-12120-8_4

liberty (usually addressed in the language of rights) and *agency* (usually addressed in the language of capability) (Beauchamp and Childress 2009, p. 100).

However, their strategy to respond to the critical objections is only partially successful. By focusing on the conceptual structural compatibility between individual autonomy and the authority of the communities or social relationships the patient is embedded in, their reply bypasses a deeper concern addressed by the criticism from a relational perspective, that is, whether it is a feasible ideal for a patient to play the role of a capable agent to the full extent, insofar as to be cast in the position of the sick person is to be in an essentially vulnerable and dependent situation. After all, "it is incapacity that qualifies a person as a patient" (MacIntyre 1977, p. 205).[1] From the patient's perspective, illness is not simply a collection of physical symptoms, but is primarily experienced as a distinct way of being in the world, a way of being that is characterized as a loss of wholeness and bodily integrity, a loss of certainty, a loss of control, a loss of freedom to act, and a loss of the familiar world (See Toombs 1992, p. 90). Moreover, it is the loss of connection with the familiar world—with the way of being in the world one used to be in—that most threatens the autonomous selfhood.[2] In this respect, connections to family play a vital role in maintaining the integrity of the patient's autonomy, because these connections provide not only caring support but also a network of shared memories and rituals which help sustain a sense of a unitary self through the crisis of alienation.[3] The first part of my paper is devoted to an analysis of a real case of an end-of-life decision, in order to show that under the dominant model of individual autonomy the patient is left with a broken, troubled voice, rather than an active agent, in pursuing his/her own life plan.

Following this line of thought, my paper aims to give a systematic reflection on the inadequateness and inappropriateness of the principle of respect for autonomy in bioethics, by referring to the Confucian tradition for those necessary yet missing concepts, i.e., goods, roles, and relationships which constitute medicine as an ordered form of human practice. In the first place, these Confucian considerations of health care help crystallize the moral uneasiness about individual autonomy encountered in a real case study that I will present here. Further, I will argue that compared to the dominant western model giving priority to individual's autonomous choice, the Confucian model of medical practice is more persuasive in that it takes seriously the basic human condition of embodied vulnerability in general, and the patient's incapability in particular, by putting the integrity of the family at the center of a caring relationship, including health care. The final goal of this paper is to show the need to reconfigure the concept of autonomy and the related medical

[1] One might question this claim by pointing to many patients who are capable of making their own decisions even though they are 'sick.' However, such objections miss the point. By using the term 'incapacity,' which I borrow from MacIntyre, I focus on patients' existential situation which distinguishes the physician-patient relationship from other contractual relationships. Therefore, the question at issue is not whether patients are capable of making autonomous decisions, but whether the individualist understanding of autonomy in medical practice is adequate for promoting the proper good of patients.

[2] As S. Kay Toombs (1992, p. 95) observes, it is extremely hard to be "clear-headed" and view clinical choices with equanimity when they relate directly to one's own uncertain future.

[3] For details see Donchin (2000, p. 248).

decision-making model so as to be more sensitive to relation of care, interdependence and mutual support that serve as the necessary conditions for the patient to be an autonomous agent, if possible.

4.2 A Real Case: The Broken Voice of the Patient

Recently, a controversial case about an end-of-life decision went to court in New York, and brought the fierce yet usually repressed conflict between family integrity and the individualist framework of the patient's autonomy to the public's attention. Grace Sung Eun Lee, a 28-year-old financial manager who is on life support with terminal brain cancer, comes from a Korean Christian family that immigrated to American years ago. She has been paralyzed from the neck down since Sep., 2012, unable to eat or breathe on her own. Although she could not speak, she is able to shake her head, move her lips and blink "yes" or "no" to communicate with others. Doctors at North Shore University Hospital in Manhasset, N.Y., testified that Ms. Lee told them that she wanted to die, and that they were prepared to grant her wish after having a psychiatrist assessment of Lee to show that she is mentally competent to make medical choices. Lee's family had been excluded from the process of decision-making until the night of Sept. 23, when Lee's mother overheard a nurse talking with her daughter, making plans to remove the breathing tube the next morning. Lee's family went to court to petition to become her legal guardian, which could give them more power over her health care decisions. By then, the heartbreaking family tragedy had been turned into a right-to-die legal battle,[4] or as the New York Times put it in the heading "Daughter's Right to Die Is Weighed Against Family's Wish to Keep Her Alive."[5] On October 6, a New York Court of Appeals upheld Ms. Lee's right to make a decision of removing her life-supporting machines. After taking the victory, however, she dramatically decided the next day to live out the rest of her life and designated her father as her healthcare guardian.

In terms of rights, Lee's story is very simple except for the unexpected end. However, the language of rights—which dominates the individualist concept of the patient's autonomy—overlooks the most intriguing and inspiring part: what is the patient's true intention in the whole process? When this question is addressed, we find ourselves faced with a Rashomon[6] tale of the patient's wish. Not only did the

[4] Pearson, E. Paralyzed bank manager terminally ill with brain cancer fights for her right to die. *New York Daily News*, October 4, 2012. http://articles.nydailynews.com/2012-10-04/news/34243320_1_tumor-tube-doctors.

[5] Hartocollis, A. Daughter's right to die is weighed against family's wish to keep her alive. *The New York Times*, October 4, 2012. http://www.nytimes.com/2012/10/05/nyregion/in-sungeun-grace-lee-case-right-to-die-is-weighed-against-a-familys-wishes.html.

[6] In "Rashomon," a famous Japanese film, the same violent event is convincingly yet divergently related by the three protagonists: a bandit, a woman who has been violated by the bandit, and her husband who is found to lie dead in a grove. Each one has his/her own narrative of the tragedy. But the movie never tells us which version to believe—or whether to believe any of them.

family and the hospital hold irreconcilable accounts of the patient's wishes, but even the patient's own voice sounded troubled, broken, and ambiguous, far from being an indisputable standard for judgment. Contrary to the liberal individualist myth that one's moods and intentions are truly transparent to oneself and largely opaque to others, there is a deep ambivalence inherent in all of Lee's expressions of her wishes for the future, which to some extent foreshadows Lee's later change of mind.

As an example, as her physician testified, "She is very tearful when she thinks about dying, but she consistently asks that the breathing tube be removed and she begs us to do that."[7] The hospital took such expressions as a firm, clear claim of the right to death, and was ready to comply with it without even notifying Lee's family. However, if one listens to these words with care, it is readable if not evident that the message Lee essentially delivered was no more than "I can not bear to go on living in the way I am currently forced to live."

In those words, usually accompanied with tears and intense emotions, she was not so much motivated by an autonomous decision about her future, as by the despair about the present disability and disease she is caught in. One crucial impact of severe illness on the patient's life is to disrupt the usual connections of the patient with the world: the patient feels herself aliened from the familiar world she used to live in and confronted with a future which is suddenly rendered impotent and inaccessible. From the patient's perspective, the meaning of illness is not only grasped in pathological terms, but first and foremost in a particular biographical situation. Thus the most significant challenge facing the patient comes from a deeper, personal level, which involves a task of merging the lost past, the alienated present, and the altered future in a narrative identity of her whole life. Not surprisingly, it is a dynamic ongoing process for the patient to create new personal meaning out of the experience of disease and disability, and incorporate these experiences into a coherent narrative of her life.

For our discussion, this narrative perspective is important in that it sheds light on the continuity between those apparently self-contradicting expressions Lee made of her wishes. On the one hand, Lee told the physicians, the newspaper and even her mother she wanted to die; on the other hand, Lee made an explicit expression to her cousin in a video, which was made and updated to YouTube by her family before the court's final judgment was announced, that she would like her father to be her medical proxy and wanted to go out of the hospital immediately. Although they might give people an impression of incoherence and contradiction, these expressions actually conveyed the same message: that the patient struggled to get out of the present impossible situation in any way and at any cost. In other words, these contradictory expressions can only make sense in the context of the whole process of adapting to impairment and striving to restore her self. In this way, the fact that Lee changed her mind in the end is not an haphazard, but rather a final decision of a process towards preserving the integrity of the self; at that point the patient no lon-

[7] Pearson, E. Paralyzed bank manager terminally ill with brain cancer fights for her right to die. *New York Daily News*, October 4, 2012. http://articles.nydailynews.com/2012-10-04/news/34243320_1_tumor-tube-doctors.

ger remained fixed on the present but was able to look to the past and to anticipate a future: as Lee explained the reasons for her changing of mind, "she was doing it to make peace with her parents and to make peace with God."[8] In other words, she is trying to look to those personal commitments which define who she was in past and continue to orient herself towards the future.

Most comments on this case tend to explain the conflict between the family and the hospital as a collision of religious belief with secular medical ethics, or a culture clash between the West and the East. However, if we take the whole process of adapting to impairment and striving to restore self into account, we can see that the conflict between the family and the hospital first and foremost is the consequence of the contradictory approaches they respectively take to assess the patient's wish.

The hospital took an impersonal, distant attitude towards to Lee's wishes and tended to assess her decisions in a formal, legal fashion. As Terry Lynam, a spokesman for the hospital, said, "We just want to comply with her wishes, *whatever they might be*."[9] Or as Beauchamp and Childress write, "The fundamental requirement is to respect a particular person's autonomous choices, *whatever they may be*" (2009, p. 107).

The family, by contrast has taken a future-orientation attitude, of a partially teleological character, towards the patient's wishes. For the family, a meaningful medical decision about one family member should be assessed according to a coherent narrative of her life. This narrative brings a unity to her life, by orienting all her decisions and actions towards some core commitments which determine who she is as a person. Accordingly, her personal identity in large part depends on how well she lives out that unity and brings it to completion.

It is the concern with the integrity of Lee's self-identity that brought Lee's family to the court. The aim of pursuing the court option was not to pit the family's claim against the patient's right—since there was no chance of success for the family to win the legal war—but to strive to gain more time for one of the family members in order to help her to continue her own narrative intelligibly in conformity with the common narrative of the family, in which the individual's narrative is largely embedded. As MacIntyre (2007, p. 221) powerfully argues, the self has to find its identity in and through its membership in communities such as the family, through the sense that my story is part of the story of other family members, just as their story is part of mine. Conversely speaking, these intertwined narratives shared by members of a family over time provide an inside perspective from which the family can judge medical decisions concerning one family member in a teleological framework which is partly based on their common past. To some extent, the family is in a position to judge whether a certain medical decision is consistent with the patient's

[8] Cf. Hartocollis, A. Changing her mind, a Queens woman decides to remain on life support. *New York Times*, October 7, 2012. http://www.nytimes.com/2012/10/07/nyregion/sungeun-grace-lee-changes-her-mind-and-decides-to-stay-on-life-support.html?_r=0.

[9] Hartocollis, A. Changing her mind, a Queens woman decides to remain on life support. *New York Times*, October 7, 2012. http://www.nytimes.com/2012/10/07/nyregion/sungeun-grace-lee-changes-her-mind-and-decides-to-stay-on-life-support.html?_r=0.

life plan, even if the patient herself might be confused due to the suffering from illness. This is why Lee's family was not surprised when Lee changed her mind and decided to stay alive; this is because for the family, the decision was a truth that they had already witnessed and lived out together with another ill family member.[10] By contrast, for the hospital, whether the patient chooses to end her life or not makes no difference: both are autonomous decisions and of the same value, as long as they satisfy the formalist requirement of liberty and agency.

It should be noted that what's most striking for me in Lee's story is not the dramatic ending when Ms. Lee finally agreed with her family, but rather how deeply connected human lives are in the process of making decisions significant for individual member as well as the family as a whole. Again, it is the concern with the integrity of Lee's life and personality that gets Lee's family involved in the situation of medical decision-making. So the key issue is not who–the individual or the family– should have the authority to make the decision, but how the decision should be made in a shared, balanced way, insofar as we are relational, social animals. Furthermore, the shared decision-making model leaves open the possibility that the individual might disagree with the family's viewpoint in the end and persuade the family instead. What is important is merely that: the family should not be excluded from the process of decision-making; rather the company of significant others contributes to individual integrity and autonomy.

In brief, what is at stake in the family's attitude is the commitment to sharing suffering in a personal way. Accordingly, another significant feature of the family's attitude is the tension between the integrity of the family and the integrity of the individual's identity, as two distinguishable yet interdependent commitments. Firstly, the integrity of the family entails the narrative integrity of each family member embodied in a single life. As Lee's father said, Lee is an only daughter, cherished by the whole family, and with the thought of losing her, "everything goes black."[11] Since the family seems doomed to lose her, what is of ultimate importance for the family is to enable Lee to come to terms with her fate in a coherent and meaningful way, which is not only crucial for the integrity of her personality, but also for the integrity of the family as a whole. Secondly, the family plays a vital role in maintaining the full extent of the patient's autonomy. As shown above, the presence of the family proves to be essential in helping the patient to maintain a sense of self in the maze of disorientation—more than often the patient felt a radical existential

[10] Lee's brother said the family was not surprised concerning Lee's final decision. "One of Lee's brothers, Paul, told the court that not only was suicide against the religious principles of his family and church, but it was also against his sister's personal beliefs. He remembered that when he had personal troubles and thought of killing himself, "It was Grace that told me not to give up." Cf. Hartocollis, A. Daughter's right to die is weighed against family's wish to keep her alive. *New York Times*, October 4, 2012. http://www.nytimes.com/2012/10/05/nyregion/in-sungeun-grace-lee-case-right-to-die-is-weighed-against-a-familys-wishes.html?_r=0.

[11] Hartocollis, A. Daughter's right to die is weighed against family's wish to keep her alive. *The New York Times*, October 4, 2012. *New York Times*, October 4, 2012. http://www.nytimes.com/2012/10/05/nyregion/in-sungeun-grace-lee-case-right-to-die-is-weighed-against-a-familys-wishes.html?_r=0.

rupture and the world was perceived as changing and hostile— by serving various duties, e.g., protecting the patient from the realities and responsibilities of the outside world, and helping to fill the gap in the narrative of the patient's life.

The dominant use of autonomy in contemporary bioethics implies the image of self-rule, analogous to a helmsman who steer the ship within the context of a wide variety of "external" considerations such as currents or weather, while not being determined by these "external" considerations (May 2005, p. 307). This image presupposes an uncorrupted rational subject or a moral mind *per se*. However, as MacIntyre points out, "to become an effective independent practical reasoner is an achievement, but it is always one to which others have made essential contributions" (1999, p. 82). In other words, the capacity of practical rationality presupposes a caring network of interdependence and attachment—the paradigm of which is the family—as its foundation. More importantly, as MacIntyre argues, "there is no point then in our development towards and in our exercise of practical reasoning at which we cease altogether to be dependent on particular others" (1999, p. 97). In the predicament of serious illness, the essential relevance of dependence to independence becomes even more explicit and dominant, as we observe in Lee's case. Generally, to become an independent practical reasoner is a lifetime enterprise which can only be engaged in through participation in a set of relationships to certain particular others. It follows that the concept of autonomy should be recast in relational terms.

In sum, the story of Lee and her family reveals that if only we pay due attention to the subtle yet complicated relationship between individual integrity and family integrity, we can approach the notion of 'autonomy' in the full sense. The term 'individual integrity,' indicates that personal identity rests on narrative configurations and reconfigurations within the cohesion of a lifetime. An authentic autonomous decision should come to terms with the agent's life plan as it is understood in his/her life history and life world. As a crucial part of this background narrative, the family plays an indispensable role in the cultivation and maintenance of individual autonomy. In this way, family integrity in general—including the family's stability, wholeness, prosperity—proves to be the condition of individual integrity. From this perspective, the prevalent model of individual-oriented autonomy in medical practice is problematic in two respects.

First, by regarding decision makers as isolated from significant relations with the surrounding others, it undermines the patient's capability for making autonomous decisions in accord with his/her character and life plan. Its abstract criterions of autonomy—liberty and agency—deprive patients of the company of significant others, the only ones through and with whom he/she can retain the integrity of self. In this way, the patient's broken voice we observed in Lee's case is not a unique phenomenon, but a symptom of a general defect in the individual-centered practice of autonomy in medicine. While individualistic autonomy enshrines the right of the individual to make his/her own choices, it can only be institutionalized in a very formal way and is therefore incapable of—as well as disinterested in—specifying a means of evaluating the contents of those decisions.

Second, on a higher level, the individualist model of autonomy is morally defective in that it underplays the interdependence between individual integrity and family integrity, and treats the family unfairly. Through the worship of individual autonomy, modern bioethics is often empty of all contents but *that of choice*, while insisting that physicians should respect the patient as an unencumbered individual who is assumed to be independent of all possible attachments.[12] As a result, the role of family is marginalized in the process of medical decision-making, while the good unique to the family goes to oblivion. The family suffers from a special aphasia in that it cannot refer to its own integrity to make its claim, although it is only because the patient is part of the family that her decision has so much influence on the whole family.

We are now at a point to acknowledge that Lee's case raises some difficulties caused by the individual-directed model of autonomy, but cannot be addressed in its own conceptual framework as long as it still takes the *autonomous person as such* (i.e., treating the patient as an unencumbered, self-encompassing agent) as the staring point. It follows that we need to reframe our conceptual schema so that it might incorporate an understanding of autonomy that stimulates health-care professionals to respond appropriately to the social situation of patients and their families. The rest of my paper is devoted to proposing a reasonable alternative view of moral autonomy that gives priority to family integrity and stability.

Two points should be noted beforehand about the nature of our inquiry. Firstly, what is at stake is not the meaning of autonomy as a moral principle in 'the literature,' but its social uses and social influence, or as Callahan puts it, "what happens to ideas out in the streets" (1984, p. 41). Secondly, bioethical principles need to be appreciated and justified in a broader context of communal human life and related practices. Following MacIntyre, I will adopt a communitarian approach, by asserting in medical ethics "that rules are less fundamental than roles and relationships and that it is the context which roles and relationships provides which alone makes sense of rules" (1977, p. 44). Accordingly, the problem with the prevalent individual autonomy consists in the lack of a proper understanding of the characteristics of medicine as a practice: *goods*, *roles*, and *relationships* are *the missing concepts* without which the notion of agent choice and rule become morally powerless. In this way, the broken voice of Ms. Lee should have been viewed as a symptom of deep disempowerment. The sheer choice celebrated by individual autonomy is incapable of providing the wise and insightful moral companion which one badly needs in fashioning one's moral life and moral goals.

Keeping both points in mind, in the following section, I will draw heavily on the Confucian tradition for some important insights concerning the roles, relationships and goods proper to medicine as an ordered form of human practice. In this way, the following discussion is not aimed to provide a cultural-specific version of autonomy. Rather its purpose is to engage in an internal dialogue between two traditions, i.e., to go back to the foundation of the principle of respect of autonomy, and re-open some questions precluded by those foundations, by way of a detour into another tradition.

---

[12] Cf. Beauchamp and Childress emphasizes that health professionals should "never assume that because a patient belongs to a particular community or culture, he or she affirms that community's worldview and values" (2009, p. 107).

4.3 The Confucian View of Medical Practice: The Art of *Ren*

Let us begin with the most foundational question: the question of what grounds and justifies conclusions in medical ethics. Among the various answers to this question, Beauchamp (2001) proposes the internal/external distinction as the basic framework for reasoning about the good of medicine, the role of physicians, and the patient-physician relationship. The internal perspective defends an ethics derived from the internal goals of medicine or from professional integrity. The external perspective maintains that the precepts in medical ethics rely upon and require justification by external standards, and in this way, medicine is treated as part of general social life rather than a self-serving profession. Certainly there also exist some variants of a mixed internal/external account.

However, the Confucian view of health care seems at odds with the internal/external distinction. On the one hand, for Confucians, medicine is definitely an ordered form of human practice, a kind of an art, or *techne* in the Aristotelian sense.[13] On the other hand, the Confucian tradition emphasizes that the meaning of medicine can only be completed within a network of mutual caring, the paradigm of which is family. Lacking a counterpart to the Aristotelian distinction between *poiesis* and *praxis*, between *techne* and *phronesis*, Confucianism tends to propose that medicine is a practice dealing with human life in general and not solely with the condition of the body: "The best physician cures the state prior to people, that is what a physician should do" (Zuo 2008, Chap. 8). In other words, in the Confucian view, the craft aspect of medicine should be absorbed and appreciated in the communal pursuit of the good life. This is why traditionally medicine is termed as the "Art of *Ren*" (*Renshu*, 仁术).

The concept of *ren* is a key term embedded in Confucianism,[14] and it is polysemous at its root. There are two senses in which the Confucian usage of *ren* are directly relevant to our discussion here.

First, *ren* stands for the tender aspect of human feelings as well as a peculiar way of life based on this particular human feature. As Confucius says, "*ren* is to love others" (*Analects* 12:22, Cf. Chan 1963, p. 40). Confucius further emphasizes that *ren* could only begin with the love for intimate family members. If we are learn to care for others, we must be first learn to care for those we find ourselves joined to

[13] Ancient Greeks routinely speak of medicine. Aristotle distinguishes three kinds of human activity: the activity of *praxis*, the activity of *theoria*, and the activity of *poiesis*. Unlike *theoria*, which produces nothing beyond itself and leaves everything as it is, *praxeis* are, like *poieseis*, useful and effective, concerned with what is human and changeable. But *poiesis* is different from *praxis* because the good of *poiesis* consists in the product rather than in the producer, i.e., aiming at an external good beyond production, while praxis aims at the action for its own sake, aiming at internal goods. The distinction between *poiesis* and *praxis* extends to be the distinction between *techne* and *phronesis*. *Techne* is concerned with reasoning about production rather than action, while *phronesis* involves deliberation about "living well in general," or about the things which manifest our happiness in "living well as a whole"(cf. Knight 2007, pp. 6–17; Waring 2000, p. 142).

[14] The Analects Confucius mentioned '*ren*' as many as 105 times, without giving a formal definition.

by the accident of birth, because we are never born into a totally alien world, but into a set of particular relationships with those who have already accepted and thus are ready to respond to the newborn.[15] In this sense, *ren* prescribes a general way of human flourishing by giving priority to the family as an intimate community of mutual caring which is sustained by deep commitments among family members as well as the cultivation of relevant virtues in the family. As Confucius put it, "when the root is firmly established, the moral life (*Dao*) will grow. Filial piety and brotherly respect are the root of *ren*" (*Analects* 1:2, Cf. Chan 1963, p. 18). In this picture, the meaning of the family can never get fully defined in terms of the individual; it rather constitutes a *sui generis* reality, irreducible to individuals, while entailing a progress of 'becoming one flesh' with other family members. The image of "one flesh" dominates the Confucian imagination and discourse on family, and bestows a great moral significance on the relations of care, interdependence and mutual support that define family life, as shown in the following paragraph from Confucian classics:

> Relations of parents to children or children to parents are like two parts of a single body or the same breath/vital energy separately breathed. [...] even if they are in different places, they remain linked. Hidden intents reach from one to the other, *they rescue one another from pain or suffering, and they are moved by the other's worries and longings*. [...]This is called 'the closeness of bone and flesh' (骨肉之親). (Lü 1991, Chap. 9, translation mine).

In sum, the rationale behind this use of *ren* as love is a deliberation about the good life, which recognizes that the circumstances of human flourishing first and in general occur within families, insofar as vulnerability and dependency are essential aspects of the human condition. To put it in a Aristotelian framework of *poiesis* and *praxis*, *ren* counts as a form of *praxis* for its own sake, a practice "actualized" by pursing the flourishing of the mutual caring communities—the foundation of which is the family—as internal goods.

Second, etymologically, the character of *ren* indicates by its structure—it consists of a simple ideogram of a human figure and two horizontal strokes suggesting human relations—a relational understanding of self. For Confucians, we exist as limited, embodied, and particular. These features locate us in the world at a particular place and time and with some particular others, whom we have not chosen but to whom we find ourselves tied. Our relation to them constitutes our concept of the self to a large extent, and thus underpins the capability of autonomy.

For example, being part of the family in an authentic way—in a way the Confucian terms as "the closeness of bone and flesh"—means one is joined with someone else in an undivided situation to the extent that one suffers from the suffering of another, that one gains his/her freedom precisely in the act of love rather than in vain autonomy, abstracted from concrete relations and particulars. The commitments to others who are united with oneself in a mutual caring community are an essential way of finding one's own identity, and the commitments of other family members also play a crucial role in helping sustain and restore personal identity in difficult

[15] Even a Hobbesian individual—who is supposed to be at war with all others—needs to be cared for by someone not at war with him in some inevitable dependent situations like infancy or childhood (Cf. Groenhout 2004, p. 26).

times. In this way, the central social significance the family plays is to provide a necessary narrative context[16] in which the members find themselves already commonly situated to "begin with," to be motivated by ends beyond their individual choices. If an individual sets out without the family story, he/she easily gets caught by accidental internal desires, as well as the external depersonalizing bureaucracy, which is exactly the dilemma that Lee was stuck with.

The Confucian understanding of medicine as the 'Art of *Ren*' implies two moral requirements which might seems exotic for a Western mind. On the one hand, it implies that the care of the sick member first and foremost is a family matter. Confucians even emphasizes that "if your parents or children are sick in bed, and you can do nothing but let their lives be decided by mediocre physicians, it's not so different from just not loving them. Accordingly, those who love their family shouldn't be ignorant of medicine" (Cheng and Cheng 1992, vol. 12, translation mine). On the other hand, the 'Art of *Ren*' entails the capability of the physician to put herself as a necessary condition for the practice medicine. One author on Chinese traditional medicine put it in this way, "If I have a disease, how eagerly would I hope a physician could save my life! If one in my family has a disease, how eagerly would I hope a physician could save his or her life! If we adopt such a perspective, the greed in our hearts will fade out, and conscience will appear to call our reverence" (Fei 1987, p. 1, translation mine). Traditional medicine stressed this idea as a basic standard, thinking that it is the only way to "draw an outsider into the situation, and feel what the patient as well as the patient's family feel" (Fei 1987, p. 1, translation mine).

In short, on the Confucian view, healthcare as 'The Art of *Ren*' is a practice involving the patient, family members, and physicians in pursuing the flourishing of a caring, loving unity as family, as a internal good.

4.4 The Internal/External Perspective of Medical Ethics vs. the Perspective of *Ren*

A comparative study between the perspective of *ren* and the internal/external perspective will help us further grasp the features of the characterization of 'Art of *Ren*.' This comparison will set up the stage on which we can further examine the practical implications of the Confucian view for medical decision-making.

Different from the internal perspective, Confucians emphasize, besides the health of the patient and the excellence of medicine as a craft,[17] that there is another

[16] For a similar argument about the family as a "narrative context" for action, see Stanley Hauerwas (1981, p. 165). He views the family as the bearer of history and hope, which is the primary means to bind people throughout time.

[17] MacIntyre gives serious consideration to the character of medicine as a practice, and mentions that "Medicine is after all an ordered form of human practice and it involves the pursuit of at least two kinds of good. There is the good of the patient whose health, life, and general well-being are at stake. And there are those goods achieved by that extension of human creative powers which the history of medicine embodies" (1978, p. 44).

important good worth pursuing, i.e., a flourishing network of mutual caring within which every individual has always already been located in some place. H. Tristram Engelhardt illuminates Hippocrates' tradition by stating that "the physician and patient are not alone in the patient-healer relationship…in addition to the physician and the patient in their confrontation with disease, there is also the art" (1996, p. 291). Confucians would add that the physician and patient are not alone in the patient-healer relationship, but first and foremost involved in an intimate caring community—on the Confucian view, the physician is invited to join the family to help care for the patient. For Confucians, just the notion of profession is inadequate to lay the moral ground for a successful practice of medicine, but has to been completed by moral deliberation about the general way of human flourishing, i.e., *ren*. Compared to the internal perspective, the perspective of *ren* seems more persuasive at two points. Firstly, as Engelhardt put it, part of the task of medicine is to "induct the patient into the lifeworld of health care," because for patients, "to accept a diagnosis is often to committed to reordering one's very life in terms of the treatments and preventive regimens that the diagnosis warrants" (1996, pp. 295–296). However, as we have observed in Lee's case, the family plays an important role in helping to maintain the patient's personal integrity while she/he has to adapt to impairment. In this respect, the presence of the family plays a vital role in helping moderate the alienation the patient feels in his/her encounter with the physician in a totally strange environment. Confucians would agree with Engelhardt in his characterization of the patient as "a stranger in a strange land" (Engelhardt 1996, p. 295), but emphasizing this difficulty indicates the necessity of locating the care for the patient primarily in a caring community like the family rather than merely within the scope of the profession. Secondly, one main objection to the internal perspective of medical morality is that internal standards are non-self-sufficient, but in some cases "may be shallow, expendable," and open to the social changes and social values which are initiated and motivated by concerns outside the practice of medicine (Beauchamp 2001, p. 606). In contrast, the strength of the position of *ren* is that it sets up a common, general goal, i.e., the flourishing of the caring community, which enjoys kind of self-sufficiency in face of social change while giving due regard to the specific features of health care practice.

On the other hand, the phrase of 'Art of *Ren*' also represents a sharp divergence from the external perspective in that it puts the concrete, particular relational nature of human life at the center of moral deliberation, whereas the external perspective appeals to "universal moral principles that are valid independent of the perspectives of particular communities and traditions of medical practice and ethics" (Beauchamp 2001, p. 612).

In sum, the doctrine of the 'Art of *Ren*' takes vulnerability and related dependence as central to the human condition, and thus demands incorporating the practice of health care into a common way of human flourishing which relies on the stable function of the family as the primary nexus of mutual caring. Insofar as it sheds light on the crucial relevance of humans as embodied, vulnerable, and dependent to our moral life—an issue which has been marginalized and even obscured from either the internal or external standards for medical morality—the Confucian position deserves a serious consideration for its practical implications. This essay is confined to examining what difference it can make to our understanding of the patient's autonomy if health care should be first and foremost construed as a practice which pursues the flourishing of a mutual caring community as one of the internal goals.

In light of the 'Art of *Ren*,' the individual-oriented model of autonomy is morally defective and powerless, because it fails to recognize a good which is crucial for personal flourishing under our condition as particular, embodied and limited, especially in the fields of medicine and health care, where the issues turn on embodiment.

The dominant concept of autonomy in health care "rests on a picture of the person as a separate being, with a distinctive personal point of view and an interest in being able securely to pursue his or her own conception of good" (Blustein 1993, p. 10). In this picture, the ultimate good that the patient can seek in health care is no more than the freedom of choice and accordingly, the patient-physician relationship is constructed by contracts, since the only moral obligations we can owe to autonomous moral agents are those formally agreed to. However, the contractual model obscures and diminishes a specific end (or good)—the healing of the patient—so that the physician-patient relationship is entered into and maintained with this end in sight. What distinguishes the physician-patient relationship from other contractual relationships is the *incapacity* of the patient. "Whereas incapacity of a certain kind disqualifies me from being a customer in a restaurant (if I am too drunk or too poor, they throw me out) or a lawyer's client (once it is established that I lack either mental capacity or the appropriate standing in a case, no lawyer will even advise me), *it is incapacity that qualifies me as a patient*" (MacIntyre 1977, p. 205, emphasis mine). Part of the good of the patient depends on having someone committed to caring for him/her, in a way deeper than those one voluntarily undertakes in some contractual model.[18] With similar concerns, Kay Toombs (1992) proposes that health care should be characterized as the healing of a patient rather than merely curing a disease. The healing of the patient entails developing a "face-to-face" relationship which is grounded in sharing the patient's suffering with an aim to restoring his or her integrity against a biographical background (Toombs 1992, p. 111). In this respect, the individualist understanding of autonomy seems to lead to negative effects instead, by encouraging physicians to treat patients impersonally and distantly, respecting our autonomy but nothing else.

But how can we construct a form of physician-relationship that is more than and different from the contractual model? The friend model is often used as an alternative to the contractual model (MacIntyre 1978; Engelhardt 1996). Engelhardt and MacIntyre both discuss the feasibility of the friend model in a medical context. To a friend you are tied by your concern for goods that are both your and his concern

[18] To illustrate this difference, MacIntyre gives a very interesting example, "But if, on entering the butcher's shop as an habitual customer I find him collapsing from a heart attack, and I merely remark 'Ah! Not in a position to sell my meat today, I see,' and proceed immediately to his competitor's store to complete my purchase, I will have obviously and grossly damaged my *whole* relationship to him, including my economic relationship, although I will have done nothing contrary to the norms of the market" (1999, p. 117).

for the same good. "The more physicians and patients share a common view of the good life, the more plausible it will be that they meet as moral friends" (Engelhardt 1996, p. 297). MacIntyre further specifies two conditions which must be satisfied for the patient and the physician to be friends (1978, p. 45). First, the physician should stand in a long-term relationship with the patient and his family, a presence at births and deaths. Second, the physician and patient should share essentially the same moral point of view. Obviously modern medical practice and its bureaucratic organization, as well as the increasing cultural and religious diversity, make it more and more difficult for the physician to play the role of the patient's friend. MacIntyre and Engelhardt both propose the dichotomy of the roles the physician could play in his/her relationship to the patient: either as a friend or as a stranger. And both raise the issue with a pessimistic tone: although the encounter between the physician and the patient as friends is a desirable ideal, it can only be realized under very limited circumstances and is generally not feasible in the context of contemporary health care.

However, the Confucian view of the 'Art of *Ren*' offers a way out of this dichotomy by elucidating the central role the family plays in dealing with the frailties of human life and related dependence which are typical of the patient's situation. Consistent with the friend model, the view of the 'Art of *Ren*' also argues that the primary good of the patient consists in having somebody committed to his/her goods in a way beyond autonomous choice. But the Confucian model goes even deeper than the friend model, because friendship is ultimately based on the individuals,[19] whereas the significance of the family can only make full sense in these relations. Family love should be understood not simply as attachment to those who are like us, but also as regard and respect for those whom we have not chosen but to whom we find ourselves tied. The familial relation is distinguished from other relations in that the family members are committed to share each other's fate from birth to death, and this initial commitment is *unconditional* in important respects. For example, friendship can be justified on the expectation of self-fulfillment,[20] whereas the family cherishes the mutual commitment unconditionally, merely for the sake of relationships *per se*. As a traditional Chinese pledge states, "However things turn out, I will be there for you." In this way, the well-functioning family embodies a network of unconditional giving and taking, which is characterized by a systematic refusal to treat the care receiver in a way that is proportional to his/her qualities and aptitudes.[21] For Confucians, only this model can serve as the norm for caring for the patient, because what the patient needs in his/her role of incapacity is

[19] As Aristotle puts it, "in loving their friend they love what is good for themselves" (Aristotle 2011, NE 1157b33). Further one "is related to his friend as he is to himself, since the friend is another himself" (Aristotle 2011, NE 1166a31). As May Sim (2003, p. 451) comments on these quotations, "One rational person can wish for the good for another rational person, and for her own sake. But the good is the excellent functioning of the individual; the relationship is 'incidental.'"

[20] We are encouraged to view the friendship primarily as the institution of personal fulfilment that is necessary for us to attain the good proper to ourselves.

[21] The teacher may give up his student when the latter is too retarded to be worth teaching. And one may put an end to a friendship when the other party does not remain the same (cf. *Nicomachean*

to always get someone to respond to his/her disease and assist him/her in continuing to live life to the fullest extent possible, no matter how seriously his/her qualities and aptitudes are impaired by the disease.

Again, the crucial point of the Confucian view is that health care, as an ordered form of social practice concerned with the frailty of the embodied human being, should be located and appreciated in the network of unconditional mutual caring. On the Confucian view, physicians do have obligations to the patient beyond those mutually and autonomously consented to, on the ground that they are invited to participate in the mutual caring community. Far from something to be wary of or a problem to cure,[22] family care provides the general context as well as the concept of a common good which unites physicians, family members, and the patient in the common pursuit of restoring the patient's integrity to the fullest extent possible. Furthermore, as we observe in Lee's case, the patient's integrity is inseparable from the integrity of the whole family. On the practical level, family involvement in the medical decision-making process also significantly mitigates the gap between the patient and the physician which was aggravated by the bureaucratic organization of health care, because the family is capable of creating and sustaining a space in which family members find themselves commonly situated to pull together in face of crisis. All that shared history, those memories and commitments, put the family in the best position to inform and reason with the physicians to get them to be actively involved in caring for the patient. In other words, the involvement of the family creates a locus that helps bring the physicians into a closer, more committed, more 'biographically based' relationship with the patient, against the background of highly specialized medical practice. Thus the Confucian view of the 'Art of *Ren*' provides an approach to overcome the dichotomy of the relationships the physician stands to the patient: either as friends or as strangers.

At this point it is perhaps worth summarizing what has been said so far about the Confucian view of the 'Art of *Ren*', and its relevance to the contemporary practice of the patient's autonomy.

Firstly, when Confucianism understands medical practice as the 'Art of *Ren*,' it sheds new light on the nature and status of health care as an ordered human practice—rather than merely a profession—and nests its social and political significance in the network of unconditional giving and taking, which is centered around the family care. This Confucian view is especially revealing at two points. First, moral reasoning about bioethical issues should pay due attention to the frailties and dependencies of human life, i.e., it should start from the particular situation of the patients, rather than importing a constellation of values inspired by the concerns outside the field of health care.[23] Second, the condition of humans as embodied, vulnerable, and dependent prescribes a common way for human flourishing, that is,

Ethics, p. 1165b). But good parents would never give up their child, no matter what kind of distress happens to this child, no matter how disabled this child is.

[22] As Chap. 14 summarizes, this attitude is typical in contemporary bioethics.

[23] As Daniel Callahan summarizes, "As a familiar constellation it encompasses a high place for autonomy, for biomedical progress with few constraints, for procedural rather than substantive

to be included and cared for in a network of unconditional taking and giving, which to some extent constitutes the common good for every one. Both lead to significant changes in framing bioethical issues, for example, they give priority to family integrity and stability in recognition of the circumstance that human flourishing first, and in general foremost, occurs within families.[24]

Secondly, as far as the issue of autonomy is concerned, the Confucian view provides crucial insights and imagination to conceive an alternative view of moral autonomy and a proper model of medical decision-making. Again, the point of the 'Art of *Ren*' is that the physician is invited to join the family to help care for the patient to the extent that the integrity of the family as a unit of mutual caring is also the internal good of medical practice. On the institutional level, it requires respecting and giving freedom to the family to carry out its indispensable tasks. It should be the entire family that takes the burden of listening to and discussing with the physician, communicating with the patient, consulting the other family members, and finally signing the consent form. Only this familist model of medical decision-making[25] can reflect this deep moral presupposition of the family: that the family is in its nature the most important nexus for human beings, by which the members find themselves mutually indebted and morally engaged to share each other's fate. Furthermore, the physician is also bestowed an active role in the pursuit of the integrity of the family as a primordial unity of caring. It implies that physicians are obligated to inform the patient directly or bring in the extra-familial authority, like a hospital ethics committee, if they find that the reasonable interests of the patient are jeopardized by a deteriorated familial relationship. It is important to note that although the Confucian model respects family authority, the family itself is not an authority on moral standards that guide these decisions. Rather it is *ren* that binds the patient, his/her family members, and physicians together in the same practice. In this way, "[t]he model is best termed not only by family-based but also harmony-oriented because it invites physicians, patients, and their family members to regard critical medical decisions as located in families, structured by ritual practices and guided by Confucian virtue in order to achieve harmonious agreement for the health care of patients" (Chen and Fan 2010, p. 574).

4.5 Concluding Remarks

In conclusion, Confucian reflection on medical practice supports a family-based, harmony-oriented medical decision model in recognition of the circumstance that human flourishing first, and in general foremost, occurs within families.

solutions to controverted ethical problems, and for a strong antipathy to comprehensive notions of the human good" (2003, p. 498).

[24] I am indebted to Engelhardt for the phrasing and related ideas.

[25] See Fan 2010 for more detailed discussion of the familialist model of Chinese medical decision-making.

I want to make some brief remarks on the universal significances of this shared decision-making model to end this essay.

Firstly, this essay argues for reconfiguring the notion of autonomy rather than getting rid of it. We live in a world prone to tyranny, and in the company of a medicine more and more like a bureaucracy in a large-scale nation.[26] Insofar as the value of autonomy is an effective means to empower patients with a full partnership with physicians, autonomy is an indispensable good in today's medical context. The issue, then, is how to understand autonomy in a way that can do justice to the vulnerability and interdependence of human existence. The point of the Confucian view is that relatedness is not the antithesis of autonomy, but a literal precondition of autonomy. As Confucians understand it, authentic autonomy is inseparable from an experienced harmonious state—a combination of bodily harmony, mental harmony, and life-process harmony—which can only be fostered and sustained in relationships, the most foundational of which is the family.

Secondly, what is at stake in the debate between the Confucian view and the dominant bioethics in the West is not so much a cultural war as the conflict between different forms of life, or different understandings of what it is to be human. The dominant bioethical theory in the West, underpinned by the Enlightenment moral project and political liberal individualism, depends on the category of 'personhood.' Persons are a universal category abstracted from particularity, from those significant relationships and roles that in key part define who I am. On the contrary, Confucians insist that moral reasoning should begin with a reflection on the limited, embodied particularity of human being, which also prescribes a universal way of framing ethical issues in general, and bioethical issues in particular. That is, as long as we want to reflect on the character of the moral life, it gains content from being located first in a contextually rich locus, such as that of family. In contrast, individual autonomy when evacuated of any particular content does not have sufficient content to determine the character of moral life. As Callahan describes it, "My autonomy is an inarticulate bore, good as bodyguard against moral bullies, but useless and vapid as a friendly, wise, and insightful companion" (1984, p. 42).

In this way, the Confucian view takes issue with the individualist notion of autonomy dominant in the West, not aiming to provide a culture-specific alternative, but inspired by the longing for another kind of universality, which a global bioethics constructed in abstract terms can not satisfy. The attractiveness of Confucian morality as well as the bioethics it supports exactly lies in the way it brings together both the particularistic and universalist dimensions of the moral life.[27] When directed at this level, a confrontation between Confucian morality and Contemporary Western morality on bioethical issues could be a serious and fruitful communication. This paper is a very preliminary attempt at this project.

[26] See Callahan (1984, p. 42) and Engelhardt (1996, p. 298).

[27] As far as I know, Engelhardt is the first one to make this point clear. See H.T. Engelhardt, "Towards a Chinese bioethics: Reconsidering medical morality after foundations" (Lecture presented at the conference, 'Constructing Chinese Bioethics,' Hong Kong Baptist University, Hong Kong, June, 2012).

References

Aristotle. 2011. *Nicomachean ethics*. Trans: R. Bartlett and S. Collins. Chicago: The University of Chicago Press.

Beauchamp, T. L. 2001. Internal and external standards for medical morality. *Journal of Medicine and Philosophy* 26:601–619.

Beauchamp, T. L., and J. F. Childress. 2009. *Principles of biomedical ethics*. 6th ed. New York: Oxford University Press.

Blustein, J. 1993. The family in medical decision making. *Hastings Centre Report* 23:6–13.

Callahan, D. 1984. Autonomy: A moral good, not a moral obsession. *Hastings Center Report* 14:40–42.

Callahan, D. 2003. Individual good and common good: A communitarian approach to bioethics. *Perspective in Biology and Medicine* 46 (4): 496–507.

Chan, W-T. 1963. *A source book in Chinese philosophy*. Princeton: Princeton University Press.

Chen, X., and R. Fan. 2010. The family and harmonious medical decision making: Cherishing an appropriate Confucian moral balance. *Journal of Medicine and Philosophy* 35:573–586.

Cheng, H., and Y. Cheng. 1992. *Er-Cheng-Yi-Shu & Er-Cheng-Wai-Shu* [The posthumous works of brothers Cheng.]. 12 vols. Shanghai: Shanghai Classics Press.

Donchin, A. 2000. Autonomy and interdependence: Quandaries in genetic decision Making. In *Relational autonomy: Feminist perspectives on autonomy, agency and the social self*, ed. C. Mackenzie and N. Stoljar, 236–259. New York: Oxford University Press.

Engelhardt, H. T. 1996. *The foundations of bioethics*. New York: Oxford University Press.

Fan, R. 2010. *Reconstructionist confucianism: Rethinking morality after the west*. Dordrecht: Springer.

Fei, B. 1987. *Yi-Fang-Lun* [On prescription]. Beijing: Chinese Medicine Classics Press.

Groenhout, R. E. 2004. *Connected lives: Human nature and an ethics of care*. Lanham: Rowman & Littlefield.

Hauerwas, S. 1981. *A community of character: Towards a constructive christian social ethic*. Notre Dame: University of Notre Dame Press.

Knight, K. 2007. *Aristotelian philosophy: Ethics and politics from Aristotle to MacIntyre*. Cambridge: Polity Press.

Lü, B. W. 1991. *Lü Shih ch'un Ch'iu*. Beijing: China Book Company.

MacIntyre, A. 1977. Patients as agents. In *Philosophical medical ethics: Its nature and significance*, ed. S. F. Spicker and H. T. Engelhardt, 197–212. Dordrecht: Springer.

MacIntyre, A. 1978. What has ethics to learn from medical ethics? *Philosophic Exchange* 2:37–47.

MacIntyre, A. 1999. *Rational dependent animals: Why Human beings need the virtues*. Chicago: Open Court Press.

MacIntyre, A. 2007. *After virtue*. 2nd ed. Notre Dame: University of Notre Dame Press.

May, T. 2005. The concept of autonomy in bioethics: An unwarranted fall from grace. In *Personal autonomy: New essays on personal autonomy and its role in contemporary moral philosophy*, ed. J. Taylor, 299–309. Cambridge: Cambridge University Press.

Sim, M. 2003. The moral self in Confucius and Aristotle. *International Philosophical Quarterly* 43 (4): 439–462.

Toombs, S. K. 1992. *The meaning of illness: A phenomenological account of the different perspectives of physician and patient*. Dordrecht: Kluwer Academic Publishers.

Veatch, R. M. 1984. Autonomy's temporary triumph. *Hastings Center Report* 14 (5): 38–40.

Waring, D. 2000. Why the practice of medicine is not a phronetic activity. *Theoretical Medicine and Bioethics* 21 (2): 139–151.

Zuo, Q. M. 2008. *The discourses of the states*. Shanghai: Chinese Classics Press.

Part III
Informed Consent: Individual-Oriented vs. Family-Oriented

Chapter 5
The Ideal of Autonomy and Its Misuse

Kyungsuk Choi

5.1 Introduction

Respecting autonomy has been an important principle for medical ethics since Beauchamp and Childress wrote their famous monograph, *Principles of Biomedical Ethics*. The principle of autonomy often seems to dominate the solutions given for bioethical issues in both Eastern and Western societies. Recently a number of Korean scholars have criticized the principle.

New approaches to autonomy have been proposed in Korea as well as the West. Some suggest a notion of "relational autonomy" which criticizes the individualist perspective. Eastern scholars frequently emphasize the value of family rather than the individual. For example, John Hardwig (1997), an American philosopher, expresses the idea that one may have a moral obligation to die for one's family member in his article, "Is There a Duty to Die?" In his article, "What About the Family," Hardwig (1990) also argues that the family should make the treatment decision when the lives of family members would be dramatically affected by the treatment decision.

While I appreciate these new approaches to autonomy and the moral complexity of human relationships they capture, I wonder if these approaches will somehow destroy or devalue autonomy as a moral value, which has played an important role in our moral foundations. Historically, autonomy has been emphasized with the development of an individualistic society. However, at our modern moral foundations, the values of autonomy and freewill cannot be devalued. What then is the proper analysis of autonomy as a moral value?

K. Choi (✉)
School of Law, Ewha Womans University, 311 Law Building, 52 Ewhayeodae-gil, Seodaemun-gu, Seoul 120-750, Korea
e-mail: choiks@ewha.ac.kr

R. Fan (ed.), *Family-Oriented Informed Consent*, Philosophy and Medicine,
DOI 10.1007/978-3-319-12120-8_5

In this paper I will distinguish the ideal of autonomy from its practical implementation. Further, I will clarify which of these has been the target of recent criticisms. I will then argue that autonomy is an important moral value, even in Asian cultures, and what we need to be weary of are misinterpretations and misuses of autonomy.

5.2 The Ideal of Autonomy

The ideal of autonomy reflects modern enlightenment thinking and its understanding of human beings. The ideology of autonomy presupposes a rational human being capable of reasoning to a conclusion if he/she is given relevant and sufficient information.

Here the notion of a rational human being is not only integral for the ideology of autonomy, but also for constructing modern societies, that is, human-centered societies divorced from the God-centered world. In the secular world, a rational human being is a member of democratic society and is a desirable person for modern societies.

The structure of modern society reflects the ideal of a rational human being. Political decision making usually follows majority rule. This rule cannot be adopted if we do not consider all persons to be equally rational. Education also sets up the ideal of a rational person. When we consider this ideal, we come to recognize the significance of critical thinking in our societies, especially the democratic ones. Economics also presupposes that consumers are rational. Even though many individuals are not in fact rational, our political, educational and economic systems presuppose that they are or at least that they should be. In this respect, rationality is the ideal we have to pursue.

Therefore, the self-determination of a rational human being is to be respected because he/she is the final authority for deciding what's in his/her best interest if he/she is competent. This idea is easily observable in the discourse of bioethics.

Beauchamp and Childress explain autonomy as follows: "We analyze autonomous action in terms of normal choosers who act (1) intentionally, (2) with understanding, and (3) without controlling influences that determine their action" (Beauchamp and Childress 2009, p. 101). The second condition, "with understanding" is related to the notion of "informed consent," which is required in biomedical contexts. I think that the importance of "informed consent" is related to our assumption that a normal chooser is a rational human being. In other words, a rational human being will determine his/her best interest if sufficient information is given. Beauchamp and Childress say, "To respect autonomous agents is to acknowledge their right to hold views, to make choices, and to take actions based on their personal values and beliefs" (Beauchamp and Childress 2009, p. 103) Thus, we can simply understand autonomy as a rational person's self-determination on the basis of personal values and beliefs when given sufficient information.

5.3 Misinterpretations and Misuses of Autonomy

From a theoretical perspective, the concept of autonomy as it is defined within biomedical ethics (including Beauchamp and Childress' account) is not the same as that set forth by Kant. Kant states, "Act always on such a maxim as thou canst at the same time will to be a universal law" (Kant 1998, p. 301). He continues, "Autonomy of the will is that property of it by which it is a law to itself (independently on any property of the objects of volition). The principle of autonomy then is: Always so to choose that the same volition shall comprehend the maxims of our choice as a universal law" (Kant 1998, p. 303). Here, autonomy is explained as a universal property of the will. Accordingly, autonomy is not mere self-determination. When we discuss autonomy in bioethical contexts, it seems we do not have in mind universalizability. Hardwig outlines the gap between Kant's understanding of autonomy and the general use of autonomy in medical ethics as follows:

> Because medical ethics has ignored patient responsibilities, we have come to interpret "autonomy" in a sense very different form Kant's original use of the term. It has come to mean simply the patient's freedom or right to choose treatment he believes is best for himself (Hardwig 1990, p. 8).

In this way, autonomy in biomedical ethics is better understood as Millian rather than Kantian. Beauchamp and Childress exposit Mill's account in *On Liberty* as follows:

> Mill concerned himself primarily with the "individuality" of autonomous agents. He argued that society should permit individuals to develop according to their own convictions, as long as they do not interfere with a like expression of freedom by others or unjustifiably harm others; but he also insisted that we sometimes have an obligation to persuade others when they have false or ill-considered views (Beauchamp and Childress 2009, p. 103).

From the above citation we can understand autonomy as mere self-determination.

Autonomy is often understood as synonymous with freedom. However, it is wrong to identify an autonomous decision simply as a free one. When the subject of self-determination is understood as individualistic, his/her decision demonstrates just his/her preference among the other options available. I do not think this decision is necessarily the same as the one a rational person would make after he/she has reviewed the relevant information with moral reflection and consideration.

Autonomy is not freedom or mere self-determination. Autonomy is ethical self-regulation as is evident in the Greek etymology of autonomy: "autos" (self) and "nomos" (rule, governance or law). Thus, one's autonomous decision is not a mere free choice in accord with one's best interest among other options. Hardwig is right to strongly criticize the individualist perspective and to emphasize the responsibility an individual has to and for his/her family. Accordingly, moral reflection should include moral consideration for the lives and wellbeing of others when making moral decisions in bioethical contexts. From a practical perspective, the ideal of autonomy has not been implemented in accord with its theoretical foundations.

Emphasis has often been placed on acquiring a signature to document informed consent rather than, more importantly, considering how the decision is reached.

This reflects our shallow understanding of autonomy. Accordingly, the ideals of autonomy and the rational human being are not realized in a desirable manner. The presence of one's signature does not ensure that an autonomous decision was made. Consequently, practical guidelines need to be developed in order to move closer to realizing the ideal of an autonomous decision made by a rational human being. To realize this goal, efforts should be made to study and develop informed consent practices, how to provide relevant information for consent, and how to identify autonomous decisions.

5.4 Is Family-Oriented Decision Making an Alternative?

As mentioned above, Hardwig criticizes the notion of individual autonomy. Alternatively, he suggests that one's family should be seriously considered in the informed consent process (Hardwig 1997, pp. 36–39). This discussion may be valuable for those Western societies in which an individual's free choice has enjoyed absolute priority even though it may not be a genuine form of autonomous decision making.

However, Hardwig's view is not novel to Eastern cultures. In Korea especially, older generations feel a moral obligation to their offspring. When elders refuse life-sustaining treatment, they do so because they understand it as a waste of money that could be better used for their offspring. This rationale is also not just an economically based compromise between family members, it is also reflective of the view that human life is not endless.

Most Koreans do not subscribe to the individualist perspective. However, some Koreans may feel the burden of tight relationships among family members[1] and feel they are wrongly forced to sacrifice their life plan for family members. Older generations, who lived after the Korean War, often had to sacrifice for older sons or younger siblings by forgoing an education to make money for the family. In this regard, Hardwig's argument for moral obligations to the family might not fit some Eastern societies where the value of autonomy is not fully developed.

[1] Here I use "family" to refer to a group of people who are related to each other by marriage and/or blood ties, such as grand-parents, parents, their children, the siblings of parents and their children, etc. However, the scope of "family" may vary depending on an individual's perception. According to Article 779 of the Korean Civil Act, family is legally defined as follows: "(1) Family members shall consist of the following persons: 1. The spouse, lineal blood relatives, and brothers and sisters; and 2. Spouses of the lineal blood relatives, lineal blood relatives of the spouse, and brothers and sisters of the spouse. (2) In the case of paragraph (1), subparagraph 2, it shall be limited to those cases where they share living accommodations." Here "lineal blood relatives" means "father, mother, and their father and mother" and "son, daughter, and their son and daughter." Unlike the Act, most Korean people seem to think that family members are persons mentioned in paragraph (1), ignoring paragraph (2). I will use "family" consisting of persons mentioned in the paragraph (1). For reference, Article 777 defines "relatives" as "1. Blood relatives within the eighth degree of relationship. 2. Affinity relatives within the fourth degree of relationship." Here the first degree of relationship is one between parents and their offspring whereas the second is between siblings.

Further, Hardwig's argument that the family should participate in individual decision making requires consideration. He states:

> Considerations of fairness and, paradoxically, of autonomy therefore indicate that the family should make the treatment decision, with all competent family members whose lives will be affected participating (Hardwig 1990, p. 9).

This point also does not quite fit Eastern societies because a family is already involved in its members' individual decision making. Too much involvement of the family is problematic.

According to Ruiping Fan, the family plays an important role in decision making in Eastern societies. He states,

> It is not a sick family member him/herself but the entire family that has real authority in clinical decision making. Western people might be concerned about this claim of family-sovereignty. But the family under this notion can be viewed as an autonomous social unit from the physician and the state, analogous to the autonomous individual in the West (Fan 1997, p. 317).

Fan is right to notice this difference between Western and Eastern cultures. However, does this difference justify "the East Asian principle of autonomy" Fan (1997, p. 315) presents? Fan's principle appears to be a "principle of family autonomy" because it views the family as an autonomous social unit that should be regarded as such by physicians and the state.

I agree that the family should participate in the process of decision making, but the decision of a family is not the same as that of an individual person in some respects. The latter cannot be reduced to the former for the following reasons.

First, a decision is fundamentally made at the individual level. A family in itself is not a legitimate subject to make a decision. Even though a group makes a decision following a majority rule vote in a democratic society, it is hard to apply this idea to a family. A majority rule vote presupposes autonomous individuals who are equal. This rule may be the best way to resolve a problem when there are conflicts of opinions among equal individuals, but a family is not comprised of equal individuals. Thus, it is not easy to say that a family may adopt a majority rule practice for decision making.

Second, Eastern culture, especially Koreans, may adopt a principle of best interest for family decision making. This principle works best when we do not know the wishes of an individual and others know what is the best for the individual. We may accept a principle of best interest when a decision needs to be made on behalf of an incompetent patient whose wishes we do not know. In such a case, the conditions listed above are met.

But what about family decision making? There is no problem when a proxy decision is a family decision. In this case, a family or its representative plays the role of a proxy. But what if a family knows the individual's wishes? Should we say that a family decision is still better in this case? I would argue that it is not. The principle of family autonomy cannot be considered the same as or similar to that of individual autonomy.

As previously stated, the family should participate in the decision making process and the individual should seriously consider the values of one's family and how their lives and wellbeing might be affected by one's decision. However, the individual should ultimately make the decision if he/she is competent. This view is found in Amitai Etzioni's article, "On a Communitarian Approach to Bioethics," which also addresses Hardwig's point. Etzioni states,

> Hardwig leans somewhat in the authoritarian direction when at one point he claims that "considerations of fairness and, paradoxically, of autonomy therefore indicate that the *family* should make the treatment decision, with all competent family members whose lives will be affected participating." Thus, a less authoritarian position would suggest that, for instance, if nine out of ten family members agree that treatment should be stopped for a given member, but the member—who is competent—rejects this conclusion, the family's wishes should not carry. However, the person does owe the family members a careful consideration of their values, reasons, and needs (Etzioni 2011, p. 367).

I agree with Etzioni's position. An individual should appreciate the value of family and take into account the values and needs of family members. However, the final decision should be made by an individual if he/she is competent.

The notion of group autonomy cannot be established if it lacks a formal process for decision making, like the established procedures of a committee or a congress. As such, there is no formal procedure for family decision making. Further, it should be noted that conflicts of interest might be present in family decision making because, in most cases, the family bears the burden of paying medical costs.

Consequently, we should ask, what is the best understanding of autonomy that preserves the ideal of rational human beings as decision makers while also honoring the individual's relationship to other family members?

5.5 A Constructive Understanding of Autonomy

How can we maintain the ideal of autonomy and its value while also reflecting the non-individualistic value of the family or community? Problems with the principle of autonomy do not come from the notion of autonomy in itself, but from misunderstanding the concept of autonomy and/or how it implemented. As previously mentioned, in bioethics, the concept of autonomy is conflated with notion of freedom, even though autonomy is not synonymous with freedom. In practice, there are no practical procedures to ensure a decision is made autonomously.

The ideal of autonomous decision making may in fact be internal and personal, causing problems for the principle of autonomy. Even though this ideal may be difficult to realize, this does not necessarily mean we should give it up. This is also true of other human values, such as justice, liberty, and love. I contend, we should still hold to the ideal of autonomy and have reason to pursue it. In its pursuit, we must educate individuals about the true meaning of autonomy and urge them to seriously reconsider what constitutes an autonomous decision.

Bruce Miller's discussion of autonomy is informative for constructing a reasonable theory of autonomous decision making. He explains different types of autonomy as follows.

> There are at least four senses of the concept as it is used in medical ethics: autonomy as free action, autonomy as authenticity, autonomy as effective deliberation, and autonomy as moral reflection...*Autonomy as free action* means an action that is voluntary and intentional...*Autonomy as authenticity* means that an action is consistent with the person's attitude, values, dispositions, and life plans...*Autonomy as effective deliberation* means action taken where a person believed that he or she was in a situation calling for a decision, was aware of the alternatives and the consequences of the alternatives, evaluated both, and chose an action based on that evaluation... *Autonomy as moral reflection* means acceptance of the moral values one acts on. The values can be those one was dealt in the socialization process, or they can differ in small or large measure. In any case, one has reflected on these values and now accepts them as one's own (Miller 1981, pp. 24–25, italics mine).

Miller's third and fourth descriptions of autonomy appear to be the most useful for understanding the ideas of "relational autonomy" and "family-oriented consent."

In the third description, autonomy as effective deliberation, one can include consideration of the wellbeing of one's family and community when one considers the consequences of the alternatives. In this way, Hardwig's worries are overcome when one seriously considers the full consequences of one's decision.

In the fourth description, autonomy as moral reflection, one may include the value of family and give consideration to the relationships between one's self and others, including family members or one's community. In addition, one should consider what the desirable and reasonable ways of life are for moral reflection. The individualist perspective Hardwig criticizes is derived from a misunderstanding of human life. It is just a fact of the world that we share our lives with others, including our family members and others in our community. Consideration for the lives of others, especially family members, and their well being should be included in one's moral reflection.

The meaning of "family-oriented consent" is well captured by Miller's description of moral reflection. Most Koreans appreciate the value of family. If one's moral view is that family is valuable to one's life, sacrifices for one's family do not compromise one's integrity. Additionally, if one wants a family member to make a decision in their place, this also does not compromise one's integrity. However, this does not mean we can replace the value of autonomy with others or that we can underestimate its value. In the above cases, the choice of a surrogate decision maker or one's sacrifice for the family must be the outcome of one's own deliberations and the values one accepts.

Daniel Callahan sets forth a communitarian interpretation of autonomy in his statement that "autonomy should be broadened to encompass an analysis of what constitutes morally good and bad free choice. The claim that so-called private choices should be exempt from moral analysis is the death of ethics" (Callahan 2003, p. 505). One should interpret autonomy such that it includes some aspects of Miller's individual approach as well as characteristics of Callahan's communitarian theory.

Further, we can learn from responsive communitarianism, that is, the relationship between autonomy and the common good of a society. Etzioni states,

> Although responsive communitarianism's starting point is the recognition that the tense relationship between autonomy and the common good must be worked out rather than assuming a priori that one of these core values trumps the other, it expects treatment to differ from one society to another and among different historical periods. Thus, in totalitarian societies and theocracies, such as in Singapore and Iran, those who advocate the balance that responsive communitarianism favors would need to promote autonomy, while in societies in which individualism is rampant, such as the US was in the 1980s, the advocates of responsive communitarianism would need to promote more attention to the common good. That is, societies often need to move in opposite directions from another to achieve the same end balance (Etzioni 2011, p. 364).

Etzioni's argument holds for the relationship between autonomy and the good of the family. The value of the family, more concretely, the good of family members, should be balanced with the value of autonomy.

At this point, I will argue that autonomy is a universal moral value. What needs to be revised is the individualistic interpretation of autonomy, not autonomy itself or a revised version of "family autonomy." For the Korean way of life, autonomy may still reflect a desirable notion of the moral agent in which family or other members of the community are just as important as the individual. Korean culture emphasizes harmony between an individual and his/her family members. However, harmony should be balanced with the value of autonomy; one cannot replace the other. This is true of the value of the family.

We have a number of values that contingently conflict with one another. The difficulty of ethics lies in balancing the various values we pursue. Consequently, the values of an individual and his/her family should also be balanced. If the value of the family is underestimated, we have to emphasize its importance as Hardwig suggests. If the value of the individual is underestimated, as it is in Korean society, we have to increase its value by emphasizing autonomy. In the East, especially in Korean culture, autonomy is still an important value in need of development.[2]

The individual (the self) and the family (a community) should be balanced. The family can be considered a community in a basic sense. From a traditional Eastern perspective, the family, rather than the individual, has been the basic unit for society and the state. However, it was not long before Korean society began to recognize an individual as having autonomy. We cannot disvalue modernity. In this regard, the value of autonomy should continue to be emphasized, but it must be balanced with other traditional values.

---

[2] The 2011 Organ Transplantation Act in Korea allows a family's refusal of organ donation even if a donor has expressed his/her explicit wishes to donate his/her organ(s). This shows that one's autonomous decision is thoroughly disvalued.

5.6 Some Practical Suggestions

In order to improve the balance of the relationship between an individual and one's family, or between autonomy and the value of family, I would like to suggest the following practices.

First, sufficient time for decision making should be given. This will help allow patients and research participants the necessary time to weigh and consider one's values in addition to how the values and wellbeing of the relevant family members will be affected.

Second, consultation services providing medical information, psychological comfort, and the like should be provided. Such services will help patients or research participants make their decisions through the processes of effective deliberation and moral reflection.

Third, the legal requirement to ask patients or research participants whether to discuss decisions with his/her family members or to consider their expected responses should be reviewed. Asking this question will allow patients or research participants the chance to consider multiple perspectives, including the values and needs of one's family members.

Finally, society should try to cultivate a social environment in which autonomous decisions can be made. For example, a decent health insurance system is necessary before advance directives can be introduced.

References

Beauchamp, T. L., and J. F. Childress. 2009. *Principles of biomedical ethics*. 6th ed. Oxford: Oxford University Press.

Beauchamp, T. L., and J. F. Childress. 2013. *Principles of biomedical ethics*. 7th ed. Oxford: Oxford University Press.

Callahan, D. 2003. Individual good and common good: A communitarian approach to bioethics. *Perspectives in Biology and Medicine* 46 (4): 496–507.

Etzioni, A. 2011. On a communitarian approach to bioethics. *Theoretical Medicine and Bioethics* 32:363–374

Fan, R. 1997. Self-determination vs. family-determination: Two incommensurable principles of autonomy. *Bioethics* 11:309–322.

Hardwig, J. 1990. What about the family? *Hastings Center Report* 20 (2): 5–10.

Hardwig, J. 1997. Is there a duty to die? *Hastings Center Report* 21(2): 34–42.

Kant, I. (1873) 1998. *Fundamental principles of the metaphysic of morals*. Trans: T. K. Abbott. In *Ethics: History, theory, and contemporary issues*, ed. S. M. Cahn and P. Markie, 275–318. Oxford: Oxford University Press.

Miller, B. L. 1981. Autonomy and the refusal of lifesaving treatment: Four cases, four senses of autonomy. *The Hastings Center Report* 11:22–28.

Chapter 6
The Confucian Alternative to the Individual-Oriented Model of Informed Consent: Family and Beyond

Kam-Por Yu

6.1 Confucian Ethics as an Alternative

In recent years, Confucian ethics has been considered as an alternative to the individual-oriented model of medical decision-making that is dominating in the modern West. As pointed out by Ruiping Fan,

> [T]he Western pattern of medical decision making has been *individualistic*, and the Chinese pattern *familistic*. … [T]hese two different patterns of medical decision making represent two different overall moral perspectives on human life and relations. (Fan 2002, p. 347)

Fan explained the Confucian alternative in this way:

> In contrast, Confucianism holds that the ideal human relation is characterized by the principle of humanity. In the Confucian account, humanity constitutes the basic virtue of individual humans and is the foundational principle of human society... Confucianism emphasizes that humanity requires one to practice love in the light of five human relations: father and son, ruler and minister, husband and wife, elder and younger, and friend and friend. Confucians understand that these five relations represent all the important natural types of human relation and form the basic patterns of human society. For Confucians, humanity requires one to give love, yet in giving love one must differentiate, follow a prescribed order, and attend to relative importance according to the different types of human relations. This is because different types of relations convey different moral significance in relation to the requirements of humanity. Confucianism requires that love be manifested in specific virtues for particular human relations. (Fan 2002, p. 356)

I would like to take the above representation of the Confucian alternative as my reference point in my discussion below. I shall argue that while it is right to say that Confucianism regards human relationships as the basis of human morality, and puts a lot of emphasis on the importance of family in a person's moral life, there is a gap between such features of Confucianism and the characterization of Confucianism as familistic in nature.

K.-P. Yu (✉)
General Education Centre, The Hong Kong Polytechnic University,
Hung Hom, Kowloon, Hong Kong, China
e-mail: kam.por.yu@polyu.edu.hk

R. Fan (ed.), *Family-Oriented Informed Consent*, Philosophy and Medicine,
DOI 10.1007/978-3-319-12120-8_6

What is the foundation of human morality? The Confucian answer is simple: human morality has its foundation in human relationships, which in turn have their source in human nature and human sentiments (Yu and Tao 2012). We find ourselves in some relationships with other people, such as the relationship with our parents and siblings. According to Confucian ethics, such natural human relationships form the most elementary basis of human morality. As it is said in the *Book of Changes*:

> Heaven and earth existing, all (material) things then got their existence. All (material) things having existence, afterwards there came male and female. From the existence of male and female there came afterwards husband and wife. From husband and wife there came father and son. From father and son there came ruler and minister. From ruler and minister there came high and low. When (the distinction of) high and low had existence, afterwards came the arrangements of propriety and righteousness. (Legge 1963, pp. 435–436)[1]

On the other hand, there are also some relationships that we enter into, such as by taking up the role of a government official or a medical doctor. Such roles and relationships are the basis of our ethical obligations. The following passage implies that responsibility comes with a certain professional role that one takes up:

> The Master said, "Do not concern yourself with matters of government unless they are the responsibility of the office you are in." Zengzi commented, "The gentleman does not allow his thoughts to go beyond his office." (*Analects* 14.26; Lau 1992, p. 141)

Confucius was once a minor official in charge of stores. He said, "All I have to do is to keep correct records." He was once a minor official in charge of sheep and cattle. He said, "All I have to do is to see to it that the sheep and cattle grow up to be strong and healthy" (*Mencius* 5B5; Lau 1984, vol. 2, p. 213).

The traditional Chinese conception of morality is *renlun* (人倫 or human relationship). Morality is regarded as based on human relationships. There are different kinds of human relationships, and different kinds of human relationships give rise to different kinds of moral obligation. By contrast, Western conceptions of morality for the most part regard impartiality as the essence of morality.[2] A *lun* or human relationship is defined with reference to a self. The human relationships referred to here are relationships between individuals, but never relationships between an individual and a collective (be it a family or a state). As noted by a number of scholars, the emphasis is put neither on the individual nor on the collective, but on the relationships. As a result, neither individualism nor collectivism is advocated in Confucianism (Liang 1989; King 1985).

The moral duty one has to another person is determined by the kind of human relationship one has with the other person, and the position one has in that human relationship. It is said in the *Analects*: "Let the ruler be a ruler, the subject a subject, the father a father, the son a son"(*Analects* 12.11; Lau 1992, p. 113). What is meant is that each person should do well with his role. The role for each person is different, and depends on the position the person occupies in his various human relationships.

---

[1] This paragraph from *The Book of Changes* is summarized in Ban Gu (32–92 A.D.) the *History of Han Dynasty* as "The *Book of Changes* said, 'As there are husband and wife, father and son, ruler and minister, superior and junior, propriety and rightness have a basis." In this summary it is made very clear that human relationships are the source of human morality.

[2] This is a characterization made by Thomas Nagel in his book *Equality and Partiality* (1991). Western philosophers such as Kant, Hare and Rawls fit this characterization very well.

Five major human relationships are commonly enumerated in the Confucian classics. In the *Mencius*, for example, the duties for different kinds of human relationships are spelt out: "love between father and son, duty between ruler and subject, distinction between husband and wife, precedence of the old over the young, and faith between friends" (*Mencius* 3A4; Lau 1984, vol. 1, p 105). This passage identifies five crucial human relationships (*lun*), and the principles (*li*) of these five human relationships.[3]

Out of the five major human relationships, three of them (i.e. father and son, husband and wife, elder brother and younger brother) reside in the family. This may give people the impression that the relation-based ethics in Confucianism has a lot to do with the family. On the other hand, it is also very clear that the family is regarded as an important domain of moral practice in Confucianism. For example, it is said in *Mencius*:

> Mencius said, 'There is a common expression, "The Empire, the state, the family." The Empire has its basis in the state, the state in the family, and the family in one's own self.' (*Mencius* 3A5; Lau 1984, vol. 1, p. 141)

It seems apparently right to say that Confucianism provides an alternative model that is different from the individual-oriented model, as it puts the emphasis on human relationships and gives a more important role to the family. However, do the above observations warrant us to conclude that Confucianism is not just relation-based but also familistic in nature? My answer is no.

6.2 How Confucian Ethics is Different from Familism

It is important to distinguish between Confucian ethics and familism. The two can be distinguished in a number of ways.

6.2.1 Multiple Value vs. Family Value

First, family is just one value among a number of core values recognized by Confucianism. It is neither the highest value nor the only value. It is a value to be considered and balanced rather than to be unequivocally upheld as sacrosanct and to override other values whenever there is conflict. The position of Confucianism is more accurately described as upholding multiple values, rather than singling out the family as having paramount moral status. For Confucianism, the real moral challenge is to balance different values, instead of arranging different values in a

[3] Some writers refer to the Confucian morality as a role-based morality. See Hansen (1993, 72). In discussing the Confucian concept of the person, Henry Rosemont emphasizes the importance of roles: "If we reject the view of human beings as free, autonomous, rights-bearing individuals, what are the alternatives?... For the early Confucians there can be no me in isolation, to be considered only abstractly: I am the totality of the roles I live in relation to specific others" (1991, 71–72).

hierarchical order (Yu 2010). According to this perspective, ethics is not just about distinguishing right from wrong, or good from bad, but also about making judgments and decisions when there is a conflict between different kinds of good. The key is not to choose between opposite values, but to balance and moderate.

According to the historian Wang Guowei (1877–1927 CE), the institutions of the Zhou dynasty (1111–249 BCE), which were designed by the Duke of Zhou (b. circa 1100 BCE), who was much admired by Confucius, and left everlasting imprints on Chinese culture and civilization, were based on three core values: *zunzun* 尊尊 (respect those who have high status); *qinqin* 親親 (maintaining good relationships with those who are related to oneself); and *xianxian* 賢賢 (appointing those who are virtuous and capable) (Wang 1973). Each of these three values is of crucial importance in itself, but each one also has serious limitations. Respecting those who have high status is essential for maintaining stability and order. Maintaining good relationships with those who are related to oneself is essential for fostering social cohesion. Appointing those who are good and capable is essential for promoting good governance. All the three values have important functions to serve, and so they should all be preserved at least to a reasonable extent. As a result, the important question to answer is not which of these three values is more important and which of these three should override the others, but rather: how can all three of them be achieved as far as possible, or how to strike a good balance between all these three areas of concern.

Likewise, in the Confucian classic *The Doctrine of the Mean*, good governance consists in a set of nine canonic principles (*jiu jing* 九經). They are: cultivate one's person; honor the good and capable; be devoted to one's kin; respect the senior ministers; understand the difficulties of the various officials; love the common people; attract the various artisans; give preferential treatment to people from afar; pacify the feudal lords (Legge 1971, pp. 408–411). The real challenge is not to do each as well as possible, but to do all of them as far as possible at the same time.

What exactly are the Confucian multiple values? Are there three or nine such values? How do we strike the appropriate balance when there is a conflict among the multiple values? These are important questions, but they are not questions that we have to answer here. It is sufficient for our current purpose to realize that family is just one aspect of concern in the Confucian scheme of values, and it is oversimplifying and misleading to characterize Confucian ethics as a kind of familism.

6.2.2 Relation-Based vs. Family-Based

It is also important to note that even if individualism is rejected by Confucianism, it does not imply that Confucianism must be associated with some form of collectivism, such as communitarianism or familism.

As pointed out by outstanding scholars of Chinese society and Chinese ethos such as Liang Shuming (1893–1988 CE) and Fei Xiaotong (1910–2005 CE), Chinese society is neither individual-based nor collective-based, but relation-based. Neither the rights of the individual nor the benefit of the collective is regarded as having priority.

In the words of Fei Xiaotong, the Chinese pattern is "self-centered," and the *self* is connected to different people with different types and different degrees of relationships:

> According to the Western pattern, all members in an organization are equivalent, just as all straws in a bundle are alike. This is quite different from the Chinese pattern. Social relationships in China possess a self-centered quality. Like the ripples formed from a stone thrown into a lake, each circle spreading out from the center becomes more distant and at the same time more insignificant. With this pattern, we are faced with the basic characteristic of Chinese social structure, or what the Confucian school has called *renlun* (human relationships). What is *lun*? To me, insofar as it is used to describe Chinese social relationships, the term signifies the ripplelike effect created from circles of relationships that spread out from the self, an effect that produces a pattern of discrete circles. ...... *Lun* stresses differentiation. In the *Book of Rites* (*Liji*), the "ten relationships" form a discontinuous classification. Gods and ghosts, monarchs and subjects, fathers and sons, the noble and the base, the intimate and the unconnected, the rewarded and the punished, husbands and wives, public affairs and private affairs, senior and juniors, and superiors and inferiors—these are principal types of human relationships. "Everyone should stay in his place" (*bu shi qi lun*); thereby, fathers are differentiated from sons, those remote from those close, those who are intimate from those who are not. (Fei 1992, p. 65)

Liang Shuming (1893–1988), in his book *Essence of Chinese Culture* (1989), contrasted Chinese society with both the individual-based and the society-based models. He argued that both the individual-based and the society-based models are Western ways of organization—the former best represented by Britain and United States, and the latter by the then USSR. He also clearly denied that Chinese society is family-based, and argued that Chinese society is relation-based. Neither the self nor the collective is given priority, but rather the different kinds of relationship a person finds himself in. In a relation-based ethics, the key question is not what is the best for an individual or the collective, but what is the best for the relationship. The key question is: what should I do given my role in the relationship. For example: what should a good son do? What should a good husband do? What should a good friend do?

> In the Confucian tradition, morality was particularistic. One's moral obligations to others were defined by their positions within one's network of personal affiliations. In Confucian moral discourse, the configuration of Chinese society was, as the great Chinese anthropologist Fei Xiaotong expressed it, "like the rings of successive ripples that are propelled outward on the surface when you throw a stone into water. Each individual is the center of the rings emanating from his social influence. Wherever the ripples reach, affiliations occur. The rings used by each person at any given time or place are not necessarily the same...." In such particularistic configuration, the nature of one's moral obligations to another depended on the precise nature of one's relationship to the other; for instance, a son's obligations to his father were different from a father's obligations to his son. (Madsen 1984, pp. 54–55)

6.2.3 Role-Based Perspective vs. Holistic Family Perspective

Even within the family, there are different kinds of relations, for example, father and son, younger and elder brothers, husband and wife. The ethics involved is based on the specific relationships as well as one's place in the relationship. This approach to ethics recognizes the important roles played by family members, but it is not familism, or specific kinds of familism such as family utilitarianism or family republicanism, which assume that there is some collective good, such as the best interest of the family, that all members of the family should try to promote, or that there is some kind of General Will, such that an unselfish true member of the family will act in a certain way for the sake of the family. In the Confucian relation-based ethics, however, different people in the family have different roles to play. What they are supposed to do is not the same. What is right for the son to do to the father may not be right for the father to do to the son. Likewise, what is right for the father to do to the son may not be right for the son to do to the father.

As there are different perspectives within the family, it is unjustified to reduce all such perspectives into a perspective called "the family's perspective." A relation-based ethics requires people to take up the perspective pertinent to their role in the relationship, instead of requiring everyone in the same collective unit to take up the same perspective.

The relation-based principles are stated in a number of Confucian classics.[4] In the *Book of Rites*, for example, it is said that there are five basic human relationships, and as there are two roles in each relationship, there are ten basic kinds of patterns of moral behavior.

> What is human morality? The father should be concerned with the well-being of the son. The son should be filial to the father. The elder brother should set an example for the younger brother. The younger brother should respect the elder brother. The husband should have life-long commitment to the wife. The wife should follow and support the husband. The senior should give benefits to the junior. The junior should be cooperative with the senior. The ruler should be actively concerned with the well-being of the minister. The minister should be loyal to the ruler. These ten things are called human morality.[5]

Let us for the moment forget about the larger society and just focus on the family. It is important to notice that even within a family, what each person is supposed to do in the family may not be the same. Let us use the parent-child relationship as an illustration: the parent should take care of the long-term well-being of the child, and the child should respect the will of the parent. Such a differentiation of roles has important practical implications. For example, as the parent is responsible for the long-term well-being of the child, it is justified for the parent to coerce the child to have healthy lifestyle, such as to have a balanced diet or to quit smoking. The parent is not doing his or her duty if the parent just lets the child makes his or her own choice. On the other hand, it would be wrong for the child to coerce the parent to have healthy lifestyle. The child can persuade the parent, but not coerce him or her.

We can put it this way: the priorities of values are different for the parent and child. For the parent, in dealing with the child, the care of the long-term interest of

---

[4] See for example the chapter "*Liyun*" in *The Book of Rites*, and Year 26 of the Duke of Zhao as well as Year 3 of the Duke of Yin in *Zuozhuan*. For a translation, see Legge 1885, vol. 1, 379–380; Legge 1960, vol. 5, 13, 718.

[5] "*Liyun*," *The Book of Rites*. My translation. For Legge's translation, see Legge 1885, vol. 1, 379–380.

the child should come before respect for freedom of choice. For the child, in dealing with the parent, the respect for freedom of choice of the parent should come before the care of the long-term interest of the parent.

The elder brother and the younger brother also have different roles to play. The elder brother should be the role model for the younger brother, and the younger brother should respect the elder brother. This implies the younger brother can adopt a more individualistic ethics, and the elder brother has to adopt a more collectivist or altruistic ethics. The younger brother just has to take care of himself, and give the elder brother due respect, but the elder brother has to bear in mind the influence that he might have on the younger brother in deliberating about what he should do. For example, in considering whether to donate a kidney to someone outside the family, the younger brother can take the decision more like a personal decision, but the elder brother has to consider whether this behavior sets a good or bad example for the younger brother. The parents, on the other hand, have a different role, and their primary duty is to ensure that the choice made by the child will not cause any significant harm to the child's long-term interest.

Unlike familism, the crucial question raised in Confucian ethics is not "What is the best for my family?" but rather "What should I do, given my role in the relationship?" The questions raised would be "What should a caring father do?" "What should a filial son do?" "What should a committed husband do?" "What should a supportive wife do?". This ethics of multiple perspectives is not adequately represented by the model of familism.

6.3 How Important is Informed Consent?

How important is informed consent? In answering this question, we will do well in taking the Confucian advice: "The gentleman devotes his efforts to the roots." (*Analects* 1.2; Lau 1992, p. 3) What is the purpose of informed consent? Why is it important? What function does it serve? To answer the question "what kind of informed consent is justified" or "how should informed consent be done?" we must first answer the question "what is the real meaning of informed consent?"

I would like to argue that (1) informed consent has remedial value; (2) informed consent has derivative value; (3) informed consent has limited value; and (4) informed consent contains composite values and internal conflicts.

"Informed consent" emerged as a remedial measure. Some practices are regarded as clearly unacceptable, and informed consent is devised as a way to prevent such kind of malpractices to happen. "Informed consent" originated as a means to protect research subjects. Nazi experiments during the Second World War were unethical because they were done without the consent of the subjects. The verdict of the Nuremberg trial listed ten principles that all research involving human subjects must observe, and one of which is "informed consent:"

> The voluntary consent of the human subject is absolutely essential. This means that the person involved ... should be so situated as to be able to exercise free power of choice without

> the intervention of any element of force, fraud, deceit, duress, overreaching or any other ulterior form of constraint or coercion. (Nuremberg Tribunal 1947)

Two problems, however, were left unsolved by the Nuremberg Code: (1) If the patient is in coma or in emergency situation, how can the patient give informed consent? (2) If a new treatment is likely to benefit a patient in coma or in emergency situation, but the treatment cannot be given to the patient because informed consent cannot be obtained, does this contradict the original intention of protecting the patient from harm?

In order to make sure that medical practitioners will not fall into the traps of violating the patients' informed consent, in recruiting them as research subjects or in including them in clinical trials, it is necessary to work out a version of "informed consent" that is workable for the practitioners. The "Declaration of Helsinki" was approved by the 18th World Medical Assembly held in Helsinki in 1964 (World Medical Organization 1964). It was less restrictive than the Nuremberg Code, and it was drafted by the Medical Association to meet the practical needs of the medical professionals. It introduced the concept of proxy or surrogate consent—when a person (e.g. a baby, a patient in coma) cannot give informed consent, a proxy can give consent on behalf of the person concerned. It also distinguished between therapeutic research and non-therapeutic research—the proxy can give consent to participate in therapeutic research (but not to participate in non-therapeutic research).

Both Beecher's 1966 paper[6] and the Tuskegee scandal made widely known in 1972 showed that the abuse of people as research subjects is not just a war time crime but can be a systematic misconduct of academic or clinical research. As a result, scientific researchers and medical practitioners today have to take informed consent very seriously, or else they would be accused of involving in unethical research or practice.

"Informed consent" is designed as an effective measure to protect the subject in a hostile environment. As such, it is an instrumental value, and its value is derivative from other more fundamental goods, such as the protection of the vulnerable or the respect of individual choice. However, as it is developed today, obtaining informed consent is more a means to protect the researchers from being accused of using the subjects inappropriately.

Let's consider the supposed values of protecting the interest and realizing the autonomy of the subject. How effective is informed consent as a means to achieve these purposes? Informed consent is an instrumental value. Its real value depends on how far it can promote the goal value. There seem to be two common justifications: (1) the protection justification: protection of people from being abused as research subjects; (2) the "autonomy" justification: Every human being of adult age and sound mind has the right to determine what will be done to his body.

According to the protection justification, the value of informed consent is derivative. It is, however, based on two dubious assumptions: (1) the subject knows

---

[6] In his paper, Beecher gave more than twenty real examples of unethical research whose results have been published in established journals.

what is good/bad for himself and will choose accordingly; (2) the subject has all the relevant information and is able to understand and make use of it intelligently.

The second justification regards informed consent as more fundamental in nature. However, the view that a person has a right to determine what can happen to his body does not seem to be a sound fundamental value, as it justifies selling organs and disallows the prevention of suicide. Respect for person can be recognized as a fundamental value, but there is a gap between respect of a person and respect of the person's choice, as well as a gap between respect of the person's choice and obtaining the person's consent. Respecting the choice of a person is just one way to respect the person, and there can be other legitimate ways.

Moreover, "informed consent" is not one single coherent value, but a composite value. "Informed consent" has three components: voluntariness, information, and comprehension. For there to be genuine consent, three conditions have to be met. First, the subject has to agree to participate without coercion or undue influence. Second, the subject has to be given information about the nature of the activity or participation, foreseeable risks, benefits, alternatives, etc. Finally, the subject has to understand the information given to him or her. There is tension and competition within these three components of informed consent. For example, there is tension between information and comprehension. The information component requires that information be given as complete and accurate as possible. The comprehension component requires the information given to be comprehensible and manageable. These are different concerns, and may lead to different attitudes and practices. For the sake of comprehension, it is however, very difficult to meet all the three conditions satisfactorily. In spite of all the efforts, the difference that can be made may be very small, especially when the stake of participation is very low.

6.4 The Confucian Response to Informed Consent

From the Confucian perspective, the rules and arrangements for informed consent are the peripherals. The nature of the doctor-patient relationship is the crux of the matter. What exactly is the doctor-patient relationship? What is the role of the doctor, and what are the doctors expected to do? Can the doctor be trusted with a certain task, or is the patient on his own take care of his own self-interest?

The Confucian role-based ethics carries an important ethical message: ethics is a collective enterprise. It is not the work of one person, but has to rely on the cooperation of a number of persons. Instead of requiring everyone to aim at the highest goal or to follow an ultimate principle, it is better to design roles according to the pursuit of a worthwhile goal, and then require everyone to do their role.

> Confucius said, "It is a better strategy to act according to one's role than to act directly according to the Way." (Legge 1960, vol. 5, p. 684)[7]

[7] *Zuozhuan*, Year 20 of the Duke of Zhao.

> Mencius said, "One who holds an office should resign it if he is unable to discharge his duties. One whose responsibility is to give advice should resign if he is unable to give it." (*Mencius* 2B5; Lau 1984, vol. 1, p. 79)

The contemporary emphasis on informed consent in all kinds of human interactions fails to recognize the special relation between the doctor and the patient, and the special role that a doctor is expected and required to play. Informed consent is important in interaction between strangers, but people who have a loving or trusting relationship do not have to take so much effort to safeguard themselves from each other. The crucial question that has to be answered is: what exactly is the nature of doctor-patient relationship? How much trust can the patient put on the doctor, and when can it be said that the doctor has betrayed the trust of the patient?

Should the doctor-patient relationship be understood as a relationship between service provider and customer? If that is the case, then the major task of the doctor is to satisfy the preference of the client, and the major ethical principle is "*Caveat emptor*" (let the buyer beware). If the doctor-patient relationship should be understood as a trusting relationship, then the doctor is committed to the task entrusted (i.e. health of the patient), and the major ethical principle to follow is "*Credat emptor*" (Let the buyer trust).

The Confucians distinguish between two kinds of professions. (1) Noble professions: A person serving in the noble profession is called *shi*, which can be translated as "gentleman," "knight," or "scholar."[8] The main purpose of a noble profession is to fulfill a noble mission. Examples of this kind of noble profession include: government official, doctor, and teacher. (2) Common professions: the main purpose of common professions is just to earn a living. Examples of such profession include merchant, farmer, and worker. Mencius said that it is alright to take up the second kind of profession just to earn a living, but it is wrong to regard the first kind of profession just as a means to earn a living. Certain strict moral codes have to be followed in taking up the noble profession.

> Mencius said, "Poverty does not constitute the grounds for taking office, but there are times when a man takes office because of poverty…. A man who takes office because of poverty chooses a low office in preference to a high one, an office with a small salary to one with a large salary. In such a case, what would be a suitable position to choose? That of a gatekeeper or a watchman. … It is shameful to take one's place at the court of a prince without putting the Way into effect."(*Mencius* 5B5: Lau 1984, vol. 2, p. 213)

In Chinese history, medicine has long been regarded as a noble profession. Here are some passages expressing such a view:

> A great physician must have a tranquil mind and be free from desires. He must first be moved by the heart of compassion, and vow to relieve the suffering of all that have a soul. When those who are sick and painful come to ask for help, no attempt should be made to find out whether they are noble or mean, rich or poor. There should be no distinction with

[8] The following are some descriptions of *shi* (gentleman): "A gentleman must be strong and resolute, for his burden is heavy and the road is long" (*Analects* 8.7; Lau 1992, 71); "What is the business of a gentleman? To set his mind on his ambition" (*Mencius*, 7A33; Lau 1984, vol. 2, 279).

> regards to age or beauty. Friends or enemies, Chinese or foreigners, clever or stupid, should be treated alike—just like one's closest family members. (Sun 1996, p. 615)[9]
> Doctors have to meet very high ethical standards. Those who are not benevolent cannot be relied on. Those who are not wise and learned cannot be used. Those who do not have perfect personal integrity cannot be trusted. This is why since ancient time people choose doctors from famous families, who have generations of proven records of good ethical practice.[10]
> It is wrong to study medicine for the sake of earning a living. Medicine is an art which reveals the secret of heaven and earth, life and death. It is an art which can intervene with the course of nature, and save people from death. Only the best type of scholars with highest level of morality should be allowed to learn this art.[11]

6.5 Confucian Virtue Ethics at Work

Let us now consider some cases.

6.5.1 Case A

> Mr. A's father has fallen into a coma and will never regain consciousness. Mr. A can afford the time, energy, and money to take care of his unconscious father, but such expenses will substantially take away from his care of other members in his family. The hospital is now asking for his consent to turn off the life-sustaining equipments. If he himself was in such condition, he would not hesitate to turn off the machine for himself. But he worries that he would be criticized as being an unfilial son if he gives consent to turn off the life-sustaining equipment. What should he do?

If he can provide better care to his family as a whole by letting his father die, should he do so? From the perspective of familism, this may be the right thing to do. But will a truly filial son sacrifice his father for the greater benefit of the family? On the other hand, what does the son really care about—to be a filial son or to be *regarded* as a filial son? I think the above case can show very well the blind-spots of both familism and conventionalism.

In Taiwan, as in other Chinese societies, family members play a very important role in health care. Surveys have been done to find out whether people would support some form of euthanasia under some specific conditions for themselves as well as for their family members (Yang 2003). It was found that most people would prefer ending their life under some specific conditions (such as in vegetative state, or in a painful dying process) but they would not prefer the same for their parents. There seems to be some apparent contradiction here. If they prefer what they regard as the best for themselves, why would they want to have something else for their parents?

[9] Sun Simiao (581–682), *Essential Prescriptions Worth a Thousand Pieces of Gold*, 1.2, "Sincere Attitudes of a Great Physician." My translation.

[10] Yang Quan of the Southern Qi period (479–503), *Wuli lun*, Quoted in Lin 1993, 34.

[11] Xu Dachun 徐大椿 (1693–1771), *Yixue Yuanliu Lun*. Cf. Xu 2002, vol. 2, 57.

The major reason given was that people were influenced a lot by the idea of filial piety. They would like to be regarded as filial, and withdrawal of treatment may be interpreted as saving resources and not caring their parent enough.

When the son is considering whether he should give consent to terminate futile treatment for his parent, what is the right ethical question that he should ask himself? In Confucian ethics, the right question for the son to ask is none of the following:

1. What should I do to promote the best interest of my family?
2. What should I do to be regarded as a filial son?

But rather:

3. What should a person with the heart-mind of a truly filial son do?

Asking question (1) may lead him to give up treatment while asking question (2) may lead him to continue the treatment regardless of its effectiveness. By asking question (3), he cannot jump to the conclusion so quickly, but has to take into consideration the well-being and the will of his parent. The question he has to answer is: what should he do to truly care about his parent and respect his or her will? The parent should not be seen as just a utility-bearer that contributes to the well-being of the family as a whole. In role-based virtue ethics, the son's duty to the parent is based on their relationship and his role as a son, and is not conditional on the parent being a vehicle constituting and contributing to the so-called benefit of the family.

The following historical story is illuminating. In the year 1393, in the early Ming dynasty period, a man called Jiang Bo-er had a mother who was not able to recover from sickness after a long time. He heard that using their son's flesh to make soup could help to cure the disease of one's parent, so he cut off a piece of his flesh, and made "spare-rib soup" for his mother. His mother did not recover. He pleaded to the gods in the temple, and he promised to kill his son to make sacrifice to the gods if his mother could recover. His mother did recover, and he killed his three-year-old son to thank the gods. The local official reported the case to the Imperial Court, and suggested commending the filial son. The Emperor, who was well advised by his scholar-ministers, gave the following edict (Xiong 2007, pp. 35–38):

> The relation between father and son is a most important natural relationship. The person concerned is so foolish to kill his son. This is total annihilation of human relationship. How ridiculous it is to suggest commending such behavior![12]

Jiang Bo-er's behavior can be justified by familism as well as conventionalism. As one's mother is regarded as a more important member in the family than one's three-year-old son, it can be argued that killing the son to save the mother is to the greater interest of the family. Jiang's behavior was also regarded as filial according to the popular opinion at that time.

However, from the perspective of role-based Confucian virtue ethics, Jiang has a dual role to his mother as well as to his son. Killing his son to save his mother is wrong, as the father-son relationship is also a primary relationship that he has, and this role also has its firmest basis in human nature and human sentiments. Jiang's

[12] Translation mine.

thinking, however, was utilitarian in nature—he gave highest weighting to his mother, lower weighting to himself, and very low weighting to his son. This is a utilitarian rather than a Confucian scheme of thinking.

6.5.2 *Case B*

> Doctor B is a family doctor. He is at the same time doing medical research. Right now he would like to recruit a number of subjects. The research he is doing cannot benefit the subjects, and it may cause some discomfort and small risk. But he is quite sure that he can go through the procedures properly and obtain informed consent from his patients if he invites them to participate in his research. Is there any ethical problem if he can obtain his patients' informed consent to participate in his own research?

As a family doctor, he has a specific role to play, and he owes special obligations to his patients. The right question that he should ask is not:

1. What procedures or rules I should follow to avoid complaints and punishments?

or

2. What agreement can I reach with my patient?

But rather:

3. What should a doctor with the heart and mind of a truly good and caring doctor do?

Of course the doctor can do his own research, and he is free to recruit subjects to participate in his research. But there is a role conflict in him recruiting his own patients as the subjects of his research. As the patients' family doctor, he is entrusted with the well-being of the patients, so far as their health is concerned. He has the obligation to protect their health interests. Recruiting them in research that will not bring them any good and may even bring them some harm contradicts his role of caring. Other people can recruit the patients in other settings, but not through the family doctor's capacity, since he has an ethical relation with the patients.

6.6 Conclusion

Seeking prior informed consent of a person whenever something of substantial impact is done to the person is appropriate and necessary in human interactions without a special relationship. When there is a relationship, informed consent is not a primary issue that has to be handled first. The nature of the relationship is the primary issue, which provides the frame of reference for deciding how relevant informed consent is, and to what extent informed consent is necessary or appropriate. In the area of health care, the doctor-patient relationship is of primary importance, not procedural issues like obtaining prior informed consent from the patients concerned.

Confucian ethics is unique in giving human relationships a central place in ethical problems. Confucian ethics is more accurately described as relation-based, role-oriented, virtue ethics, than as familism. Giving human relationships a central place in ethics justifies not just giving family members a more important role to play in the medical decision-making process of a patient by his or her family members, but it likewise provides support to give a special role to the doctor, as the doctor-patient relationship is also an important and ethically relevant relationship.

Reference

Beecher, H. 1966. Ethics and clinical research. *New England Journal of Medicine* 274 (24): 1354–1360.

Fan, R. P. 2002. Reconstructing surrogate decision making: Aristotelianism and Confucianism on ideal human relations. *Philosophy East and West* 52 (3): 346–372.

Fei, X. 1992. *From the soil: The foundations of Chinese society. Trans. G. G. Hamilton and W. Zheng*. Berkeley: University of California Press.

Hansen, C. 1993. Classical Chinese ethics. In *A companion to ethics,* ed. P. Singer, 69–81. Cambridge: Cambridge University Press.

King, A. Y. C. 1985. The individual and group in Confucianism: A relational perspective. In *Individualism and holism: Studies in confucian and taoist values,* ed. D. Munro, 57–70. Ann Arbor: Center for Chinese Studies, The University of Michigan.

Lau, D. C., trans. 1984. *Mencius*. 2 vols. Hong Kong: The Chinese University Press.

Lau, D. C. 1992. *The analects*. Hong Kong: The Chinese University Press.

Legge, J., trans. 1885. *Li Chi [The book of rites]*. 2 vols. Oxford: Oxford University Press.

Legge, J. 1960. *The Chinese classics*. 5 vols. Hong Kong: Hong Kong University Press.

Legge, J. 1963. *The I Ching: The book of changes*. New York: Dover Publications.

Legge, J. 1971. *Confucian analects, the great learning, and the doctrine of the mean*. New York: Dover Publications.

Liang, S. 1989. *Zhongguo Wenhua Yaoyi [The essence of Chinese culture]*. Hong Kong: Sanlian Shudian.

Lin, Y. 1993. *Rujia Wenhua Yu Zhong Yixue*. Fuzhoushi: Fujian Kexue Jishu Chubanshe.

Madsen, R. 1984. *Morality and power in a chinese village*. Berkeley: University of California Press.

Nagel, T. 1991. *Equality and partiality*. New York: Oxford University Press.

Nuremberg Tribunal. 1947. Nuremberg Code. Reprinted in *British Medical Journal* 313 (7070): 1448–1449

Rosemont, H. 1991. *A Chinese mirror: Moral reflections on political economy and society*. La Salle: Open Court.

Sun, S. 1996. *Sun Zhenren Qianjin Fang*. Beijing: Renmin Weisheng Chunbanshe.

Wang, G. 1973. Yin zhou zhidu lun. In *Guantang Jilin,* ed. G. Wang, 451–480. Hong Kong: Zhonghua Book Co.

World Medical Organization. 1964. Declaration of Helsinki. Reprinted in *British Medical Journal* 313 (7070): 1448–1449

Xiong, Z. 2007. *Kan Liao Mingchao Bu Mingbai*. Hong Kong: Sanlian chubanshe.

Xu, D. 2002. *Yixue Yuanliu Lun*. Hong Kong: Dizhi Wenhua Chuban Youxian Gongsi.

Yang, H. I. 2003. Bad living than good death? A cultural analysis of family paternalism in death and dying in Taiwan. In *Asian bioethics in the 21st Century,* ed. S. Y. Song, Y. M. Koo, and D. R. J. Macer, 99–107. Christchurch: Eubios Ethics Institute.

Yu, K. P. 2010. The handling of multiple values in confucian ethics. In *Taking confucian ethics seriously,* ed. K-P. Yu, J. Tao, and P. J. Ivanhoe, 27–51. Albany: State University of New York Press.

Yu, K. P., and J. Tao 2012. Confucianism. In *Encyclopedia of applied ethics.* 2nd ed. 4 vols., ed. R. Chadwick, 1: 578–586 San Diego: Academic.

Chapter 7
The East Asian Family-Oriented Principle and the Concept of Autonomy

Lawrence Y. Y. Yung

7.1 Introduction

I shall begin with a brief sketch of the East Asian principle of autonomy first proposed in Ruiping Fan's (1997) "Self-determination vs. Family-determination: Two Incommensurable Principles of Autonomy."[1] The East Asian principle of autonomy supports family-sovereignty in bioethics as it regards the family and the patient together as an autonomous unit comprising the final authority in clinical decision making. Moreover, it assumes an objective conception of the good collectively understood by family members and it upholds harmonious dependence within the family.

There has been a persistent concern in bioethics about the role of the family in clinical decision making. A major worry is that the family's involvement in an adult patient's clinical decision making may interfere with the patient's autonomy and his best interests. Therefore, while the family-oriented principle is subject to critical scrutiny, the East Asian principle of autonomy is criticized for being "a principle of family autonomy" and hence a misimplementation of the ideal of autonomy, like the Millsian individualistic notion of autonomy as freedom to choose treatment in bioethical contexts.[2]

In this chapter I will discuss Kyungsuk Choi's criticisms of the East Asian principle of autonomy and his objections to the family-oriented principle in East Asian bioethics. In addition to Choi's objections, I will also address two moral issues

[1] Fan argues elsewhere that the Western and Chinese models of medical decision making manifest two different perspectives of human life and human relations. See Fan 2002.

[2] See Chap. 5.

L. Y. Y. Yung (✉)
Department of Public Policy, City University of Hong Kong, Tat Chee Avenue, Kowloon, Hong Kong, China
e-mail: sayung@cityu.edu.hk

R. Fan (ed.), *Family-Oriented Informed Consent*, Philosophy and Medicine,
DOI 10.1007/978-3-319-12120-8_7

pertaining to the ethics and practice of the family-oriented principle. The first issue is that the family-oriented principle may give rise to the problems of family-oriented paternalism in bioethics and withholding information from patients. The second issue is that under undue pressure from family members, a patient may succumb to familial interests and sacrifice his own interests. I argue that these issues pose serious problems for the East Asian family-oriented principle as a viable alternative to individual-oriented decision making in bioethics.

Contrary to Fan, Kam-por Yu argues that the Confucian model of informed consent is not a familial model.[3] I will examine Yu's position and complete my discussion by reflecting on the primacy of autonomy in bioethics against the backdrop of Fan's account of the East Asian principle of autonomy and Choi's call for a return to the Kantian non-individualistic understanding of moral autonomy as the ideal conceptualization of patient autonomy that provides the ethical foundation for informed consent.

7.2 The East Asian Principle of Autonomy and Family-Sovereignty

Fan argues that East Asian bioethics and Western bioethics can be differentiated not only in Tom Beauchamp's broad sense of morality but also in his narrow sense of morality.[4] In particular, East Asian bioethics and Western bioethics differ in the narrow sense with regard to the minimal substantive content of the principle of autonomy, which bears on clinical decision making. Accordingly, the East Asian principle of autonomy and the Western principle of autonomy are incommensurable on three issues: (1) who holds the final authority of decision-making, (2) what is the basis for the decision made, and (3) what is the major value upheld by the principle of autonomy. On Fan's view, "the East Asian principle of autonomy contains the minimal substantive content of family-sovereignty, an objective conception of the good and harmonious dependence, while the Western principle of autonomy carries the minimal substantive content of self-sovereignty, a subjective conception of the good and individual independence" (Fan 1997, p. 319). According to Fan the Western principle of autonomy is not a universal moral precept because East Asian bioethics is informed by its own principle of autonomy.[5] With its claim of family-sovereignty, the East Asian principle of autonomy appears to be drastically different

[3] See Chap. 6.

[4] See Fan 1997. According to Beauchamp's distinction, morality in the narrow sense is constituted by moral precepts that are vague, indeterminate, and yet universally binding whilst morality in the broad sense is constituted by moral precepts that are subject to further culturally and theoretically conditioned interpretations. Hence, no ethical precept in the broad sense of morality is absolute; only ethical precepts in the narrow sense of morality are universally binding. See Beauchamp 1996.

[5] Fan anticipates the objection that the East Asian principle of autonomy he proposes is not a principle of autonomy at all because it goes beyond the basic scope of the concept of autonomy

from the Western principle of autonomy, which regards the individual as the locus of autonomy.[6] In the absence of a standing formulation of the East Asian principle of autonomy, Fan has proposed the following:

> Positive Formulation: Every agent should be able to make his or her decisions and actions harmoniously in cooperation with other relevant persons.[7]
>
> Negative Formulation: No harmoniously made decisions and actions should be subjected to controlling constraints by others (Fan 1997, p. 316).

In contrast to the Western principle of autonomy, the East Asian principle is a less known and less established concept in bioethics.[8] Indeed, the East Asian principle of autonomy is "yet to be clearly articulated" (Fan 1997, p. 313). Unlike the Western principle of autonomy, which is an individual-oriented principle, the East Asian principle is a family-oriented principle. In the West, a competent patient is given final authority over his own body and well-being in bioethical contexts while his family can at best play a secondary role.[9] The patient can accept or refuse medical treatment according to his wishes and his private conception of the good, even if his family has a different view. In the West, the family is obligated to respect the patient's self-sovereignty and his capacity for self-determination. In East Asia, however, family-sovereignty is the norm, i.e., the family and the patient, together as an autonomous unit, share final authority when making clinical decisions. As Fan notes, "it is not a sick family member him/herself but the entire family that has real authority in clinical decision making. Western people might be concerned about

---

as it is understood in the West. See Fan 1997, p. 321. I discuss the concept of autonomy in the last section of this chapter.

[6] Unfortunately bioethicists have not been able to resolve the question regarding what we are to respect in respecting autonomy. Different suggestions include: autonomous choice, autonomous execution of a choice, autonomous agency, and the will of an agent. The conceptual enquiry into the notion of autonomy goes beyond bioethics and becomes metaphysical.

[7] On Fan's account, the other relevant persons in the clinical context would include family members, like the spouse, parents and adult children, in addition to the physician. Fan's discussion makes no reference to people who are emotionally or psychologically close to one another, e.g., adopted families and other domestic and intimate relationships.

[8] Here I am using "the Western principle of autonomy" as a loose term, without specifying any particular formulation of the principle of autonomy in Western bioethics. The principle of autonomy takes a central place in Western bioethics as it is assigned the task of grounding informed consent. However, despite the primacy of the principle of autonomy, bioethicists have not been successful in formulating a universal principle of autonomy acceptable to all. In the final section of this chapter I discuss Beauchamp's dissatisfaction with the unsuccessful attempts of bioethicists to establish a theory of autonomy that clearly "spells out its nature, its moral implications, its limits" (Beauchamp 2004, p. 214).

[9] In fact, even an incompetent patient in the West can exercise this final authority through arrangements like surrogate decision-making, living wills and advance directives. In general, the family is marginalized in clinical decision making in the West. As Mark J. Cherry notes, "The family is placed within a hermeneutic of suspicion, as failing to protect the best interests of its members" See Chap. 3 for an exegesis of the decline of the family under the individualistic character of the social-democratic egalitarian ideology that underlines the dominant approach to Western bioethics.

this claim of family-sovereignty. But the family under this notion can be viewed as an autonomous social unit from the physician and the state, analogous to the autonomous individual in the West" (Fan 1997, p. 317).

7.3 Family-Oriented Principle: Challenges and Responses

Kyungsuk Choi raises three problems concerning the claim of family-sovereignty introduced by Fan's account of the East Asian principle of autonomy. The first problem refers to the lack of a formal procedure for clinical decision making in the family. In Choi's own words,

> Majority rule presupposes autonomous individuals who are equal. Majority rule may be the best way to solve a problem when there are conflicts of opinions among equal individuals. But a family is not composed of equal individuals. Thus it is not so easy to say that a family may adopt a majority rule (See Chap. 5).

The second problem has to do with the burden of determining an incompetent patient's best interests and respecting his wishes. Suppose a patient is deemed incompetent because he is in a coma and, in accord with family-sovereignty, the family plays the role of a proxy decision maker on his behalf. Being incompetent, the patient cannot take part in the decision making process and thus cannot defend his own wishes. Consequently, his family may ignore his wishes, especially if his family has no knowledge of them and, therefore, cannot respect them. As a result, the family can only make a decision on the basis of the patient's best interests. Even if the family is aware of the patient's wishes, they may knowingly violate them on the grounds that they should act in accord with the best interest of the patient. At any rate, under family-sovereignty, one's family may disregard an incompetent patient's wishes in an attempt to act in the patient's best interest. This problem arises because family members are usually responsible for paying medical fees and the financial burden may constrain their clinical decisions.

However, the problems Choi raises are not restricted to the family-oriented model. To begin, the lack of a formal procedure for decision making can also be found in individual-oriented models. Generally speaking, illness may diminish a patient's ability to exercise the critical reflection required for patient autonomy. Due to deteriorating health conditions and the emotional turmoil caused by illness, a patient's cognitive capacity to understand the diagnosis and prognosis of his conditions, as well as his medical options, may have been compromised. His cognitive capacity to make autonomous decisions based on the information given to him may be greatly undermined. Moreover, despite the institutional implementation of informed consent procedures, a patient may still fail to critically reflect upon and evaluate his preferences and options for treatment before coming to a decision. In fact, the presumption that the patient is an atomistic, autonomous decision maker capable of making autonomous choices in his own best interest has been challenged by feminists and communitarians who conceive of autonomy as a relational concept and define selfhood as deeply imbedded within one's social relations with others.[10]

Family involvement in bioethics can help a patient in different ways. Self-interpretation and interpretation of the self by others are interrelated. As Charles Taylor puts it "we define our identity always in dialogue with, sometimes in struggle against, the things our significant others want to see in us" (Taylor 1994, p. 28). A patient's self-interpretation is important to his capacity for autonomy. His autonomous decision may be undermined by some internal weakness or psychological impediment such as irrational preferences, false consciousness, a belief in oppressive norms, or deformed desires. In these cases a patient would need the help of others in order to make a truly autonomous decision. Family-oriented decision-making can provide a much needed platform for patients to discuss their conditions with family members before making a decision. In the process of making a decision with his family, the patient may gain a better understanding of his conditions, his options, and even his genuine wishes.

Family-oriented decision-making is a shared decision making process that upholds an objective conception of the good accepted by the family. In dealing with the internal weaknesses and psychological impediments of a patient, family members may help the patient realize that his current wishes and preferences are not truly in his best interests or they do not fit with the objective good. As Fan points out, for the family-oriented principle "what is important is not that one's clinical decisions must be made by oneself in conformity with one's present desires. It is more important that they are made for one's long-term good impersonally understood" (Fan 1997, p. 317). However, according to Fan, under the East Asian principle of autonomy the family can significantly discount a patient's currently held desires, preferences or expectations if they do not agree with the objective good.[11] Does this mean that the East Asian principle of autonomy supports family-oriented paternalism in bioethics? Does the East Asian principle of autonomy allow the family to interfere with the rational preferences and genuine wishes of a competent patient? I discuss family-oriented paternalism in the next section.

Family involvement may be particularly helpful when the institutional implementation of informed consent procedures fails to encourage the patient to engage

[10] See Stoljar 2011 for a recent example. Drawing on Charles Taylor's social theory of agency and his communitarian arguments against the atomistic conception of the self, Stoljar argues that informed consent (in the sense of information plus noninterference) is not sufficient for patient autonomy partly because patients, as agents, "are situated in historical, social, class, race and gender contexts" (Stoljar 2011, p. 376). As such, this social situation has an impact on identity, self-conception, and the nature and development of important capacities like autonomy. In sum, autonomy is intrinsically social and relational because it is conditioned and limited by the social situation of human existence.In contrast, Jules Holroyd argues that the conditions for autonomous agency are not constitutively relational though the conditions for autonomous choice or autonomous action may be relational. See Holroyd 2009.

[11] Fan's example is, "if a patient refuses treatment because he judges his life is no longer worth living, while the relevant others do not think so in terms of the objective conception of the good, the patient's wish would not be followed, whether or not the patient is competent" (Fan 1997, p. 318).

in the critical reflection required for patient autonomy. A lamentable characteristic of modern medicine is that "the patient-doctor relationship has become highly constrained, formalized, and regulated instead of being a personal and communicative relation of interpersonal respect" (Rehbock 2011, p. 526).[12] Therefore, a family platform may provide the kind of caring and emotional support a patient needs most during illness. It may also empower the patient when dealing with difficulties from his illness and/or a highly constrained, formalized and regulated relationship with his physician. Aside from helping a patient endure the healthcare experience and providing long-term care after discharge, family involvement may also empower the patient by providing the assistance he needs for autonomous decision making.

Finally, Choi's concern about conflicts of interest could be a legitimate and practical worry. However, on the other hand, the family may also provide the financial aid a patient needs in order to cover medical costs. A more significant moral worry is that family involvement in the decision making of an adult patient may conflict with his best interests and may cause the patient, under undue pressure, to succumb to the interests of his family. I discuss this problem in the next section.

7.4 The Problem of Family-Oriented Paternalism in Bioethics

Among the three issues Choi raises, the burden of promoting the patient's best interests and respecting his wishes is a serious problem for family-oriented decision-making. Choi's worry is that under the claim of family-sovereignty, an incompetent patient's wishes are vulnerable to the discretion of the family who, in acting as proxy decision makers, may choose to promote the patient's best interests at the expense of his wishes. However, Choi's discussion does not spell out the full extent of this worry.

I will now examine the worst-case scenario arising from an endorsement of family-sovereignty. In determining the best interests of the patient, family members play the role of interpreters. Consequently, they may impose their conception of the good and their values upon the patient who may have very different values and preferences. However, family-sovereignty allows it on the grounds that family members are meant to act and decide on the best interests of the patient. This seems rather paternalistic.

Family-oriented paternalism is morally objectionable because it allows family members to substitute their judgments for those of the patient as if they knew better what was conducive to the interests and wellbeing of the patient. In addition, family-sovereignty may leave no room even for a competent patient to take part in

[12] Rehbock depicts a bleak situation for the patient as a solo decision maker: "If one only has to read and sign a long paper with a good deal of information and is otherwise left alone with the final decision, one does not get the feeling that one's autonomy is being respected" (Rehbock 2011, p. 526).

the decision making process because "generally, the Chinese take for granted that the entire family makes medical decisions for a patient, whether the patient is competent or not" (Fan 2000, p. 90). When the family takes charge completely it may lead to concealing illness even from a competent patient.[13] Therefore, the practice of family-sovereignty may lead to the worst case scenario of family-oriented paternalism in which, even if competent, the patient's condition is concealed from him and he is isolated from the clinical decision making process while family members violate his expressed wishes by acting in what they take to be in his best interests.

Obviously, the worry that family-sovereignty may become family-oriented paternalism raises a genuine moral concern for the East Asian principle of autonomy, rendering it an unlikely, if not a completely unjustifiable, principle of autonomy for Western people. The individualistic approach to bioethics in the West and the familial approach to bioethics in East Asia represent two opposing attitudes regarding the role of the family in bioethics. In general, the Western approach has taken great care to prevent or withhold the family from interfering with an individual's decision making, while the East Asian approach looks upon the family as an important source of help and support rather than a source of paternalistic interference. These opposing attitudes to the role of the family in bioethics suggest that East Asian bioethics and Western bioethics have different conceptions of the nature of the family and of individuals.

According to Fan, the ethics and practice of family-sovereignty in East Asian societies are shaped by the Confucian understanding of the family and individuals. Due to the Confucian moral requirements that "one should take one's family as an autonomous unit from the rest of society, flourishing or suffering as a whole" and, in medical contexts, that "the injury, disease or disability of one family member must be taken as a problem of the entire family" (Fan 1997, p. 317), it is only natural that it takes the entire family instead of just the patient alone to make medical decisions. In general, family-sovereignty is deep-rooted in the daily lives of East Asian societies. For instance, making a medical decision in accord with family-sovereignty is like making a decision to relocate the family.[14] People in East Asian societies have a Confucian understanding of the family and individuals and they generally adopt a familial approach to any issues that concern the family as a whole.

In contrast, although people in the West may approve of a familial approach to relocation, they generally do not welcome a familial approach to decision making in bioethics. In relocating a family, all family members in the household are stakeholders and the family should decide on the relocation by taking into account each member's needs for living, education and work. However, this is not an argument by analogy for the normative claim that insofar as a patient's illness concerns his family, other family members should take a more active role as stakeholders in his decision making. After all, relocation and medical decisions are different in

[13] See Fan 2000, 2004 for the implications of the East Asian principle of autonomy on the practices of truth-telling and on the Confucian view on truth telling in medicine. See also Chap. 15.

[14] I thank Eirik Lang Harris for the example of relocation.

important ways. The disanalogies are obvious. Relocating a family usually causes substantial change in the lives of everyone in the household whereas a patient's illness may only mildly affect other family members. For instance, suppose a patient and his siblings live in different cities and they only see each other a few times over the years and have very independent lives. In this case the patient's illness may not affect the siblings' lives other than causing some distress and concern. In the case of medical decision making, it is about what shall or shall not be done to the patient. It can be a matter of life and death. People in the West do not favor a familial approach to bioethics as they advocate respect for the patient's autonomy.

Additionally, in planning for relocation, a family may try to balance the needs of its members. However, it is inevitable that some needs may precede others and thus some needs may be compromised. For instance, if the parents have to change jobs and relocate to somewhere far away from where they live, the parents' need to relocate would certainly precede their children's need to stay in the community, peer-groups and schools with which they identify. And of course some parents would relocate their family in order for their children to receive better schooling. In any case, when relocating a family each member is a stakeholder and their needs will be weighted and considered differently depending on the circumstances.

In contrast, when the family makes a medical decision by exercising family-sovereignty, the patient's needs and wellbeing are usually given overriding priority so that other family members sacrifice themselves for the sake of the patient. For example, family members may take leave from work in order to take turns looking after the patient.Nevertheless, it is not uncommon for a patient to refuse treatment because he wants to relieve his family of the financial burden of his medical treatment. For instance, elderly patients usually choose to sacrifice themselves in order to save family resources for younger generations. In some of these cases they may succumb to familial interests because of undue pressure from other family members. However, in normal circumstances the ethics of family-sovereignty can have the potential merit of prioritizing the patient's best interests above those of other family members.

Gerald Dworkin famously argues that paternalism could be justified in those "pure cases" in which "the class of persons whose freedom is restricted is identical with the class of persons whose benefit is intended to be promoted by such restriction" (Dworkin 2008, p. 283). Such cases are justifiable because in retrospect the person whose freedom is restricted would consent to the interventions if he possessed the relevant information. Likewise, family-oriented paternalism can be justified if the patient would consent to the interventions if he possessed the relevant information for the decision made. In such cases the patient would come to endorse the good brought about by the intervention as serving his best interests and wellbeing. However, there are two unsettling challenges for the family-oriented principle. First, it is not clear how the family can protect the best interests of a patient when family members disagree and when the patient is subject to undue pressure to succumb to family interests. Second, it remains to be seen what a morally appropriate response for the family would be when the rational choice and genuine wishes of a competent patient do not fit the objective conception of the good supported by the family.

7.5 The Family-Oriented Principle: Confucianism and Familism

Yu argues that because of three crucial differences between Confucianism and familism the Confucian alternative to the individual-oriented model of informed consent is not a familistic model, albeit it supports a form of "family engagement" in medical decision making. Yu suggests that the first crucial difference between Confucianism and familism lies in a contrast of values. In his own words:

> family is just one value among a number of major values recognized by Confucianism. It is neither the highest value nor the only value...The position of Confucianism is more accurately described as upholding multiple values, rather than singling out the family as having paramount moral status. For Confucianism, the real moral challenge is to balance different values, instead of arranging different values in a hierarchical order... (See Chap. 6).

Confucianism on Yu's view upholds multiple values including the "family value" and others like the three core values (*zunzun*, *qinqin*, and *xianxian*) and the nine canonic principles for good governance (*jiu jing*).[15] Therefore, Yu concludes, "family is just one aspect of concern in the Confucian scheme of values, and it is an oversimplification and misleading to characterize Confucian ethics as a kind of familism" (See Chap. 6).

This argument appears elusive for a number of reasons. A confusing misnomer occurs in this argument as it seems to conceive of a "family value" and then conflates it with "family" as a concern for Confucianism. It may make sense to say that the family is "just one aspect of concern in the Confucian scheme of values." For Confucianism is also concerned with the state and the individual. But this constitutes a rather weak argument and it does not give much support to the conclusion that Confucianism is not familistic. Yu seems to reason that since family is just one aspect of concern for Confucianism, whereas family is the only concern for familism, Confucianism is therefore not familism. But this is problematic. After all, Confucianism can still be familistic if its concern for the state and the individual is extended or derived from its primary concern for the family. In other words, Confucianism is familistic if it supports a familistic conception of politics and takes the family as the ultimate foundation of human life on which one's social and political life is to be modeled.

Moreover, it is puzzling why the premise that the family is a Confucian value among others (i.e., the three core values and the nine canonic principles) leads to the conclusion that Confucianism is not familistic. Yu seems to believe that the "family value" is familial whereas the three cores values and the nine canonic principles for good governance are mostly nonfamilial, and, therefore, Confucianism is not familistic.

---

[15] The three core values are first articulated by Wang Guowei (1877–1927) in his study of the Zhou dynasty. Basically, *zunzun* means to respect those who have high status, *qinqin* means to maintain good relationships with one's relatives, and *xianxian* means to appoint those who are virtuous and capable. The nine canonic principles for good governance are found in *The Doctrines of the Mean*.

But this argument is muddled because it simply fails to define or explain the "Confucian family value." Presumably this 'Confucian family value' must be a familial value. It is true that the majority of the nine canonic principles are nonfamilial as they are related to good governance while only one (i.e. *xianxian*) of the three core values is nonfamilial.[16] So it seems that the three core values and the nine canonic principles could be used to support the claim that Confucianism upholds multiple nonfamilial values, therefore it is not a kind of familism.

However, this argument does not work because it is tainted with preconceptions. The argument mentions mainly Confucian nonfamilial values and is so unreasonably selective with Confucian values that it risks underrepresenting Confucianism. It is puzzling why this argument omits the two fundamental Confucian values, i.e., *xiao* and *di* (both are familial), which are of the utmost importance to Confucianism. The omission of *xiao* is particularly problematic. It is the most fundamental of all other Confucian values and it is the root of *ren* and many other values (*Analects 1.2*).[17] Therefore, the fact that the father and son relationship comes first in the five cardinal human relationships (*wulun*) is not without reason. In fact, those nonfamilial relations (i.e., the ruler and subject relationship as well as the friend and friend relationship) are modeled on, or at least in continuum with, the familial relationships. This familistic reduction of human relationships in Confucianism greatly weakens Yu's claim that the family is "just one aspect of concern in the Confucian scheme of values."

The second contrast is between relation-based and family-based. Yu draws on Fei Xiatong's view that "Chinese society is neither individual-based nor collective-based, but relation-based… the Chinese pattern is "self-centered," and the *self* is connected to different people with different types and degrees of relationship" (See Chap. 6). Yu then appeals to the work of Liang Shuming who "denied that Chinese society is family-based, and argued that Chinese society is relation-based. Neither the self nor the collective is given priority, but rather the different kinds of relationships a person finds himself in" (See Chap. 6).

The first problem with this contrast is that Yu does not support it with arguments. The contrast is presented more like a summary of the ideas of Fei and Liang. Obviously the lack of argument is problematic. For instance, to judge the passage alone, we do not know on what ground Liang "denied Chinese society is family-based" and why he "argued that Chinese society is relation-based." It reads such that we are supposed to take the ideas of Fei and Liang as absolute truth. A related problem is that some tension exists between Yu's quotations of Fei's view on *renlun* (human relationships) and the Chinese pattern of "self-center" (i.e., the claim that the individual is the center of the social relationships that circle around him), and Liang's view that neither the self nor the collective is given priority. If we judge the passage alone we cannot determine how this "Chinese pattern of self-center" is different from the liberal Western self. In other words, Yu appears to have risked

[16] See Chap. 6, for his account of the nine canonic principles.

[17] In the *Analects* there are several sections on *xiao,* see 1.9, 1.11, 2.5–8, 4.18–21, 13.18, 17.21, 19.18

supporting his account of Confucianism with some form of individualism, making it difficult to identify what remains uniquely Confucian in his account.[18] At any rate, Yu's point is simply that both Fei and Liang support the view that Chinese society is relation-based instead of family-based.

However, I think it is an exaggeration, if not a gross distortion, to insist that Chinese society is relation-based instead of family-based. The second contrast, as Yu presents it, seems to exaggerate the relational dimension of the Confucian self while ignoring the fact that Confucianism puts more emphasis on familial relations and non-familial human relations are modeled on familial relationships.

The final contrast is between role-based perspectives and the holistic family perspective. On Yu's view, familism assumes either "some collective good, such as the best interest of the family, that all members of the family should try to promote" or "some kind of general will, such that an unselfish true member of the family would act in a certain way for the sake of the family" (See Chap. 6). By contrast, Confucianism, as Yu argues, supports the role-based perspective, which "requires people to take up the perspective pertinent to their role in the relationship" (See Chap. 6). Therefore, Yu concludes, the Confucian "ethics of multiple perspectives is not adequately represented by the model of familism" (See Chap. 6).

I want to discuss the implications of this final contrast for bioethics. As Yu summarizes:

> The priorities of value are different for parents and children. For the parent, in dealing with a child, care for the long-term interest of the child should come before respect for freedom of choice. For the child, in dealing with a parent, respect for the parent's freedom of choice should come before care for the long-term interest of the parent (See Chap. 6).

This summary seems to suggest that despite all the perspectives held by family members, medical decisions are eventually made in accordance with some pre-established priorities of value. Curiously, if that is the case, to what extent do multiple perspectives contribute to the final decision? Given the priorities of value, what roles do relation-based perspectives play in clinical decision making? Can an adult child disregard the priorities of value by favoring the long-term interest of his ill father instead of respecting his freedom of choice? Moreover, are there any justifications for the priorities of value?

To conclude, Yu's attempt to differentiate Confucianism and familism is not satisfactory. Confucianism has been closely identified with familism for a long time. The "standard interpretation" is that classical Confucianism was the ideology of the *zongfa* system, which gave rise to the problems of submissiveness, traditionalism, and nepotism in Chinese society. It also led to "the problem of stranger" in nonfamilial impersonal contexts. Lastly, it conflicted with the rule of law and prevented the growth of public virtues or spirit in Chinese culture. Confucianism supports partiality as its familial and clan ethics regulate human relations and conduct according to familial and intra-familial bonds. Consequently, the ties between unrelated people are undermined, making social networking and civil cooperation

---

[18] A full explanation of this point is beyond the scope of this chapter.

difficult in Chinese society.[19] How Confucianism and familism have influenced each other is open to debate,[20] although I cannot go any further into this dispute here. However, my previous discussion should suffice to show that Yu's view, that the Confucian model of informed consent is not familistic, does not hold. Yu's charge that familism assumes "the best interest of the family" or "the holistic family perspective" may be valid for familism *per se*, but it not valid against Fan's family-oriented principle. Fan's family-oriented principle is familistic in that it regards the family and the patient together as the final authority in bioethical contexts. Under the family-oriented principle, clinical decisions are made on the basis of the patient's long-term good objectively understood by family members. The process does not invoke "the holistic family perspective" and it does not serve the best interest of the family.

7.6 The Primacy of Autonomy

Ruth Faden and Tom Beauchamp attest to the primacy of autonomy over beneficence and justice in the growing literature on informed consent.[21] However, they also note that given its conceptual uncertainty, respect for autonomy can be elusive when it is applied in bioethics. They warn that "major confusion can emerge over the precise analysis of autonomy if we move beyond the core idea that the autonomous person is not bound by controlling constraints and is in control of personal affairs" (Faden and Beauchamp 1986, p. 8). After actively working within the flourishing bioethics literature for decades, Beauchamp begins to question moral philosophers' success at "bring[ing] ethical theories and methods to bear on problems of practice," i.e., "the actual moral difficulties and issues presented in health policy and the health professions when decisions must be made about a proper action or policy" (Beauchamp 2004, p. 209). Beauchamp complains, "no clear connection has been made between conceptual analysis of autonomy and topics in bioethics" (Beauchamp

[19] However, contrary to the "standard interpretation," Joseph Chan argues that "classical Confucianism does contain important nonfamilial principles and values that can balance its familial ethics and constrain its tendency to cast everything in that light" (Chan 2004, p. 62).

A different approach can be found in Kim 2010. Instead of taking Confucian familism as "the critical obstacle to civil society", Kim Sungmoon proposes "to reconstruct a civil society by creatively repossessing Confucian familism…and to present it as an alternative to liberal civil society, which is predicated on moral individualism and asserts its unique mode of civility ("socialiablity") and citizenship ("strangership")—tenets that are largely unpalatable to the Confucian moral sensibility" (Kim 2010, p. 477). On Kim's view, a characteristically modern yet Confucian civil society can be founded on a moral individualism cultivated in Confucian familism.

[20] See, for instance, a recent chapter of this dispute between Liu (2007) and Guo (2007). Liu criticizes Confucianism as "consanguinitism in essence" and Guo responds to Liu.

[21] Faden and Beauchamp state, "respect for autonomy is the most frequently mentioned moral principle in the literature on informed consent, where it is conceived as a principle rooted in the liberal Western tradition of the importance of individual freedom and choice, both for political life and for personal development" (Faden and Beauchamp 1986, p. 7).

2004, p. 214). The second problem is that moral philosophers' conceptualization of autonomy seems to be leading nowhere.

The distinction between the metaphysical and moral concepts of autonomy has fostered confusing views and uncertainty currently surrounds the meaning of "autonomy," its relationship to the concept of *persons*, the descriptive or normative character of these concepts, and the connection between these notions and that of *respect* for autonomy (respect for persons). What we are to respect about autonomy remains unclear, and it remains obscure what "respect" means. Most obscured is how practice is affected by a theory of autonomy (Beauchamp 2004, p. 214).[22]

Beauchamp considers Gerald Dworkin's theory of autonomy more successful and penetrating than other formulations of autonomy although he still finds "many problems" with Dworkin's theory. In short, Beauchamp is pessimistic about the future of ethical theory in bioethics and he is deeply concerned that, despite the primacy of autonomy in bioethics, the various attempts at conceptualizing it have failed to produce a satisfactory account of respect for autonomy and, above all, the various formulations offered have no practical implications.

Beauchamp's deep concern is revealing. As some contest, Fan's account of the East Asian principle of autonomy may not be a proper principle of autonomy. For instance, in Choi's view, the East Asian principle of autonomy is "a principle of family autonomy" that cannot be a viable principle of patient autonomy because it risks revising and devaluing the ideal of autonomy in bioethics. Nevertheless, Fan's articulation of the East Asian principle has the merit of bringing the East Asian perspective to bioethics, especially its family-oriented principle, and thereby helps define the difference between Western and East Asian approaches to bioethics. Choi believes that family-oriented decision making cannot be a viable alternative to individual-oriented decision making in bioethics because his objections claim "the decision of a family is not the same as that of an individual person" (See Chap. 5). However, I have defended the family-oriented principle against these objections and have also laid out some important challenges for the family-oriented principle.

It should be noted that Choi's objections to family-sovereignty are only intended to regulate the role of the family in bioethics because he also holds that the family should be involved in one's clinical decision making. Drawing on John Hardwig, Choi calls for a return to the Kantian, non-individualistic understanding of moral autonomy as the proper ideal of autonomy in bioethics (See Chap. 5). His proposal is that the ideal notion of autonomy should reflect the patient's relationship with other family members and it should include moral consideration of others' lives and wellbeing.[23] By and large the ideal autonomous decision in bioethics would refer to

[22] Things actually begin to change. For instance, the primacy of autonomy is challenged by James Taylor, who argues that "the ethical foundation of informed consent is really concern for human well-being" (Taylor 2005, p. 384).

[23] Choi suggests that Bruce Miller's four senses of autonomy in bioethics (namely autonomy as free action, autonomy as authenticity, autonomy as effective deliberation, and autonomy as moral reflection) can be expanded to reflect the ideas of relational autonomy and family-oriented consent. His proposal is to incorporate consideration for other family members, as well as the community's

the decision of a rational person after reviewing all relevant information concerning not just one's own life and wellbeing but also the lives and wellbeing of others.

Similar proposals exist. Like Choi, Theda Rehbock also holds that "the problem is not the principle of autonomy itself, but the conceptual understanding of it as well as the concrete realization and institutional implementation of informed consent procedures under conditions of modern medicine" (Rhebock 2011, p. 526). On Rehbock's view, we need "a more precise, broader, and more differentiated understanding of the concept of autonomy and of its relation to other ethical principles" (Rehbock 2011, p. 524) if we are to see the principle of autonomy as unconditional and universal. Accordingly, "the principle of autonomy should not be restricted or abandoned in favor of other ethical principles such as beneficence or public interest" (Rehbock 2011, p. 524). Rehbock's proposal is to combine the non-individualistic Kantian account of autonomy as moral autonomy with an "Aristotelian account of happiness as a common good that includes justice and moral virtue (Rehbock 2011, p. 527). It is beyond the scope of this chapter for me to judge whether a Kantian formulation of autonomy can really answer all of the problems concerning respect for autonomy in bioethics, but I do maintain that any satisfactory account of autonomy has to address Beauchamp's concerns: it should have practical implications and it should indicate to us what it is about autonomy that we are to respect.

7.7 Conclusion

In this chapter I have argued that the East Asian family-oriented principle is defensible in theory as a shared decision making model and is thus a viable alternative to individual-oriented decision making in bioethics. I have responded to some misconceptions concerning the familistic nature of the family-oriented principle. Moreover, I have also discussed whether the family-oriented principle is workable in practice. I identified and considered two crucial problems in employing the family-oriented principle, i.e., family-oriented paternalism and conflicts of interests between a patient and his family. These obstacles may appear formidable but they are not insurmountable in practice.

Acknowledgments I would like to thank the editor, Ruiping Fan, for his encouragement and patience. This chapter has benefited from valuable comments and constructive suggestions from Philip, J. Ivanhoe, Ho-Mun Chan, Sungmoon Kim, Eirik Lang Harris, Hektor King-tak Yan, Hsin-wen Lee, and Richard Kim during a monthly talk session in March 2013 at the Center for East Asian and Comparative Philosophy (CEACOP) at the City University of Hong Kong.

---

wellbeing, into autonomy as effective deliberation and to add the value of family and consideration of sound relationships between the self and others to autonomy as moral reflection. (See Chap. 5).

References

Beauchamp, T. L. 1996. Comparative studies: Japan and America. In *Japanese and western bioethics,* ed. K. Hoshino, 25–47. Dordrecht: Kluwer Academic Publishers.

Beauchamp, T. L. 2004. Does ethical theory have a future in bioeth*ics? The Journal of Law, Medicine* and Ethics 32(2): 209–217.

Chan, J. 2004. Exploring the nonfamilial in Confucian political philosophy. In *The politics of affective relations: East Asia and beyond,* ed. H. Chaihark and D. A. Bell, 61–74. Lanham: Lexington Books.

Cherry, M. J. 2014. Individually directed informed consent and the decline of the family in the West. In *Family-oriented informed consent: East Asian & American perspectives,* ed. R. Fan, 43–62. Dordrecht: Springer.

Choi, K. 2014. The ideal of autonomy and its misimplementation. In Family-oriented informed consent: East Asian and American perspectives, ed. R. Fan, 83–92. Dordrecht: Springer.

Dowrkin, G. 2008. Paternalism. In *Philosophy of law.* 8th ed., ed. J. Feinberg and J. Coleman, 281–291. Belmont: Wadsworth Publishing.

Faden, R. R., and T. L. Beauchamp. 1986. *A history and theory of informed consent.* New York: Oxford University Press.

Fan, R. 1997. Self-determination vs. family-determination: Two incommensurable principles of autonomy. *Bioethics* 11:309–322.

Fan, R. 2000. International perspective: Informed consent and truth telling: The Chinese Confucian moral perspective. *HEC Forum* 12 (1): 87–95.

Fan, R. 2002. Reconsidering surrogate decision-making: Aristotelianism and Confucianism on ideal human relations. *Philosophical East and West* 52:346–372.

Guo, Q. 2007. Is Confucian ethics a "Consanguinism"? *Dao* 6:21–37.

Holroyd, J. 2009. Relational autonomy and paternalistic interventions. *Res Publica* 15:321–336.

Kim, S. 2010. Beyond liberal civil society: Confucian familism and relational strangership. *Philosophical East and West* 60 (4): 476–498.

Liu, Q. 2007. Confucianism and corruption: An analysis of Shun's two actions described by Mencius. *Dao* 6:1–19.

RehBock, T. 2011. Limits of autonomy in biomedical ethics? Conceptual clarifications. *Cambridge Quarterly of Healthcare Ethics* 20:524–532.

Stoljar, N. 2011. Informed consent and relational conceptions of autonomy. *Journal of Medicine and Philosophy* 36:375–384.

Sutrop, M. 2011. Changing ethical frameworks: From individual rights to the common good? *Cambridge Quarterly of Healthcare Ethics* 20:533–545.

Taylor, C. 1994. The politics of recognition. In *Multiculturalism: Examining the politics of recognition,* ed. A. Guttman, 25–74. Princeton: Princeton University Press.

Taylor, J. 2005. Autonomy and informed consent: A much misunderstood relationship. *The Journal of Value Inquiry* 38:383–391.

Yu, K. P. 2014. The Confucian alternative to the individual-oriented model of informed consent: Family and beyond. In Family-oriented informed consent: East Asian & American perspectives, ed. R. Fan, 93–106. Dordrecht: Springer.

Part IV
Family Consent in End-of-Life Decision Making

Chapter 8
Family Consent in Medical Decision-Making in Taiwan: The Implications of the New Revisions of the Hospice Palliative Care Act

Shui Chuen Lee

8.1 Introduction

The second revision of the Hospice Palliative Care Act added a new criterion that allows for a patient's family to petition for the withdrawal of life sustaining treatment if a terminal, incompetent patient's wishes are unknown and the patient satisfies DNR criteria. In the third and most recent revision, the requirement of consulting an ethics committee has been removed. These changes reflect the long established practice of family consent in medical decision-making in Taiwan as well as in Confucian cultural contexts. This paper presents some of the key features of this practice and the attendant rationale for family decision-making in hospice care. It argues how and why family decision making is beneficial for both the patient and the family as well as for medical professionals. Some short-comings and difficulties are discussed and solutions proposed. This paper also examines how family consent and the mediation of medical professionals can enforce and protect the best interest of the patient.

This paper ultimately proposes a Confucian model for medical consultation and argues for the cooperation of the patient, the patient's family and relevant medical professionals in making decisions that best reflect the will of the patient and reduce the suffering of the patient and family members. Further, it will be argued that family participation should be extended to and used as a model in other bioethical contexts. To conclude, the underlying core Confucian values and ideas for family involvement and medical consultation will be highlighted and developed by contemporary Neo-Confucianism.[1]

[1] The contemporary Neo-Confucianism referred to in this paper is the so-called third generation of Confucianism, that is, the successor of the pre-Chin Confucianism of Confucius, Mencius and Xunzi, and the Song-Ming Neo-Confucianism. The main figures include the late Professors Tang

S. C. Lee (✉)
Graduate Institute of Philosophy, National Central University, Chungli, Taiwan
e-mail: shuiclee@cc.ncu.edu.tw

R. Fan (ed.), *Family-Oriented Informed Consent*, Philosophy and Medicine,
DOI 10.1007/978-3-319-12120-8_8

8.2 The Development of the Hospice Palliative Care Act in Taiwan

As medical technology and treatment have greatly improved in Taiwan, more and more lives have been saved and some have been sustained or prolonged in vegetative or semi-vegetative states. During the late 1990's, only about one-tenth of terminal cancer patients obtained hospice care while the rest occupied a large portion of intensive care facilities, resulting not only in a great demand for medical care, but also causing needless harm to patients and family members. Since laws in Taiwan require saving the patient at all costs, natural death is uncommon and end-of-life treatment becomes a last torture rather than a comfortable farewell to family and friends. Consequently, it has become more and more urgent for Taiwan to establish clear legal procedures for guiding end-of-life treatment. After years of vigorous debate, "do not resuscitate" (DNR) orders, as part of the Hospice Palliative Care Act, were finally legislated into law on June 7, 2000. The law's title suggests that its content provides for palliative care rather than directly addressing the ending of life for terminal patients. Reflecting the taboo surrounding this issue, some medical professionals will not mention "letting die" or "death" when referring to the law. However, the law addresses allowing terminal patients to die with their consent, DNRs, and withholding of other related invasive treatments. This is the first law to break the engrained medical tradition of saving patients at all costs, providing a first step towards solving some of the grave problems facing terminal patients, their families and medical professionals.

However, as it originally stood, medical professionals found the law to be almost useless because it only addressed the legal process for withholding life-sustaining treatments. When a patient was rushed in, in most cases, it was unclear whether the patient had a DNR order and it was rare to have two doctors available to certify the patient fit the legal criteria. At the same time, other medical laws required the immediate provision of life-sustaining treatment. Additionally, no law provided for the withdrawal of life-sustaining treatment once it had begun. Thus, the law was practically useless.

After two years revisions were clearly necessary. The second version of the Act allowed for the withdrawal of life-sustaining treatment if the patient's condition satisfied the requirements for withholding such treatment. This revision did make the law workable. Further application of the law meant further difficulties because patients usually entered a coma or became incompetent. According to the Act, if a patient did not have a DNR order in place prior to entering an incompetent state, life-sustaining treatment could not be withdrawn. This led to further revisions and protracted debate due to its introduction of family surrogacy when the patient's wishes are not clear. The debate reflected the deep-seated differences between the traditional Confucian acceptance of family involvement in medical decision-making and the disposition of some medical professional to respect patient autonomy.

Chun-i, Mou Tsung-san and Hsu Fu-kuan, who have written influential classics on Chinese and Confucian philosophies.

8.3 The Rationale for Family Involvement

The main revision included in the Act's third version introduced family members as surrogate decision makers when the patient was in a comatose state or incapable of making known his or her will. The Act outlines how the patient's family can arrive at a decision to enact a DNR when the patient is comatose, incompetent or under the legal age. It should be noted that family involvement is permitted only when the patient loses competency. This means personal choice still remains the foundation for decision-making. The Act also specifies that a family decision cannot contradict the explicit will of the patient before entering into a state of incompetence. The Act is written with Western priorities, that is, with the spouse, adult children, grand children, parents, brothers and sisters, grandparents and so on in that order. In cases of conflict, family members given priority on this list can overrule the decisions of others. The Act also guards against abuse or neglect of the patient's interests by further requiring that in such cases an ethics committee must be consulted. Consequently, it is clear that the third version of the Act is in line with the mainstream thought of Western bioethics. It is clear that family surrogacy is introduced because the patient's wishes are unknown and needed for the purpose of end-of-life decision making. Family decision making is crucial and necessary in such cases, but it is not a mechanism capable of overruling the personal decision of the patient.

However, in practice, family consensus is the aim of the Act and medical professionals usually uphold it. Even in cases where a DNR is in place and the patient's condition fulfills the required criteria, if there are family objections, medical professionals will not take the actions prescribed by the Act. This shows that family consent is paramount. In many cases, the presence of all relevant family members is requested. For example, in determining critical medical treatment for a parent, the consent of all brothers and sisters is usually requested. Even if the spouse and elders are present, medical professionals insist that other known family members be present before the final decision is made. In non-urgent cases medical professionals usually request the family members hold meetings to arrive at a consensus decision before taking further action. There have been cases where reasonable decisions were made and later contested by absent family members who blamed the medical professionals for acting against the will of the deceased patient. Sometimes extended family members, such as uncles or cousins, participate and are involved in the final decision. It simply is not true that family members with legal priority necessarily make the final decision. It is in practice that the significance of the Confucian concept of the family is recognized as a decision-making entity.

Family involvement appears to bring to the bedside a complicated decision making procedure that sometimes leads to disaster or a delay in treatment. Some also doubt the legitimacy of family participation and see it as a violation of the patient's right to make his/her own medical decisions. However, because the family is a closely knit and sharing unit, its members do not only share household expenses, but also intimate experiences and affection. Because family members share common experiences that constitute their self-identity, family members usually have a strong identity with their family and with each other. Joy and grief are shared and not just observed like bystanders. Strong sympathetic or empathetic feelings usually exist between family members. In Taiwan, it is said that when one family member is sick, the whole family is sick. The pain of a patient is also the pain of their family members, much like how a mother very often feels more pain with her child's illness than her child. Mutual concern and intimacy derive strong mutual responsibilities that naturally develop into a kind of care and duty to each other. End-of-life decisions are especially critical and painful for the patient and the family.

In Taiwan, family relations are generally still very intimate and family involvement in all kinds of personal activities is common practice. Medical decisions are no exception. Even though family members may not be living under one roof, they still share household expenses, the joy and grief of the family, and take strong responsibility for each other. For instance, a family member usually accompanies the patient when seeing a doctor and they receive the diagnosis together. Family members very often are the main respondent to queries and instructions from health professionals. The medical law in Taiwan is even written in such a way that diagnoses and results can be disclosed either to the patient or to the family, and in some cases the diagnosis is not made known to the patient at all. Family members can provide the most needed help and trust when one is sick and vulnerable. Family members are also usually a reliable source of the patient's values and preferences, which provides the best guidance for treating the patient and grounds the medical decisions for the health professionals. In cases of terminal illness, the family's legitimate participation is critical because it is often the case that the wishes of an incompetent or comatose patient are unclear for health professionals trying to determine which alternatives to deploy. The family as a whole is usually competent to make such a decision. This may help health professionals avoid some of the hard moral dilemmas. On the other hand, family participation is also useful for preventing the abuse of the weak and lonely patient by health professionals.

It should be acknowledged that family participation sometimes makes cases more complicated and sometimes can result in a deadlock. One common problem is when family members cannot arrive at a consensus among themselves. Family members may have significant disagreements about what is the best for the patient. The need for family discussion is important. Here medical professionals and ethics consultants can assist in conducting and moderating family meetings. Medical advice is important for preventing misunderstandings and unfounded worries about the medical possibilities. Ethics consultations can also clarify and reduce mistaken conceptions of the ethical and legal responsibilities involved. The hope is that through good will family members will arrive at a reasonable and harmonious decision. Through open discussion medical professionals can learn not only about the differences expressed among family members, but also about the family's structure and the values and preferences of the patient and the family as a whole. With such understanding, in cases of deadlock, medical professionals would be better able to protect the best interest of the patient. In cases of malicious manipulations, medical professionals are usually powerful regulators and protectors of the vulnerable patient. Since family members may be far apart and untraceable, the Act specifies how a family decision can be made without all members present and also provides a priority ranking of the family members in case deadlock occurs.

There may be instances where the family makes a choice that does not benefit the patient or is against the patient's expressed will. In such cases, medical professionals need to use their best judgment in managing the final decision. In cases of apparent abuse, medical professionals need to have courage and the support of the hospital and their colleagues to fight for the patient's rights and best interest. In cases of deadlock, one might consider the family as temporarily dissolved and the individual as temporarily constituting the family—one person is one family. In such a case medical professionals should employ their best judgment to treat the patient in line with his/her best interests.

A small puzzle may be what constitutes a family in modern society. In this essay, it shall be defined as one with the basic natural family as its base, usually containing those living under the same roof. Since members of Chinese society share household expenses even if they live and work in other cities or countries, these individuals are also regarded as family members. Sometimes, people with no legal status, such as partners or friends living and sharing together like family, are also eligible for membership. In such cases it is better if the patient has properly confirmed such a partner's membership in his/her family. It is often the case that a key person in the family is someone who does not belong to the family at all. It is supposed that this key person can represent the values and preferences of the patient and, therefore, his/her opinions should be consulted and carefully considered.

8.4 The Incorporation and Elimination of the Ethics Committee

In the case of DNRs, family consensus and decision is vital. Proponents of the Act largely include families who have had a terminal relative suffering the great pain of dying and those health professionals who take care of medically hopeless and suffering patients. Health professionals and hospice services in Taiwan will not accept a decision made by one or some family members without the consensus of all known family members because they are afraid of endless protests against them if a family member was not told of the decision and would have opposed the DNR if informed.

In order to prevent the possible abuse of family members, the third version of the Act requires an ethics committee be set up to make the final decision. It is not that medical professionals favor individual autonomy and the rights of the patient, but rather it is a check to guard against family members abusing the incompetent patient. It is necessary to be careful about such issues when preserving family involvement in medical decision-making, especially concerning end-of-life cases, because there are in fact tensions between patients and family members in East Asian societies.

In the old traditional Chinese family, there is no doubt that family members were more intimate and ethically related in almost all kinds of individual and family activities. Medical decisions were usually made in the best interest of the patient through a harmonious consensus in which everyone was satisfied. However, there were abuses, exploitation, and suppression of the wishes of vulnerable family members by those with superior power in the decision making process, especially the ruling father or some other person. Although family ties are still strong in modern Asian societies, it is obvious that they have weakened and diverse personal interests have become a part of modern life. Conflicts abound between family members. In medical decision making, decisions reached by patients or family members are usually other-regarding and self-sacrificing, however, it should be acknowledged that there are obvious cases in which a conflict of interests between the patient and family members, or amongst individual family members, may lead to the unnecessary suffering of the patient and a difficult moral dilemma for the relevant medical professionals. In the Act's third version, an ethics committee is required to review the decision. This is a reflection of the concern of medical professionals and bioethicists in Taiwan that the family could abuse the Hospice Palliative Care Act. This requirement is understood as shielding medical professionals from the burden of denying a family decision as well as providing critical protection for the benefit of the patient.

However, a similar problem arises for committee members. In one case reported by ethics committee members involved in the decision, the committee refused a reasonable decision arrived at by the head physician and the major family members for a DNR because some committee members were worried about accusations by other opposing family members. Requiring an ethics committee review can be seen as a result of the growing distrust of physicians and families in such cases of life and death decision making. In fact, the traditional mutual trust between physicians and patients has eroded over time because medical professionals have begun to protect their own interests and shy away from accusations brought by patients and their families. Understood as such, the ethics committee basically serves as a veto power in the decision making process.

The composition of the ethics committee includes members of fairly diverse disciplines and different walks of life. The ethics committee is meant to be representative and more objective in evaluating cases because they are less affected by personal predilections. However, committee members are usually somewhat detached and thus cannot meet the intensive commitment required for end-of-life decision making. Since such decisions require highly emotional and committed participation, it is not easy for someone removed from the patient's suffering and death to make a proper decision. The patient's family members actually share in the suffering. They are not and could not be bystanders. They are in deep emotional as well as moral dilemmas. It runs deep for intimate family members. Hence, their decision is critical and should be respected. Medical professionals are also supposed to be as concerned as the family members. In Chinese medical tradition, medical professionals are expected to treat the patient as they would be relatives and family members. Those called Confucian doctors have great compassion for the suffering patient. Hence, even though medical professionals may not be as affected as family members, they are committed to the patient as well. However, it is different for ethics committee members. As a fourth party in the decision making process, and very often not really involved with the patient or medical matters, it is understandable that committee members are more conservative in reaching a final decision. Consequently, some of the painful decisions made by family members and the involved medical professionals are lightly vetoed by the committee. To block the decision of the patient and family in this way, the ethics committee becomes a mechanism that furthers the pain and suffering of both the family members and the health professionals. Thus, it is reasonable that this provision was recently deleted from the fourth version on December 21, 2012. Henceforth, decisions made by family members supported by the medical professionals are final. Family members usually respect the suggestions of the professionals, however, in cases of conflict, family decision has the last say both legally and in daily practice.

The role of medical professionals has changed from dominant decision maker to collaborator and facilitator in the decision-making process. In typical cases, health professionals should not only provide medical advice, but also facilitate the family decision making process. They should attempt to learn about their patient's preferences, the individual family members, and their role in family matters. In foreseeing possible moral dilemmas, the health professional should help the patient and his/her family members hold meetings to express their respective view points and their different opinions, to inform them of the possible outcomes and their respective legal rights and duties. In this way, when a final decision needs to be made, the decision can be reached smoothly and most satisfactorily for everyone involved. In the most difficult cases, health professionals have to stand by the rights and interests of the patient.

A good and reasonable decision in favor of a DNR requires the cooperation of the patient, his/her family members, and the relevant health professionals. It is unreasonable for any party to have the upper hand. If a decision becomes an insolvable case, the patient's preference should prevail, for, when this happens, we regard the patient's family as dissolved and reconstituted with him/her as the only member. Medical professionals also have to protect the best interest of the patient, even against the patient's family. This is demanded by the professional code of health professionals and need be upheld and supported both by its professional members and the public.

8.5 A Confucian Model of Medical Decision-Making and Its Ramifications

It is part of the ethos and philosophy of Confucianism that family is one of the most basic human relations. It is called an "ethical relation" to signify the somewhat intimate relationship that constitutes naturally the identity and reality of our lives[2]. Since no human child can survive its early years without family-like caring

[2] The justification for intimate relationships and relational autonomy was first proposed by feminists. My concept of Confucian ethical relational autonomy is a reflection of the Confucian emphasis on the ethical family relation. See: Lee2007; cf. MacKenzie and Stoljar 2002.

by others and because the natural family is the paradigmatic case for all, family is regarded as part and parcel of our self-identity.[3] In traditional Chinese society, social and political duties are often bound up with the family, which is regarded as an inseparable whole for both fame and blame. Nowadays, in Chinese society, family ties are still quite strong. Though family members very often do not live under the same roof anymore, much sharing, including support for parents and elders, is still part of family life. Consequently, allowing family members to join us in personal decision-making is reasonable and very often provides the strongest protection for the individual. However, we need to acknowledge a number of reservations. First, in Confucianism and the Chinese family, there are other ethical relations, for example, nation and subjects, friends and social relations, that are significant as well. They all make different claims on one's responsibility. The family relation is of course a primordial one, but that does not mean it is absolute. The involvement of other relational parties, one's nation for example, is regarded as equally important, especially when medical resources are needed that are far beyond the financial resources of a person or a family. Mencius commended that a sage king had to provide family care for those in need (*Mencius, 1B:5*). This means that public authority and support are in the background of family relations. Second, parental disobedience is morally permissible, and sometimes necessary, when a family's decision is unreasonable (*Mencius, 5A:2*). The legitimacy of parental interference is based upon the reasoning that parents should be benevolent towards their children. Confucius never recommends blind obedience to one's parent or ruler[4]. Mencius admired Shun (the Sage king) in his deeds of disobedience to his father's unreasonable commands. Furthermore, other important and related persons could form part of the family and have equal status to other family members. For example, in the tradition of Confucian doctors, physicians are understood as becoming part of the patient's family. Medical professionals are not simply outsiders. Care providers form a close relationship with the patient and a virtuous physician is to treat his/her patients as his/her own relatives. In the physician patient relationship, the professional is assumed to treat patients as a parent devoted to their children. Thus great trust is usually vested in the physician to make the best medical decision for the patient. In the Confucian tradition, the physician plays a large part in medical decisions for the patient and the patient's family because of his professional skills and virtue. Hence, medical professionals have a duty to protect the reasonable interests of the patient, as a parent would for his/her child. In other words, the Confucian model of medical decision-making is a semi-familial one. In this model, family involvement in decisions is basic. However, the patient has the final say over his/her own treatment and decision-making, but the patient has to face the blame of his/her family if his/her choice hurts the family's interests. Medical professionals have to take into consideration the opinions

[3] Christine Korsgaard would call such an identity "practical identity" or "moral identity," with the implication that it defines our obligations and values. See: Korsgaard 1996, pp. 90–130.

[4] The *Book of Rites* mentions Confucius' story of Tsengzi who sustains slight, but not life threatening, punishment from his father.

and decisions of the family as a whole and also has to have the courage to stand for the rational as well as the personal interests of the patient.

In accordance with the usual Chinese ethos of death, the dying person, and sometimes the sick, becomes the supreme decision-maker. The patient's decision commands the respect of family members even though it may not be preferable for other family members and sometimes it may not be in the best interest of the patient him/herself. This is why it was previously proposed that the patient make the final decision, as though the family had been dissolved, when there are conflicts. Hence, the involvement of the family is not only necessary, but a better way to deal with the dilemmas that arise in end-of-life situations.

Based upon the Confucian conception of the family relationship, a proposal can be made for a basic model of medical decision-making.[5] The basic element of this model consists of four steps. First, physicians should try to build a good patient-physician relationship, so that both sides can achieve some degree of mutual trust and understanding; and in the process the patient's values, preferences and choices can be noted and clarified. Second, medical professionals should observe and determine the degree of competence and the rights of the patient, and should foretell the collapse of the patient and any possible urgent situations. Third, medical professionals should conduct family meetings to clarify the medical diagnosis, possible treatment, prognosis, duties and rights of the patient and his/her family, and provide medical recommendations. Finally, medical professionals should help generate reasonable consensus regarding medical treatments. With such a process for family consent, the decision is assumed to be a harmonious consensus within the family and also reasonably satisfactory for the medical professional.

This model could be applied to all sorts of medical issues, including the DNR requests addressed by the Hospice Palliative Care Act, treatment for terminal patients, and other related bioethical issues. For instance, the law providing for legal abortion in Taiwan has been accused of being too lenient, not discriminating between different types of requests for abortion and allowing for abortion based on the subjective feelings of the pregnant woman. On the one hand, it protects the pregnant woman from any possible harm due to the pregnancy, but on the other hand, it is too lenient, causing youngsters to engage in careless sexual behavior, which leads to an exceedingly great number of abortions in Taiwan. Some conservatives have proposed a mandatory waiting period and consultation before receiving an abortion. Without mutual trust between the pregnant woman and the consultant, the consultation would be a one-sided argument against abortion and may violate the spirit of the Confucian doctor. Even though by law a pregnant woman could independently decide to have an abortion, in practice, medical professionals in Taiwan usually require the husband to witness her request form. This practice may invite objections from the husband and often involves a violation of the woman's wishes. Consequently, this has led to women seeking underground abortions. There are also many reported cases of women being forced by their families or boyfriends to have an abortion against their will. A healthy abortion consultation should give full pro-

[5] I have elaborated a Confucian model of clinical consultation in Lee 2008.

tection to the woman while providing due respect to the innocent fetus, and should aim at reaching a mutual understanding with family members.

This kind of family consultation is also proper for research involving family members. For example, research involving biobanking should require family consultation because genetic materials and information contain many common familial elements. Even by personal freedom standards, individual consent is not enough in such cases. Genetic information is in fact the common property of the family.

The Confucian model proposed for family decision-making may also provide an alternative for the West. In Taiwan, some medical laws provide for the involvement of the family. In practice, family involvement is openly permitted even though in some instances there is a balance of shared authority with due respect given to the individual preferences of the patient.

8.6 Concluding Remarks: Clarifications and Refutations

The proposed model of family participation in medical decision making is built around those core ideas and values of Confucianism most prominently expressed in the *Analects* of Confucius and the thought of Mencius. The core value is *ren,* or *the mind of ren,* for Confucius and, for Mencius, the unbearable mind of other's suffering. For Confucians, this is the foundation of morality, which determines the moral actions of human beings. Some understand the Chinese word *ren* as composed of the roots, "two" and "man," and thus regard Confucianism as an ethic of relation. A recent archaeological discovery demonstrates that the original Chinese word, *ren,* is a direct combination of the two roots, "body" and "mind." This disposes a limiting interpretation of *ren* and Confucian philosophy to a relational one. There is no doubt that human relation guides our actions in practice, but it is obvious that this is not the only way. Confucian ethics is not limited to the familial relation. In the *Analects* and *Mencius* there is also mention of moral practice involving only the self. In fact, this is the most basic and profound way to cultivate one's morals. For instance, Confucius told his best student, Yan Hui, to: "Suppress your selfish desires and return to the ritual of *ren*" (*Analects*12:1). This is pure moral cultivation of the self. One is to rid oneself of any immoral desire and follow the spirit of *li*, that is, *ren* or *the mind of ren*. Mencius also emphasized the "thinking" of the heart/mind, that is, the determination of the unbearable heart/mind of what should be done. In the two later pre-Chin classics *Zhongyun* and the *Great Learning*, the practice of self-awareness for detecting and removing immoral desires is the most prominent and important step of moral cultivation. It is from this primordial starting point that different and diverse human lives are evaluated. Thus, human relations are the ways in which human morality can be expressed and moral ideals achieved. In a more concrete way, it is codified in *li* or rituals. However, Confucius points out that *li* has to be built upon the foundation of *ren,* and *li* should always be transformed according to the command of *ren* (*The Analects*, 2:23). This means moral rules should constantly be critically reviewed by the moral mind. Confucius would likewise take filial piety as the first, natural step of the practice of *ren* because it is when and where one grows up all along, and the parent-child relationship is surely the most basic relationship. However, as has been indicated in this essay and as many records in *The Analects* show, respect for elders and parents is not the only way to practice *ren* or to be a virtuous Confucian gentleman, *junzi*. Furthermore, Confucius, as well as Mencius, condemns those who stick rigidly to the requirements of rituals, especially those who regard them as absolute moral commands. They are said to be the thief of virtue (*The Analects*, 17:13)! Not only should we be flexible in how we express moral concerns, but practice should also be updated through learning. There is an oft-quoted saying of Confucius that expresses the idea that we can always learn something through common people, which implies that there are good deeds and achievement through which we can learn to improve ourselves. Confucius said he himself never tired of learning. It is not only a devoted love for knowledge and wisdom; it is a moral command to always improve ourselves and our ability to be better. To improve our bioethical actions we must update our scientific knowledge of the world and our social and moral knowledge of human society. Self-limiting and self-exaggeration is not the attitude of Confucianism.

Though the basic concern of Confucianism is rooted in family matters, Confucians never stop here. Confucians are always concerned with others, especially their suffering. The others Confucians are concerned with expand from the family to the social, the national to all people, and from animals and plants to, ultimately, the whole universe. Confucianism is holistic. Even within the human world, it never ends without encompassing all human beings. Family is but one stage of our moral deliberation. Therefore, Confucianism cannot be properly understood as familism. Neither can it be understood as only having moral concern for the family alone, or as assigning supreme value to the family. Hence, the above analysis of family consent is based upon the underlying requirement of *ren* or the moral heart/mind for the sympathy of the suffering patient. It is through the closely knitted moral community of the family and the Confucian semi-familial relationship of the physician and the patient that we specify the necessity and significance of family participation.

There are a number of similarities between Confucianism and the ethical theories of the West. Care ethics is most closely related to Confucianism because both emphasize the importance of a shared caring relation. Nel Noddings, an American feminist, understands care as the primordial relationship and derives ethical care from natural care, forming the foundation of morality. However, Noddings limits caring as a kind of personal affection and neglects the rational aspect of moral deliberation. Noddings does not accept natural caring as itself moral and her concept of caring is limited to concrete relations. For Confucianism, caring and sharing the suffering and joy of others is a natural universal sentiment. It is more than just a sentiment. It is a moral and rational command to ourselves. The unbearable mind is by no means stuck to our personal affection, as parent to child, but in itself is a universal caring for all.

Given that Confucianism also emphasizes the importance of community, some understand Confucianism as a kind of communitarianism. This interpretation recognizes the community implications of Confucianism, but forgets that the community is just another way to realize or express the moral command that flows from the moral mind. Community itself is not sacred and always needs to be checked by our moral mind or the virtue of *ren*. A community is a moral community only if it satisfies or is constituted by the principle or command of our moral mind. Moral principles constitute a moral community. For Confucianism, it is obvious that the origin of morality stems from our moral mind, which issues directives for action: when these directions are codified, they become moral principles. It does not matter whether they are called principles of *ren, yi, li, chi, shun*, or other names, as they all come from the moral mind and are constitutive of our moral community. These principles are the origin of community authority in moral matters. Similarly, virtue is the successful achievement of moral cultivation built into our predisposition to perform moral deeds. But, again, virtue is not the starting point of morality.

Although Confucianism cannot be strongly identified with Western ethics, this does not mean that Confucianism has nothing in common with Western theories. Confucianism does contain some Western moral elements, such as care, individual freedom, justice, etc., but it attributed different weightings to these elements. The primordial core of Confucianism is the heart/mind, such that we share each other's joys and pains. Through the sharing of experience, we form the first moral community within our family and extend our moral concern to an expanding circle from family to society, and ultimately to the whole universe. It is no accident that Confucians have a cosmic feeling: feeling upheld in awe of Heaven, the grand universe. In the mundane world, we start from the near at hand, that is, from our home. For the diversified modern world, our moral sharing may seem limited, especially to family sharing, however, this does not mean that we are and must be limited. In the field of medicine in particular, Confucianism provides a wealth of philosophical resources for better and more reasonable medical activity as an art of humanity or *ren-shu*.[6]

References

Confucius. 1984. *The Analects*. In *The four books,* annotated & ed. by Zhu Xi. Taipei: Legein Publisher.

Korsgaard, C. 1996. *The sources of normativity*. Cambridge: Cambridge University Press.

Lee, S. C. 2007. On relational autonomy: From feminist critique to a Confucian model for clinical practice. In *The Family, Medical Decision-Making, and biotechnology: Critical reflections on Asian moral perspectives,* ed. S. C. Lee, 83–93. Dordrecht: Springer.

Lee, S. C. 2008. A theory and model of ethical consultation. In *Bedside consultation: Theory and practice,* ed. S.C. Lee and D.J. Tsai, 1–11. Taipei: Hu-nan Book Company.

MacKenzie, C., and N. Stoljar, eds. 2002. *Relational autonomy: Feminist perspectives on autonomy, agency, and the social self.* Oxford: Oxford University Press.

Mencius. 1984. *Mencius*. In *The four books,* annotated & ed. by Zhu Xi. Taipei: Legein Publisher.

[6] In *Mencius*, 1A:7. "*Ren-shu*" is a political term first coined by Mencius. It was then extended by Chinese physicians in the Ming Dynasty to mean the art of medicine.

Chapter 9
Filial Duty as the Moral Foundation of Caring for the Elderly: Its Possibility and Limitations

Ilhak Lee (李日學)

9.1 Introduction

Humans are "social animals": one is born into the human relationship, nurtured with the support of other human beings, and has a chance to live life in its fullest sense only when he/she remains in this relationship. The family is at the core of this relationship. There is little doubt about the importance of family in one's life. In Asian countries, where family ties are strong, the younger generation regards it as their moral duty to care for their elderly parents in return for the grace they receive in the early days of their lives. This sense of responsibility and thankfulness to one's parents also extends to one's ancestors.

This concept is expressed especially well in the Confucian social order. Confucianism has influenced Korean society in every aspect of life. It has served as the backbone of social relationships, as well as political and socio-cultural standards. Among its values, filial duty (孝) guides children in the care of their parents. Filial duty was once regarded as a natural representation of the loving and caring relation between parents and children on which the whole social relationship was built. However, due to rapid and inevitable social changes this basic value seems obsolete and even oppressive (especially against women), because current culture put emphasis on personal autonomy and independence. Making the situation worse, filial duty has been (mis)understood as an unconditional, unlimited commitment and sacrifice, meaning children should do what, in the opinion of "others," is best for their parents, not what the parents prefer. It seems in this case children have little chance of demonstrating obedience, which is the proper understanding of filial duty. They also seem to have little chance for a discussion with their parents about the treatment they prefer, or what the parent would want. If we are to preserve the family

I. Lee (✉)
Department of Medical Law and Ethics, College of Medicine, Asian Institute for Bioethics and Health Law, Yonsei University, , 408 Administration Building, Yonsei Health System, Yonsei-ro, Seadamun-gu, 120-752 Seoul, Korea
e-mail: ilhak.lee@gmail.com

R. Fan (ed.), *Family-Oriented Informed Consent,* Philosophy and Medicine,
DOI 10.1007/978-3-319-12120-8_9

as the core of Korean social life and address the undesirable tendencies mentioned above, we must renew society's understanding of filial duty and its implications.

The idea of filial duty represented in the *Xiao Jing*'s Book of Filial Duty (孝經) explores what it means to care for elderly patients. The concept of "filial duty(孝)[1]" can serve as moral justification for children's obligation to care for their parents and provides good reason to involve family members in end-of-life care decision-making. However, as family members are included as principal decision makers, there is a need to support the patient actually receiving end-of-life care. It should be remembered that family members as decision makers are another form of substitute decision making and it is better to set limits on their authority. This issue will arise when we consider how elderly parents are cared for in Korea, especially when they become critically ill.

9.2 The Concept of Filial Duty in the Confucian Literature

9.2.1 Historical Background: The Confucian Chosun (朝鮮) Dynasty

Confucianism worked as a governing ideology from the point of its introduction to Korea. The Chosun (朝鮮) Dynasty (AD 1392–1897) was the period when Confucianism, especially Neo Confucianism (性理學) dominated every aspect of life, from politics to the conventions of the lay public. The strong influence of Neo Confucianism in Korea can be explained by several factors. First, politicians recognized that Confucianism would be the best way to achieve an ideal political system and, consequently, during the fourteenth–sixteenth centuries codes of law were actively enacted in an attempt to establish a Confucian political structure in Korea. Second, political leaders and Confucian scholars vigorously encouraged the spread of Confucian conventions in Korean society generally. People's lives were framed according to the Confucian conventions of a coming-of-age ceremony, a wedding ceremony, the mourning of one's parents, and ancestral rites (冠-婚-喪-祭) and all were educated about these rules. These efforts resulted in the establishment of a Neo-Confucian Korea beginning in the seventeenth century. Even though a foreign

[1] English translation for 孝 is not a simple task. There are some translations of this concept in English, e.g., filial piety (James Legge), filial conduct (Roger T. Ames), the treatment of parents (Arthur Waley), and filial duty (Ivan Chen). There are even some plain translations like "the love between parents and children." Each translation has its strengths and weaknesses, and nuance is often missed in translation. The conventionally used term, "filial piety." is unsatisfactory in several senses; filial piety is introduced in context of ancestral ceremony with religious sense but my focus is obligation to living parents with the sense of moral obligation. So I chose less used one, duty which fits well. In this article filial duty will be used for its neutrality and moral sense.

Although I have expressed the English meanings of some Chinese characters in this text, the translations are still problematic, so I leave some of the notes below concerning translations in characters for those able to read them.

power took control of the Korean peninsula during the first half of twentieth century with an apparent discontinuation of the Confucian tradition as a result, Confucianism survived the rapid modernization that took place in twentieth century. The weight of this tradition is due to the "long and thorough" cultural traditions. Consequently, the Confucian tradition is still the working ideology in Korea. The rites of passage are still Confucian in their content and procedures, and personal character is still valued when evaluating one's suitability for a public position. Moreover, everyday personal relationships are still deeply ruled by Confucian notions, including making distinctions according to age and gender and taking a hierarchical approach to organizations. The ancestral rite is a very important mechanism for justifying and preserving these Confucian conventions. People are reminded of the identity and continuity of family ties and those distinctions made according to age and/or gender, which fortify the resulting patriarchy by tending to ceremony (Cheong 2007, pp. 193–207).

Consequently, because end-of-life decision making is related to life and death, family relations, and public policy, it is important to understand contemporary Confucianism among Koreans.

9.2.2 *Filial Duty in Confucianism*

9.2.2.1 Loyalty (義理) and Propriety (禮) as Fundamental Interpersonal Virtues

Loyalty is the fundamental virtue of Confucianism. Loyalty is based on a sense of human (and emotional) relationship to others, and a loyal person is one who shows the utmost commitment to this relationship. However, according to differences in the strength of a relationship, loyalty is distinguishable between persons. Propriety means expressing a degree of loyalty proper to the relationship. Propriety is relative to the timing, place, and person (時宜). In Confucianism, one is right only when one makes a right action according to his place (名分) within relations, more specifically, the social order. In particular, according to filial piety, children have a duty to their parents because of their relation to them and this duty is to express their gratitude and love in a proper manner.

9.2.2.2 Filial Duty in the Book of Filial Duty (孝經)

In *The Analects* (論語), filial duty is mentioned as the foundation of all human relationships: "filial piety and fraternal submission! Are they not the root of all benevolent actions?" (Confucius *Analects*, Chap. 2).[2] We can briefly examine the Confucian teachings of filial duty in the Xiao Jing's Book of Filial Duty (孝經).[3]

[2] Filial piety(filial duty) and fraternal submission!- are they not the root of all benevolent actions? (Book I. Hsio) (學而. 孝弟也者其爲仁之本與).

[3] There are several versions of 孝經. I refer to 今文孝經 when citing.

Filial Duty Begins with One's Own Body

The scope of the Confucian notion of filial duty can be understood by reading the Book of Filial Duty. The first chapter of the Book teaches that filial duty begins with revering one's own body: "Our bodies—to every hair and bit of skin—are received by us from our parents, and we must not presume to injure or wound them. This is the beginning of filial piety" (Confucius *Xiao Jing,* Bk. I).[4] This natural and even basic duty expands throughout the chapter to encompass the duty of public service, which includes serving the King and gaining societal recognition, the final goal of the duty.[5] It begins with service to one's parents; it proceeds to service of the ruler; and it is completed by the establishment of character (Confucius *Xiao Jing,* Bk. I). In the eighth book, the duty is understood as a natural responses to one's parent's love, therefore, it is unimaginable for a man to be loyal to others but not his own parents. Only when a man has been a loyal child does he deserve other social roles in government or society.[6]

The relationship between a father and a son and its attendant duties thus belong to the Heaven-conferred nature; they contain in them the principle of righteousness between ruler and subject. The son derives his life from his parents, and no greater gift could possibly be bestowed upon him. As ruler and parent, the father deals with him accordingly, and no generosity could be greater than this. Hence, he who does not love his parents, but loves other men, is called a rebel against virtue, and he who does not revere his parents, but reveres other men, is called a rebel against propriety (Confucius *Xiao Jing,* Bk. 8).

The Practices of Filial Duty

In the tenth chapter of the Xiao Jing, the five practices that fulfill children's duties are provided.[7] The services a filial son performs for his parents are as follows:

---

[4] Our bodies—to every hair and bit of skin—are received by us from our parents, and we must not presume to injure or wound them. This is the beginning of filial piety. (開宗明義: 身體髮膚, 受諸父母, 不敢毀傷, 孝之始也).

[5] It commences with the service of parents; it proceeds to the service of the ruler; it is completed by the establishment of character (始於事親, 中於事君, 終於立身).

[6] "The relation and duties between father and son, (thus belonging to) the Heaven-conferred nature, (contain in them the principle of) righteousness between ruler and subject. The son derives his life from his parents, and no greater gift could possibly be transmitted. His ruler and parent (in one), his father deals with him accordingly, and no generosity could be greater than this. Hence, he who does not love his parents, but loves other men, is called a rebel against virtue, and he who does not revere his parents, but reveres other men, is called a rebel against propriety. (聖治: 父子之道, 天性也, 君臣之義也。父母生之, 續莫大焉。君親臨之, 厚莫重焉.故不愛其親而愛他人者, 謂之悖德; 不敬其親而敬他人者, 謂之悖禮).

[7] "The service which a filial son does to his parents is as follows: In his general conduct to them, he manifests the utmost reverence. In his nourishing of them, his endeavor is to give them the utmost pleasure. When they are ill, he feels the greatest anxiety. In mourning for them (dead), he exhibits every demonstration of grief. In sacrificing to them, he displays the utmost solemnity. When a son

> In his general conduct to them, he manifests the utmost reverence. In his nourishing of them, his endeavor is to give them the utmost pleasure. When they are ill, he feels the greatest anxiety. In mourning for them (dead), he exhibits every demonstration of grief. In sacrificing to them, he displays the utmost solemnity (Xiao Jing Book 10. Trans. Legge).

Practicing this duty then includes reverence(敬), nourishing(養), caring when ill(病), mourning(喪) and sacrificing(祭)[8]. These practices are natural but it is their performance as a response to love received that is important. In familial relations reciprocity is stressed. Theoretically parents will not be abusive to their children, because it is natural to love children (it is basic human nature). Likewise, children must do their utmost with their filial duty toward the parents. Are there any limits to the requirement of filial duty? It seems that, from the Book, one possible limit is when obeying parents will lead to the disgrace of the parents. In fact the Book does set limits on the practice of the duty: A child must argue against, rather than obey, the will of his parents if it does not fulfill the requirement of righteousness.[9] "Hence, since remonstrance is required in the case of unrighteous conduct, how can (simple) obedience to the orders of a father be accounted filial piety" (Confucius *Xiao Jing,* Bk. 15)? Accordingly, while fulfilling one's duty to one's parents, reason and rationality are still to be pursued.

The Adoption of Filial Duty in Korean Society

Filial duty expands beyond familial relationships to include society. It is understood as a cornerstone of the ideal Confucian nation. The virtuous relationship between father and son(孝) can be applied to the virtuous king-servant relationship(忠). Due to the implications for society of filial piety, the Korean government published materials containing exemplar cases of filial duty (孝行) for laypeople who could not read Chinese characters.[10] Combined with an emphasis on family ties, filial duty was always the top priority for ordinary people. Younger generations lived with and cared for their elderly parents, providing them with nourishment and respect. If one was regarded as disloyal to his/her parents he/she would not live in the same com-

is complete in these five things, (he may be pronounced) able to serve his parents. (紀孝行: 孝子之事親也, 居則致其敬, 養則致其樂, 病則致其憂, 喪則致其哀, 祭則致其嚴).

[8] Ancestral sacrificing will be better understood as "ancestral sacrificing ritual". In Korea, there were 10–16 ritual ceremonies remembering/celebrating ancestors' hidden merits. (4 generations upward(father, mother—great-great grand father/mother) has his/her own ceremony, and over has several ceremonies in total. There ritual ceremonies followed Neo-Confucian courtesy, minimally modified.

[9] Therefore when a case of unrighteous conduct is concerned, a son must by no means keep from remonstrating with his father, nor a minister from remonstrating with his ruler. Hence, since remonstrance is required in the case of unrighteous conduct, how can (simple) obedience to the orders of a father be accounted filial piety? 諫諍: 故當不義, 則爭之。從父之令, 又焉得爲孝乎!

[10] For example, in the Chosun dynasty (朝鮮, 1392–1897), the government published a book for public education entitled, *The Picture Book of Deeds According the Three Bonds* (三綱行實圖), which emphasized filial duty(孝), loyalty(忠), and marital loyalty(烈) as cardinal virtues (AD 1431).

munity. The obligation of filial duty therefore extends in proportion to how much the family means to the individual.

The exemplar cases put forth were not the stories or ordinary, everyday people. They contained foolish filial duty(愚孝), foolish loyalty(愚忠) and foolish virtuous women(愚烈). The exemplars required individuals to sacrifice themselves in extreme duty with no aim and no limits. Filial duty implies reciprocity and rationality in theory, but in reality it eroded to a brute justification of a vertical hierarchy where being of a higher caste requires obedience, sacrifice and commitment from those who are younger and of a lower social position.

This distorted understanding of filial duty in combination with loyalty (忠) has survived through the current day. During the Japanese occupation period (1910–1945) and the Japanese dictatorship (1945–1992), filial duty was confused with loyalty and obligation to dictators. In Korea, the concept and practice of filial duty still plays a crucial role in social life. However, we can imagine that the practice of filial piety is also often misled by factors outside of its original meaning. As such, it is necessary to compare people's conventional understandings with the original meaning of filial duty. It is also necessary to consider whether the responsibility of care can be justified by the Confucian notion of filial duty and how this value can be saved. This discussion leads us also to consider policy implications for the aging population whose burden of care often falls on younger generations, usually their children.

9.3 Future Research for Applying Filial Duty in Medicine

Filial duty, with its virtues and vices, is still a guiding principle for assisting family members in making decisions about end-of-life care. It locates the individual within those relationships that give him aims and meaning in life. Filial duty provides a good moral foundation for surrogate decision making for elderly patients because it requires children to care for their parents while also considering the genuine will of their parents. Children who are willing to care for their elderly parents and are very attentive to their will, can be reliable surrogates. However, there are still some issues to be resolved in the interest of a better decision making process.

First, we need more empirical information about the actual surrogate decision making process: What factors affect the decision? How is consensus reached? How are the children involved in advance care planning?

Second, contextual barriers to soliciting and fulfilling the patient's care preferences need to be identified.

Third, medical professionals need to be made aware that inviting the whole family to participate in advance care planning can result in better outcomes. At present, they typically abandon one of the stakeholders, either the patient herself or the family. Consequently, the medical care setting needs to be changed; more palliative/hospice care facilities should be made available, service charges for advance care planning should be reimbursed, and professional ethics consultations for end-of-life care should be provided.

These issues come into play in the following situation in Korea.

9.4 Filial Duty in the Twenty First Century in Korea

9.4.1 How are Dying Patients Cared for in Korea?

Life in Korea, especially when one becomes ill, centers on the family relationship. One is provided social resources (education and support) by his/her family and in response, as a member of the family, one adopts the duty of family. It includes participating in important rites (e.g., wedding, mourning, and ancestral sacrificing), and taking care of his/her old and ill parents. Elderly parents typically abide by the decisions their children make. The younger generation recognizes the care of their elderly parents as a moral obligation; meanwhile they find it quite burdensome (Cho 2006). In a national survey, over seventy percent of respondents answered that caring for one's elderly parents is a family obligation in 2002, but in 2008 this number dropped to thirty-seven percent (Statistics Korea 2011).[11] Recent cases of "lone death" (孤獨死, 無緣死) in Korea confirms the concerns of society. There was also the tragic story of an abandoned elderly couple that chose to commit suicide because there was no one to turn to (Huh 2012). The traditional obligation of filial duty is losing its grasp on the younger generation of today: they find the obligation more suitable for the state (Statistics Korea 2011).

However, in matters of healthcare, the moral obligation of caring still abides in the younger generation. When a parent becomes sick and cannot care for oneself, the common understanding among Koreans is that caring for one's elderly parents is a very natural obligation (Kim and Song 2012). In one study about end-of-life expenditures, almost 80 % of the out-of-pocket fees are covered by the children of the patient, with 15 % from the spouse, and 8 % from the patient (Lee 2012).

9.4.2 Family Members as Surrogate Decision Makers

As children bear the burden of care for their parents, they also play the role of surrogate decision-maker for determining end-of-life care. Only a small percentage of patients document their treatment preferences in advance directives or living wills, leaving the family members or physician to make the decisions. Although there

[11] I am not sure what led to this change in general attitude. But this 'expressed' change is a representation of underlying changes, like the generation gap (such that baby boomers have become seniors, and their children raised in a different culture have now become the majority of the respondents) and the governmental responsibility of social welfare that has become the dominant ideology among the younger generation.

have been campaigns to increase the public awareness of such documents (Cho 2012), it may take a long time to include patients in end-of-life decision making. For example, in a palliative care hospital, only 0.7 % of patients had of advance directives. Similar results have been reported several times (Park unpublished; Heo 2009). At this point one may ask, "does it matter if patients entrust their welfare to their children and that children do their best for the good of their parents?" The problem is that end-of-care decisions are often too personal to entrust even to the patient's own children. They are often overwhelmed by the possible loss of their loved one or exhausted from the long, burdensome care that is required. Confusion, a lack of pertinent medical knowledge, and the unavailability of legal and ethical counseling can lead to tragic decisions. Family members of dying patients are often in need of help and the moral justification for the authority of one's children needs to be explored. For now we have to trust that their children have the best intentions, and filial duty is often referenced in this regard.

From the perspective of biomedical ethics, the responsibility of filial duty has a double edge. Patients can make their wishes known through their husbands and wives or their daughters and sons, who take the wellbeing of their loved-one to be a top priority. Patients become a member of the community through a strong bond and the common goal of regaining health. They are located in a network of caring people. This network is composed of people with mutual understandings of each other so that decisions can be made on the basis of their deep communications. Patients can influence decisions by stating their treatment preferences. The principle of filial duty in end-of-life care holds for children. They are to consider both the physical welfare and the wishes of the patient. This considerate approach is compatible with other principles in biomedical ethics.

However, there are many obstacles to deal with. In Korea, patients are often times not provided with basic information about their disease and instead, family members, most often the eldest son, make medical decisions for them and are responsible for those hospital charges not covered by insurance (approximately 50 % of net expenses). Koreans consider care for the elderly to be the responsibility of one's children, which is understood as filial duty (孝ko, hyo or xiāo). Worldwide, it is quite natural for children to love and care for their aging parents, but in Korea it is considered the most fundamental duty of a human being.[12] In fact, people are evaluated by how they regard their parents because filial loyalty is fundamental to all Confucian ethics. Ignoring this duty is tantamount to committing a serious sin worthy of social blame and scrutiny. The old saying goes: "Serve your parents with everything: it will be of no use to mourn them after they have passed." This cultural belief supports the idea that a good child would be willing to cut off one's own finger so that their dying parent might be revived by the vitality of the blood and their sacrifice.

As such, children feel that it is their duty to keep a parent "alive" and to provide every possible means of curative care, regardless of the likelihood of recovery. Even in cases where parents explicitly refuse aggressive medical care, children feel (or

[12] See Kim 2008, *Confucius' Analects,* 1:2,6,11; 2:5,7; 4:18.19. 21.

are expected to feel) guilty if they cease to request every possible medical treatment. In contemporary Korea, children will sacrifice their jobs, houses, and savings to pay the necessary hospital expenses in order for their parents to remain in the hospital longer. Understandably, because children seek every treatment measure available, the economic burden is unimaginable.

In the meantime, the Korean family structure is changing: most Koreans now live separately from their parents, resulting in the loosening of family ties. Children seek to escape the burdensome duty of caring for their elderly, disconnected parents. When there remains little emotional bond to one's parents, it is difficult to fulfill one's filial duty. In addition, there is another problem related to this change of family structure. As a consequence, children are becoming less reliable sources of their parent's preferences for treatment. Filial duty requires a child to think from the perspective of the welfare and happiness of one's parent, but it does not provide an objective or quantitative standard for one to apply during deliberation. Because there is no proper standard to rely on, decisions are usually made in a cautious manner, i.e., the parent remains connected to a ventilator for as long as possible. Children do not want to risk any blame or scrutiny for "killing" their parent, especially not from their acquaintances. Consequently, a patient's living will is ignored and the decision is made without consulting the patient.

9.5 Policy Implications

Fulfilling one's filial duty is a step toward personal integrity and the perfection of self. To admit of an obligation to one's parents is to recognize one's personal position within society, especially with relation to one's family, ancestors and descendants. Consequently, policy initiatives should be undertaken to encourage a more rational and patient-centered model of surrogate decision making, however, not in the conventional or medicalized sense[13].

First, palliative care for dying patients should be more accessible in terms of quality and expense. Children should be informed of alternative treatments and their attending costs and benefits: aggressive life-sustaining treatment can be critical in the care of patients. However, sometimes it is useless or even harmful for the patient. Children should understand that palliative care is not abandonment of the patient and should not be seen as a failure of filial duty; in fact, it is often times a true act of beneficence.

Second, the patient should be included in the shared decision-making process. Conventionally, the patient is excluded from information sharing and discussions of the treatment options. These practices are undesirable from the perspective of patient autonomy, however, from the perspective of filial duty it is seeking the parent's wishes and letting them conform to the will of heaven.

[13] The term 'medicalized' is adopted to reflect the attitude of trying every possible means of treatment, even with little hope.

It is difficult for clinicians, as well as family members, to break bad news. Policy initiatives to promote advance care planning and documentation through advance directives should be undertaken. The newly revised Cancer Control Act (2011)[14] mandates disclosing the patient's diagnosis as well as the goals of palliative care to the patient. During intake sessions, physicians have the chance to ask for the living wills of patients. A patient's participation in care planning can improve compliance and assist the patient and family in preparing for what's to come.

9.6 Conclusion

Filial duty is representative of the natural relation between parents and children. It can serve as a guiding principle for biomedical decision making. Consequently, we don't need to abandon our tradition to be *ethical*. However, we do need to amend our view of filial duty such that it encompasses rational deliberation rather than just an unconditional, unlimited commitment. In addition, there are further policy issues that must be addressed before asking people to change their views on one's duty to care for terminal elderly patients.

References

Cheong, S. B. 2007. *The cultural grammar of Koreans*. Seoul: Saeng-gak-ui-namoo.

Cho, M. K. 2012. Advance directives: Patient's right to decision and treatment is important. *Mediphana News*. http://www.medifonews.com/news/article.html?no=82071. Accessed 23 Nov 2013.

Cho, S. 2006. Challenges of the aging society and familial support for the elderly. *Korean Population Studies* 29 (3): 139–157.

Confucius. 1960. Confucian Analects. In *The Chinese classics: With a translation, critical and exegetical notes, progegomena, and copious indexes.* Vol. I., 137–145. Trans. J. Legge. Hong Kong: Hong Kong University Press.

Confucius. 2011. *Xiao Jing*. Trans. J. Legge. http://www.ctext.org/xiao-jing. Accessed 15 Nov 2013.

Heo, D. S. 2009. Opinion on the legalization of foregoing futile life supporting treatment. Proceedings for Legislating Death with Dignity. Korea.

Huh, J. K. 2012. I killed your demented mom: I will follow her. *Chosun Ilbo*http://news.chosun.com/site/data/html_dir/2012/10/31/2012103100230.html. Accessed 23 Nov 2013.

Kim, E. J., and J. H. Song. Increasing lonely death. *Sisain*http://www.sisainlive.com/news/articleView.html?idxno=14389. Accessed 23 Nov 2013.

Kim, Y. O. (trans.). 2008. *Confucius' analects: Korean translation with commentary*. Seoul: Tongnamu.

Kim, Y. O. 2009. *Hyokyeong: Korean translation with commentary*. Seoul: Tongnamu.

Lee, E. J. 2012. Children of patients pay the care of elderly patients. *Money Today*http://news.mt.co.kr/mtview.php?no=2012021717148286574. Accessed 23 Nov 2013.

---

[14] Korean Cancer Control Act (2011). Act No. 10465. Available: http://www.law.go.kr/lsSc.do?menuId=0&p1=&subMenu=1&nwYn=1&query=암관리법&x=-609&y=-208#liBgcolor0 (accessed November 23, 2013).

Legge, J. (trans.) 1885. The classic of Filial Piety (XiaoJing), In *The sacred books of China: The texts of Confucianism,* Pt. 3, ed., 465–488, Oxford: Clarendon Press.
Park, J. C. unpublished. Advance directive, who writes it, how? Seminar material for "Monthly Seminar of Korean Association on Medical Law" Presented March 2011. Seoul, Korea.
Statistics Korea. 2011. Social survey. http://kosis.kr. Accessed 15 Nov 2013.

Chapter 10
End-of-Life Decision Making in Hong Kong: The Appeal of the Shared Decision Making Model

Ho Mun Chan, M. W. Tse Doris, Kam Hung Wong, Julian Chuk-Ling Lai and Chun Kit Chui

10.1 Introduction

Making decisions about end-of-life care (EOL care) for the terminally ill is often difficult for patients, family members and health care professionals, as EOL decision-making poses legal, medical, ethical, religious and cultural issues that cannot be easily tackled, such as considerations of the patient's dignity and autonomy, and 'paternalism.' The case of the 41-year-old brain damaged woman, Terri Schiavo in the U.S. (who was in a vegetative state for more than 15 years) has provoked discussion of the right-to-live/right-to-die and the implementation of advance directives

H. M. Chan (✉)
Department of Public Policy, City University of Hong Kong,
Tat Chee Avenue, Kowloon, Hong Kong, China
e-mail: sachm@cityu.edu.hk

M. W. T. Doris
Hospital Chief Executive, Caritas Medical Centre, Hospital Authority, Kowloon,
Hong Kong, China
e-mail: mwtse@ha.org.hk

K. H. Wong
Department of Clinical Oncology, Queen Elizabeth Hospital, Hospital Authority,
Kowloon, Hong Kong, China
e-mail: wongkh@ha.org.hk

J. C.-L. Lai
Department of Applied Social Studies, City University of Hong Kong,
Kowloon, Hong Kong, China
e-mail: ssjulwin@cityu.edu.hk

C. K. Chui
Department of Public Policy, City University of Hong Kong, Kowloon, Hong Kong, China
e-mail: floyd.chui@gmail.com

R. Fan (ed.), *Family-Oriented Informed Consent*, Philosophy and Medicine,
DOI 10.1007/978-3-319-12120-8_10

(including living wills, health care proxies/durable power of attorney). In particular, the case also raises a crucial question: "who should be the right person to make the decision to embrace the EOL care?"

Advance directives (ADs) have been widely recognized in some Western societies—e.g. the U.S., Canada, Australia—as instruments for competent persons to express their prior treatment choices of EOL care (for example, with respect to withdrawal/withholding life-sustaining treatment) before they become incapable of making such decisions. However, in many cases, ADs have failed to guide clinical decision-making and their utility has been questioned (Biegler et al. 2000). In addition, their application to non-Western societies has also been studied (Akabayashi et al. 2003; Bito et al. 2007; Bowman and Singer 2001; Colclough and Young 2007; Kwak and Haley 2005). According to the liberal model, the establishment of ADs is rooted in the principle of respect for autonomy. The fundamental value that underpins the use of ADs is the principle of respect for autonomy. Human beings should have the right to demand their EOL care in accordance with the advance medical instructions that they have made before they become mentally incapacitated. ADs were designed to solve the dilemma of decision-making in the case of patients who were no longer capable of exercising self-determination, and to avoid paternalistic decision-making by the physician or the family (Vollmann 2001).

Aside from giving advance treatment instructions, patients can use ADs to appoint their proxies. In the absence of ADs, the EOL decision will usually be made by the family. According to the prevailing liberal model, neither the proxy nor the family is supposed to make their *own* decision (Veatch 1998). Rather, when the patient loses his/her decisional capacity, the surrogate decision-maker (the proxy or the family) is expected to follow a substituted judgment standard by imagining being in the patient's situation, and then figuring out what s/he would have chosen for him/her had s/he been mentally competent and fully understood his/her situation. When no definite choice could be derived from going through this thinking process, the surrogate decision-maker is expected to fall back on the principle of best interests. According to this principle, the surrogate decision-maker should follow what a reasonable person would have chosen to promote his/her best interests had s/he been in the circumstances of the patient. The family is supposed to make the decision for the patient but not for themselves. The interests of the family are marginalized. The decision of the family merely serves as a means for the patient to exercise his/her autonomy or protect his/her best interests.

Thompson et al. (2003) conducted a vignette study, using hypothetical cases, to compare the views of cancer patients, healthy controls and medical staff on the binding nature of advance directives, and whether instructions in advance directives should be followed because they are "the right thing to do," and not because they are legally binding. Results have shown that in cases with cancer patients, instructions laid down in the ADs are considered less binding by the participants of the study than the opinions of physicians and nurses. (Thompson et al. 2003, pp. 297–298). Recently, the debates about a culture-based approach to EOL decision making have increased (Akabayashi et al. 2003; Bito et al. 2007; Bowman and Singer 2001; Colclough and Young 2007; Kwak and Haley 2005). However, it is argued that autonomy has a transcultural value (Sanchez-Gonzalez 1997). Research findings have shown that Asians were more likely to prefer family-based decision making than other ethnic groups (Chan 2004a, b; Ohi 1998; Tuschida 1998; Tse et al. 2003; Ohara 2000).

In Hong Kong, ADs are still not commonly used in EOL care planning. In August 2006, the Law Reform Commission of Hong Kong released its report "Substitute Decision-Making and Advance Directives in Relation to Medical Treatment" (2006). The Commission recommended that the Government should play a role in promoting public awareness and understanding of the concept of ADs. Only in July 2010 did the Hospital Authority issued the first detailed guidance on ADs in adults for clinicians in Hong Kong. However, the principle of autonomy that underpins the use of ADs is rooted in the culture of Western societies, and controversies exist about its applicability to non-Western societies. Furthermore, even with an EOL decision that is made with the aid of ADs, there may still be a risk of a premature death, or a prolonged and undignified dying process with avoidable distress and suffering. This risk can be minimized if the decision is regulated and guided by a public policy framework, and guidelines are based on a plausible, ethical EOL decision-making model. Yet there is no universally accepted model across different cultures. The key issues of the controversy are concerned with the role of ADs and the problem of determining who has the principal authority to make decisions about EOL care for the terminally ill.

The study reported in this chapter aims to examine opinions on life-sustaining treatment and ADs of patients, family members, and health care providers in Hong Kong with respect to who should be the principal decision-makers for the patient in various EOL conditions, and how to resolve the conflicts among various stakeholders in the decision making process. One of the main objectives of the study is to construct an ethical model for EOL decision-making that could be practiced in the cultural context of Hong Kong, and compare it with the liberal model, which is prevalent in the US and some Western other countries.

To achieve the above aims and objectives, the study investigated the views of advanced cancer patients, family members, physicians and nurses, by questionnaire surveys and face-to-face interviews based on two vignettes, with the first one adopted from Sahm et al. (2005) study in Germany and the second from Thompson et al. (2003) study in Scotland.

10.2 Research Design, Respondents, and Methodology

There are four groups of respondents in the study: advanced cancer patients ($n=34$), family members[1] ($n=40$) and health care providers: doctors ($n=23$) and nurses ($n=42$). The respondents were recruited by the snowball method in the hospice

[1] "Family members" included 6 bereaved relatives and 34 relatives accompanied advanced cancer patients attending the outpatient clinics.

center of the Department of Clinical Oncology and the out-patient clinic of the Palliative Care Unit of two public hospitals in Hong Kong respectively. The study was conducted from November 3, 2008 to April 28, 2009.

Data was generated through a combination of quantitative and qualitative research, i.e. questionnaires and individual face-to-face interviews. The aim of the questionnaire study is to collect data about respondents' in the following four aspects:

1. Demographic data, personal attribute, and relevant background information
2. Value and attitude profiles
3. Knowledge about different aspects of end-of-life treatment and care
4. Subjective experience in regard to life sustaining treatments (LSTs) and ADs
5. Choice of EOL decision making model

Self-administrated questionnaires were provided to medical professionals beforehand. For patients and family members, their responses to the questionnaire items were codified by the interviewer. After the completion of the questionnaire, two vignettes were presented to the respondents for individual interviews. The respondents were asked to approve or disapprove of the decisions made by the doctors in the vignette. After that, they were asked to state their own preferences if they were in a similar situation. They were then asked to explain the reasons for their responses to the two vignettes.

Analyses were conducted to identify significant relationships between data collected from the questionnaire survey and respondents' judgments and preferences in response to the two vignettes. SPSS 17.0 (SPSS Inc., fall 2008) was used to process the quantitative data.

All interviews and discussions were recorded by MP3 recorders and were transcribed verbatim with the help of software Express Scribe. The thematic analysis and approach mainly followed the methods used by Ritchie and Spencer (1994). The preliminary classifications of codes were in accordance with the key issues or questions in the vignettes, as well as the categories from Thompson et al. (2003). Revision and modification of codes were carried out in the process of indexing each transcription. The findings, including translated quotations from the interviewees, were then put into charts.

10.3 Results

10.3.1 Vignette 1

A 58-year-old patient has survived a bowel cancer operation, but now has metastases in the liver and lungs, which are being treated (chemotherapy) and which are stable at the moment. Now she is reasonably well and cooks in the afternoons for her granddaughter.

When she has a heart attack, which results in cessation of the heartbeat, she is not resuscitated because her husband says that she did not want to be resuscitated.

	Patients	Family members	Nurses	Physicians	*p*
Right/wrong	20.8/79.2 (n=24)	32.4/67.7 (n=37)	44.7/55.3 (n=38)	70/30 (n=20)	<0.01
Interviewee's preference	16.7/66.7/16.7 (n=30)	22.5/67.5/10 (n=40)	58.5/36.6/4.9 (n=41)	78.3/13/8.7 (n=23)	*n.s.*

Fig. 10.1 Responses to Vignette 1 in the four groups of participants: patients, family members, nurses, and physicians

	Professionals	Non-professionals	*p*
Right/wrong	53.4/46.6 (n=58)	27.9/72.1 (n=61)	<0.005
Interviewee's preference	65.6/28.1/6.3 (n=64)	20/67.1/12.9 (n=70)	<0.001

Note: The chi-square-test is used above in comparing frequency of answer in groups and association with categorical variables. The figures shown above are given in percentage. 'Right/wrong' indicates opinions that express agreement or disagreement with doctor's decision mentioned in the vignette. 'Interviewee's preference' indicates respondents' own decision for the same circumstance and the three options are: not want to be resuscitated/ want to be resuscitated/ not want to make any definite statement.

Fig. 10.2 Responses to Vignette 1 in participants regrouped as professionals (nurses and physicians) and non-professionals (patients and family members)

The decision of the doctor was,

- *Right*—because the patient did not want to be resuscitated
- *Wrong*—because her quality of life was quite good and death was not near

Would you, in this situation, if you had been given an advance directive,

- not want to be resuscitated,
- want to be resuscitated,
- not want to make any definite statement. (Figs. 10.1, 10.2)

The results indicate that 20.8 % of patients (n=24), 32.4 % of family members (n=37), 44.7 % of nurses (n=38), and 70.0 % of physicians (n=20) approved the doctor's decision in Vignette 1 *(right—because the patient did not want to be resuscitated)*, while 79.2 % of patients, 67.7 % of family members, 55.3 % of nurses, and 30.0 % of physicians disapproved the doctor's decision in Vignette 1 (*wrong—because her quality of life was quite good and death was not imminent*). When asking the respondents' preferences in case they were in the similar condition: 16.7 % of patients (n=30), 22.5 % of family members (n=40), 58.5 % of nurses (n=41), and 78.3 % of physicians (n=23), would not want to be resuscitated; 66.7 % of patients, 67.5 % of family members, 36.6 % of nurses, and 13.0 % of physicians would want resuscitation; 16.7 % of patients, 10.0 % of family members, 4.9 % of nurses, and 8.7 % of physicians did not want to make any definite statement. In sum, if the results were categorized as two groups: professionals (nurses and physicians) and non-professionals (patients and family members), 53.4 % of *professionals* (n=58) considered that the doctor's decision in Vignette 1 was *right*, 46.6 % disapproved of the decision, while 27.9 % of *non-professionals* (n=61) approved the doctor's decision in Vignette 1, 72.1 % disapproved the decision. Of professionals (n=64), 65.6 % would not want to be resuscitated, 28.1 % would want resuscitation, and 6.3 % did not want to make any definite statement, if they were in the similar condition. For non-professionals groups (n=70), 20.0 % would not want to be resuscitated, 67.1 % would want resuscitation, and 12.9 % did not want to make any definite statement.

The phrase "death was not near" is ambiguous. It may mean "death from cardiac arrest was not imminent" or "death from her underlying cancer was not imminent". Follow-up interviews clarified that the latter was the understanding of the respondents who disapproved the doctor's decision.

10.3.2 Vignette 2

A 68-year-old retired male lives with his wife and enjoys hiking. Although it was discovered that he has terminal liver cancer, his condition is stable and his life remains normal. In his advance directive, it is clearly stated that he does not want any life-sustaining treatment under a life-threatening medical situation. A few days ago, he was infected with pneumonia and was sent to a hospital. Since his life was threatened, the doctor decided to give antibiotics to the patient in order to save his life.

The decision of the doctor was,

- Right because the patient has a fair quality of life and antibiotics are not only for sustaining-life, but also effective in treating pneumonia
- Wrong because the patient's will is being ignored

Would you, in this situation, if you had given an advance directive,

- not want to receive antibiotics
- want to receive antibiotics
- not want to make any definite statement (Figs. 10.3 and 10.4).

The results of Vignette 2 indicate that 92.0 % of patients (n=25), 90.6 % of family members (n=32), 80.5 % of nurses (n=41), and 95.2 % of physicians (n=21)

	Patients	Family members	Nurses	Physicians	*p*
Right/wrong	92/8 (n=25)	90.6/9.4 (n=32)	80.5/19.5 (n=41)	95.2/4.8 (n=21)	*n.s.*
Interviewee's preference	6.9/72.4/20.7 (n=29)	13.2/71.1/15.8 (n=38)	22/70.7/7.3 (n=41)	13/82.6/4.3 (n=23)	*n.s.*

Fig. 10.3 Responses to Vignette 2 in the four groups of participants: patients, family members, nurses, and physicians

	Professionals	Non-professionals	*p*
Right/wrong	85.5/14.5 (n=62)	91.2/8.8 (n=57)	*n.s.*
Interviewee's preference	18.8/75/6.3 (n=64)	10.4/71.6/17.9 (n=67)	*n.s.*

Note: The chi-square-test is used above in comparing frequency of answer in groups and association with categorical variables. The figures shown above are given in percentage. 'Right/wrong' indicates opinions that express agreement or disagreement with doctor's decision mentioned in the vignette. 'Interviewee's preference' indicates respondents' own decision for the same circumstance and the three options are: not want to receive antibiotics/ want to receive antibiotics/ not want to make any definite statement.

Fig. 10.4 Responses to Vignette 2 in participants regrouped as professionals (nurses and physicians) and non-professionals (patients and family members)

approved the doctor's decision *(right—because the patient's quality of life is fair and antibiotics is an effective treatment of pneumonia)*, while 8.0% of patients, 9.4% of family members, 19.5% of nurses and 4.8% of physicians disapproved (*wrong—because the patient's will is being ignored*). When asking for the respondents' preferences in case they were in a similar condition: 6.9% of patients ($n=29$), 13.2% of family members ($n=38$), 22.0% of nurses ($n=41$), and 13.0% of physicians ($n=23$) would not want to receive antibiotics; 72.4% of patients, 71.1% of family members, 70.7% of nurses, and 82.6% of physicians would want to receive it; 20.7% of patients, 15.8% of family members, 7.3% of nurses, and 4.3% of physicians did not want to make any definite statement. In sum, if the results were categorized as two groups: professionals (nurses and physicians) and non-professionals (patients and family members), 85.5% of *professionals* ($n=62$) approved of the doctor's decision, 14.5% disapproved it, while 91.2% of *non-professionals* ($n=57$) approved of the doctor's decision, and 8.8% disapproved the decision. Of professionals ($n=64$), 18.8% would not want to receive antibiotics, 75.0% would want to receive it, and 6.3% did not want to make any definite statement if they were in the similar condition. For non-professionals groups ($n=67$), 10.4% would not want to receive antibiotics, 71.6% would want to receive it, and 17.9% did not want to make any definite statement.

10.3.3 Reasons for and Against the Doctor's Decision

The responses of Vignette 1 may seem to show that non-professionals tend to disregard the wishes the patient laid down in ADs, tend to consider ADs less binding than health professionals, and lay more emphasis on the quality of life of the patient than on autonomy (Sahm et al. 2005). Yet the responses to Vignette 2 show no such difference. This result can be explained by different perceptions of treatments mentioned in the vignettes. Although both resuscitation and antibiotics are considered as life-sustaining treatments, respondents believed that the former is more invasive than the latter. Further, some respondents regarded antibiotics as medicine for curing pneumonia rather than a kind of life-sustaining treatment that can only prolong life without improving patient's quality of life. The reasons for this response towards the doctor's decision in the vignettes therefore cannot be explained simply in terms of their attitudes towards the dilemma of autonomy vs. quality of life. Sahm et al. (2005) approach was quantitative, but ours was both quantitative and qualitative. Respondents were asked to explain the reasons for their responses towards the doctor's choice in the two vignettes. The results of our qualitative study gives a picture seemingly more complicated than Sahm et al. thought. Aside from autonomy, respondents also explained their views of the doctor's decision by making reference to best interests of the patient, the value of the family, the value of medical knowledge, and professional experience. The key issue did not seem to be simply a matter of autonomy vs. quality of life. The following is a summary of the responses.

1. *Autonomy*: Some respondents argued that out of a respect for autonomy the patient's prior instruction should be followed; while some said that not following the AD did not compromise the value of autonomy because the wishes of the patient were not so clear, and so it was not so clear that the AD was applicable, hence ignoring it was not a violation of the patient's autonomy.
2. *Best Interests:* Some respondents said that following the doctor's decision of not prolonging life was in the best interest of the patient because his health condition was regarded as poor; while some said that such a decision was wrong because the patient's quality of life was perceived to be reasonable good.
3. *The Family:* As for the value of the family, some respondents said that if the family member acknowledged that the AD should be followed, then it should be so; while some said that if following the AD would prolong the poor health condition of the patient, the family might feel upset later on, and so it should not be followed if we really respect the family.
4. *Medical Knowledge and Professional Experience:* Some said that authority of the medical doctor should be duly respected because of his/her expert knowledge, and so his decision should be taken if s/he chooses to act against the patient's oral AD, even if it was not supported by the husband in Vignette 1. Yet some other respondents said that if the doctor trusted what the husband said, the decision of following the AD should be correct because it is a professional judgment of the medical doctor.

Respondents did not have a tendency to rely on a single value, such as autonomy—as in the case of the liberal model—to justify their views. Other values were widely addressed in the justification of their responses. Furthermore, complexities and contingencies are common in EOL situations. It is often difficult to ascertain whether the patient can foresee accurately what his/her situation will be when s/he becomes incompetent, and know exactly what is his/her own prior wish, even if here is an AD. Nor is it easy to determine whether the family knows exactly what the patient wants or is entirely trustworthy. Furthermore, terms used in AD, such as LST or "life-threatening," are often vague. The values that respondents drew on to justify their views, including autonomy, quality of life, the family, and so on, are not well-defined in themselves either. It is not surprising that different respondents could use the same value to justify different preferences and different values to justify the same choice. The values drawn upon by respondents therefore could not serve as predictors of their preferences.

Indeed, similar findings were generated by a questionnaire survey too. The instruments developed by Schwartz (1992) were used to measure the value profile of respondents. Four value dimensions were measured by 10 value types, which were based on 57 values. (For the mean scores and the comparison of the 10 values in different groups of participants, please refer to Table A.1 and Table A.2 in the Appendix.) The comparison of the value profile and the responses to the doctor's choice in the two vignettes shows that in Vignette 1 the respondents who agreed with doctor's decision have a statistically higher mean score (0.708) on the value of Benevolence, than those who disagreed with doctor's decision (0.4849) ($F(1,101)=5.979$, $p=0.016$). Although such a result is found, no single value is overwhelmingly decisive in predicting responses since, as stated above, respondents also consider other values, such as autonomy and best interests, in their decision-making processes. Respondents were also asked to weigh the quality of life against the sanctity (intrinsic value) of life itself. Again, no significant results were found between the weighting and respondents' responses in the questionnaire and in the two vignettes. The same applies to demographic background. Most respondents (50.7%, $n=70$) declared no religious faith. Christianity was respondents' most popular religion (37%, $n=51$), with 28.3% ($n=39$) of them being Protestant and 8.7% ($n=12$) Catholics. 8.7% ($n=12$) and 2.9% ($n=4$) of respondents were Buddhists and Taoists, and one of them (0.7%) was a Hindu. Even though a high proportion of respondents were Christians, none of the respondents said that people should fight death to the very end, even if the patient's condition was terminal and incurable. Except for being a doctor or a nurse, all the above factors are not good predictors of the responses to the doctor's choice in the two vignettes.

10.3.4 Knowledge About and Acquaintance with LST

Findings of the questionnaire survey indicate that, in the professional group (nurses and physicians), respondents who approved of the doctor's decision in Vignette 1 possess a better understanding of LST than those who disapproved. Knowledge about LST is examined by five questions on the truthfulness of statements on LST and the respondents are asked to answer by choosing among options provided: '*Yes*,' '*No*,' and '*Do not know*.' If the answer given by respondents is correct, it will be counted as '1' (in numerical sense), otherwise it will be counted as '0,' and then a mean score is generated: the higher the score, the better understanding of LST the respondent possesses. In a comparison of mean scores on knowledge about LST in relation to their responses given in the vignette study, a statistically significant difference was found among respondents who held different views about the doctor's decision, $F(1, 117)=9.283$, $p=0.003$. The mean score on knowledge about LST of those who approve doctor's decision in Vignette 1 (*right—because the patient did not want to be resuscitated*) is 0.645, which is significantly higher than those who disapproved of the doctor's decision in Vignette 1 (*wrong—because her quality of life was quite good and death was not near*), whose mean score is 0.50141. By further categorizing the respondents into two groups: professionals (nurses and physicians) and non-professionals (patients and family members), and comparing their mean scores on knowledge about LST in relation to their responses given in the vignette study, a statistically significant result is found in the group of professionals. In terms of the score on knowledge of LST, the respondents who approved the doctor's decision in Vignette 1 (*right—because the patient did not want to be resuscitated*) are significantly different from those who disapproved of the doctor's decision in Vignette 1 (*wrong—because her quality of life was quite good and death was not imminent*), $F(1, 57)=4.368$, $p=0.041$. The mean LST knowledge score for respondents who approved of the doctor's decision (0.69677) is significantly higher than the score for those who disapproved of the doctor's decision (0.58519). The difference between the means is 0.11158. The effect size d is about 0.55, which indicates a strong association between variables.

Respondents were also asked to indicate their acquaintance with LST in the questionnaire survey. There were a total of 23 respondents (including 22 patients and 1 family member) who reported that they had received LST before. The following are further details about which kinds of treatment they have received: 2 patients reported that they had received *artificial nutrition and hydration*; 4 patients had received blood products; 21 respondents (20 patients and 1 family member) had received *disease specific treatments*, and 3 patients reported that they had received *antibiotics*. No respondent from the professionals group (nurses and physicians) reported experience of being treated by LST. Due to the insufficiency of respondents' having personal experience of LST, there is no significant result between the respondents' acquaintance with LST and their responses in vignette study. However, by further investigating respondents' responses given in the in-depth interview, it is shown that experience did play a role in determining some respondents' decisions. For instance, the following three bereaved family members shared their experience at the vignette study. In her answer concerning the medical decision in Vignette 1, FM38, a family member, believed that the doctor should resuscitate the patient, since curing people is his or her duty and the patient's health condition is reasonable. But she did not want resuscitation if she were the patient in the vignette. Here is her reason given: 'I do not want to (receive resuscitation) because I myself…I think of my mother, I think she suffered a lot.' Then she started to describe how her mother was being cared at home: 'I think she was in a lot of pain. Sometimes when she took painkillers, her condition might become better. When she was brought to the park, she seemed to be very happy, seemed to be…but when she was in pain, she could not sleep at night and we would not sleep either.' This created psychological pressure among the family, especially to FM38, since 'you cannot let her (the patient) know you are crying. I may have suppressed too much, I became very weak. (…) I still cannot relieve (my emotion), so I have to visit psychologist. I know I have suppressed it for many years, trying not to be sad. However it (taking care of the family member) is really exhausting.' As a result, she decided not to be treated if she were the patient in Vignette 1. Similarly, when asking the preference of family member FM32, if she were the patient in the condition mentioned in Vignette 2, she preferred not to have antibiotics to avoid suffering from irreversible illness or medical treatments. This response may be influenced by her personal experience in decision-making for her mother: she once requested doctors and nurses to intubate the unconscious patient in order to facilitate her urination and defecation. Although this treatment was expected to be futile, medical professionals still followed her instruction. Eventually the treatment not only failed to improve the condition, it created greater suffering for the patient. This bereaved family member thought that the decision was 'silly,' and that it is better not to prolong a patient's pain and suffering by further treatment. FM33, another bereaved family member, disapproved of the decision made by the doctor in Vignette 2: 'He (the patient in Vignette 2) is now brought to the hospital due to pneumonia. To be honest, with cancer and pneumonia…no treatment can help the patient in such situation. Yes, what you (the doctor in Vignette 2) did is just prolonging patient's suffering.' He continued explaining with support from personal experience: 'It is because the patient is old … I am not knowledgeable in medical issues. But recently I was being told by a doctor that whenever an elderly, who is a patient with cancer, is brought to the hospital and infected with pneumonia, his or her condition is very dangerous. Just like my father, (having) pneumonia, coughing with sputum, fever … After taking an antipyretic, the fever comes down for a while, and later his temperature rises again, (…) The sputum in his windpipe is hard to remove and it makes breathing very difficult.' Due to these reasons, he thought that the health condition of patient in Vignette 2 is not likely to be restored, and thus the doctor's decision to treat the patient is 'unnecessary.'

10.3.5 The Preferred Model of End-of-life Decision Making

In the survey, respondents were asked to choose between different models of EOL decision-making. The most popular choice across the four groups of respondents is the model of shared decision making—the attending healthcare team and the family jointly making the decision after taking the wishes of the patient into consideration: 39.4% of patients (n=13), 65.0% of family members (n=26), 50.0% of nurses (n=21), and 65.2% of physicians (n=15) choose this model. The liberal model (that the EOL decision should be based on the patient's preference) was not as popular: 24.2% of patients (n=8), 12.5% of family members (n=5), 21.4% of nurses (n=9), and 17.4% of physicians (n=4) prefer this model. Even though the patients who opted for the shared decision making model amounts to 39.4%, this was their most popular choice in comparison with other models.

The results from other questions about EOL decision making show that ADs were regarded as expressions of patients' wishes, but that they should not be the only voice that ought to be heard in the decision making process. That patients should be treated 'in absolute accordance' with their ADs was not a popular choice across the four groups of respondents: only 6.7 % of patients (n=12), 7.5 % of family members (n=3), 42.9 % of nurses (n=18), and 13 % of physicians (n=3) chose it. When asking the respondents whether it is acceptable not to follow patient's AD, although the result seems to be diverse among different groups—48.5 % of patients (n=16) and 67.5 % of family members (n=27) thought that it is acceptable, while 47.8 % of physicians (n=11) thought that it is unacceptable and 43.9 % of nurses (n=18) express a 'neutral' opinion (χ^2: 21.9, df=6, $p<0.005$)—it does not imply that ADs were regarded as unimportant in EOL decision-making. This is seen in that, in another question, respondents were asked to express their views on how much a patient should be treated in accordance with his or her AD. The majority of the three groups of respondents below preferred that patients should be treated 'as much as possible' according to their ADs: 62.5 % of family members (n=25), 52.4 % of nurses (n=22) and 82.6 % of physicians (n=19), though 40 % of patients (n=12) regarded ADs 'just as references'.

Those respondents who thought that treatments should not be implemented absolutely in accordance with ADs were asked to answer a follow-up question: who should make the medical decision for the patient if treatments need not to be provided absolutely in accordance with his or her AD? They were also asked to state the reason(s) for their answers. Most of them, 51.9 % of patients (n=14), 69.4 % of family members (n=25), 79.2 % of nurses (n=19), and 80.0 % of physicians (n=16) thought that it should be jointly decided by family and medical professionals. Here are some major reasons given by the respondents who were choosing this option: 43 respondents (including 12 patients, 23 family members, 3 nurses and 5 physicians) thought that medical professionals should take part in the decision making process because they possess relevant knowledge and/or understanding of patients' health condition, which enables them to explain to the family members. 24 respondents (including 7 patients, 10 family members, 3 nurses and 4 physicians) believed that family members understand the patient's wishes, values or condition, so their views should take into consideration. Besides, due to their close relationship with patients, 15 respondents (including 4 patients and 11 family members) asserted that family members should decide or even have the right to decide on behalf of patients.

Similarly, when asking respondents about who should represent the wishes of the patient if the instructions of the ADs are unclear, joint decision-making by the family and medical professionals was most popular among the four groups: 41.4 % of patients (n=12), 55.0 % of family members (n=22), 60.0 % of nurses (n=24), and 78.3 % of physicians (n=18) chose this model. Even though the patients who opted for the such joint decision making amounts to 35.7 %, it was their most popular choice in comparison with other options. Respondents were also asked to explain their chosen options. The following were the major reasons given by respondents who favored joint-decision making by the family and medical professionals: 46 respondents (including 10 patients, 21 family members, 9 nurses and 6 physicians) thought that medical professionals should take part because they possess relevant knowledge and/or understand patients' health condition, while 27 respondents (including 6 patients, 7 family members, 9 nurses and 5 physicians) believed that family members should interpret patient's AD because they understand the patient's wishes, values or condition; among these 15 respondents (including 4 patients and 11 family members) also thought that family members should be involved for another reason: their close relationship with the patient entitles them to make the decision. There were also 8 respondents (6 nurses and 2 physicians) who believed that the family and medical professionals should jointly interpret the patient's AD, since both parties understand what the patients' best interest is.

Respondents were also asked to express their views on the formulation of the details of the AD. Again, most of the respondents among four groups preferred joint decision-making: 35.7% of patients (n=10), 67.5% of family members (n=27), 69.0% of nurses (n=29) and 91.3% of physicians (n=21) thought that the details of AD should be formulated by the patients, the family, and the medical professional. Even though the patients who opted for the such joint decision making amounts to 35.7%, it was their most popular choice in comparison with other options.

10.3.6 The Prevalence of Familism

The findings so far show that respondents did not in general believe that EOL decision-making should be entirely dictated by the wishes of the patient as in the liberal model, though they believed that their wishes should be taken into consideration seriously. As we have discussed in this paper, the shared decision-making model was most preferred. Medical professionals were expected to play a significant role by many respondents, due to a respect for their medical knowledge and professional experience. Family members were also expected to play a key role in the decision-making process. This shows that Hong Kong's society is still under a strong influence of familism.

Owing to the close connection and caring relationships between the patient and other family members, they often have great concern for the well-being of the patient, and so can help medical professionals to ascertain his/her wishes and their best interests. Yet according to familism, the role of the family is not confined to serve this functional purpose. The EOL decision is regarded a collective decision made by the patient and other family members, and the wishes or the interests of the patient can sometimes be overridden by the family's decision. According to familism, the patient is not conceived as an independent or isolated being but rather as a self situated in a family network. The wishes of the patient should not be dictated solely by what s/he wants for himself or herself. According to the distinction between interests *in* the self (self-centered interests) and interests *of* the self (Wicclair 1999), people are often willing to sacrifice their interests in the self, such as health, material interests, or other self-centered interests, for the sake of promoting the interests of other family members, which means that interest of the self may include the interests of other family members. According to the liberal model, one may choose not to include the interests of other family members in one's interest of the self, and make it identical with one's interest in the self. On the contrary, according to familism, the interests of other family members are inseparable from the interest of the self, and so one's interest in the self sometimes has to be sacrificed for the sake of promoting the interests of other family members.

In the follow-up interviews of the two vignettes, a majority of respondents asserted that the family should play a significant role in EOL decision-making. For Vignette 1, 12 patients, 11 family members, 25 nurses, and 17 physicians expressed such a view. The corresponding figures for Vignette 2 are: 18 patients, 28 family members, 37 nurses, and 19 physicians. In both cases, more patients assert the significant role of the family than those who deny it. Some medical professionals justified their view for some prudential reasons, such as avoiding disputes, complaints and litigations, by following the family's choice. MP65, a physician said that in Vignette 1 'if the family requests treatments (even if the patient is in a poor condition and the doctor believes that she should not receive any treatment), the attending medical professionals can still follow the family's suggestion since it can "buy more time" for communication between the family and the doctors.'

Some respondents justified the role of the family based on a concern for the well-being of the patient. They believed that the family can help medical professionals to understand the wishes of the patient and decide what is in their best interest. Seven patients, 3 family members, and 12 physicians believed that the family understands the values, feelings, and conditions of the patient, and thus can explain or provide more information about the patient's will. Three nurses (MP 34, MP46, and MP47) said, 'the family is an important source of information (for knowing more about the patient's will) because of the close family relationship.' MP 65, a physician, said that even if the patient has an AD, s/he may have changed his mind, and the family can reconfirm the validity of the AD.

Some respondents justified the involvement of the family not for instrumental reasons or for promoting the interests of the patient. They just believed that the family has a value of its own. Quite a few respondents maintained that the family's wishes should be followed insofar as it does not violate the patient's wishes, or best interests, or if the patient has not expressed his/her wish. Some respondents even held a more radical form of familism. Three family members (FM 9, FM 26 and FM37) said that 'treating a patient is not solely a medical issue but also involves other aspects, such as financial and psychological considerations for the family.' Two physicians, MP57 and MP65, and a nurse, MP41, said that medical professionals should show concern about the feelings and emotions of the family. MP41 said that the patient's refusal of treatment should be rejected if the family thinks otherwise because 'the family members may regret not saving the patient.' Three nurses, MP47, MP49, and MP53, also expressed similar views. There were also the responses that the family should be the final decision-maker, and that their decision should override the doctor's and the patient's opinion when there is no consensus. Some respondents even thought that the interest *in* the self of patient can sometimes be sacrificed, depending of the situation. MP40, a nurse, said, 'If the illness [of the patient] is treatable, medical professionals should decide. If the patient's condition is irreversible, family members should be allowed to decide for the patient, even if the treatment [decided] may cause pain and suffering to the patient.'

In sum, the findings show that a strong ethos of familism was manifested in the interviewee's responses to EOL decision-making issues discussed in the follow-up interview.

10.4 Conclusion

Hong Kong is an Asian society with a distinctive integrated culture of East and the West. The liberal model, a prevailing theoretical model in the Western literature and law, makes EOL decision-making to be exclusively dictated by the value of autonomy, though in practice the family regularly participate in shared decision making. The results of the above discussion show that though the value of autonomy is recognized in Hong Kong, it is not as highly recognized as in the liberal model, because other values are also taken seriously. On the other hand, under the influence of the Confucian ethos, health professionals are regarded as elites, and treated with deference because their knowledge and experience are highly relevant to the well-being of individuals and society. Our findings also show that knowledge about and experiences with the use of LST do play a significant role in EOL decision-making. The involvement of health professionals is therefore crucial. Confucianism also emphasizes the value of the family. Our findings show that people in Hong Kong are still under a strong influence of familism in EOL decision-making. On the contrary, according to the liberal model, it is the patient who determines the degree to which s/he wants the family and medical professionals involved. Since Hong Kong is also influenced by Western culture, the model of shared decision-making is most preferred because it is an amalgam of medical paternalism, familism and liberalism—a possible result of East meeting West.

The study shows that a certain form of pluralism prevails in Hong Kong's society. This result provides support to the conclusion that the shared decision-making model is the more suitable EOL one for Hong Kong. Our results show that people in Hong Kong tend to draw on more than one relevant value in EOL decision-making. There is a tendency to make their preferences compatible with the relevant values, and people would try to rationalize their choice by drawing on them as far as possible. As a result, their preferences cannot be explained in terms of the domination of a single value. Furthermore, due to the complexities and contingencies of EOL situations and the fact that these relevant values are not well-defined in themselves, people can interpret them in ways compatible with their preferences. So they can use the same value to justify different preferences or use different values to justify the same choice. It follows that the values that people draw upon cannot serve as predictors of their preferences. If people's preferences had generally been predicted by the single value of autonomy, the liberal model would be a better choice for Hong Kong, and the patient should be the principal EOL decision maker. However, since there is no single value to rely on in EOL decision-making, the shared decision-making model should be a more suitable model for Hong Kong. Indeed, the Hospital Authority which provides around 90% of the hospital services in Hong

Kong adopts a consensus-building approach in its two guidelines in regard to EOL decision-making (Hospital Authority 2002, 2010). According to these guidelines, the attending healthcare team should try to build up a consensus with the patient and the family in the EOL decision-making process.

Appendix

Table A.1 Means and standard deviations of the 10 values of the Schwartz Value Survey from Four Groups of Participants (Patients, Family Members, Nurses, and Physicians)

	Patients (n = 25)		Family members (n = 39)		Nurses (n = 32)		Physicians (n = 19)	
	MSD		MSD		MSD		MSD	
Achievement	−0.191	0.628	−0.165	0.702	−0.087	0.654	0.032	0.728
Hedonism	−0.173	0.861	−0.225	0.867	−0.720	0.868	−0.449	0.990
Power	−1.016	0.870	−1.186	0.732	−1.005	0.778	−1.189	1.085
Benevolence	0.405	0.292	0.576	0.482	0.534	0.406	0.838	0.532
Universalism	0.138	0.399	0.126	0.502	0.304	0.456	0.247	0.521
Conformity	0.680	0.696	0.546	0.633	0.262	0.519	0.082	0.639
Security	0.623	0.559	0.641	0.754	0.801	0.411	0.306	0.695
Tradition	0.107	0.465	−0.340	0.698	−0.293	0.482	−0.501	0.853
Self-direction	−0.040	0.441	−0.058	0.524	0.072	0.362	0.306	0.517
Stimulation	−1.932	1.499	−1.558	1.031	−1.751	0.980	−1.378	1.131

Table A.2 One-Way Analysis of Variance (ANOVA) summary table comparing four groups of participants (Patients. Family Members, Nurses, and Physicians) in the 10 values of the Schwartz Value Survey

Source	*df*	*SS*	*MS*	*F*	*p*
Achievement					
Between groups	3	0.681	0.227	0.494	0.687
Within groups	111	51.028	0.460		
Total	114	51.709			
Hedonism					
Between groups	3	5.788	1.929	2.451	0.067
Within groups	111	87.361	0.787		
Total	114	93.149			
Power					
Between groups	3	0.901	0.300	0.425	0.736
Within groups	111	78.499	0.707		
Total	114	79.400			
Benevolence					
Between groups	3	2.090	0.697	3.667	0.015

Source	*df*	*SS*	*MS*	*F*	*p*
Within groups	111	21.084	0.190		
Total	114	23.173			
Universalism					
Between groups	3	0.703	0.234	1.051	0.373
Within groups	111	24.751	0.223		
Total	114	25.454			
Conformity					
Between groups	3	5.288	1.763	4.594	0.005
Within groups	111	42.590	0.384		
Total	114	47.878			
Security					
Between groups	3	2.936	0.979	2.526	0.061
Within groups	111	43.004	0.384		
Total	114	45.940			
Tradition					
Between groups	3	4.759	1.586	4.003	0.010
Within groups	111	43.984	0.396		
Total	114	48.742			
Self-direction					
Between groups	3	1.909	0.636	2.949	0.036
Within groups	111	23.952	0.216		
Total	114	25.861			
Stimulation					
Between groups	3	4.036	1.345	1.015	0.389
Within groups	111	147.115	1.325		
Total	114	151.151			

A statistically significant difference was found among four groups of participants on the following values: Benevolence, $F(3,111)=3.667$, $p=0.015$, on Conformity, $F(3,111)=4.594$, $p=0.005$, on Tradition, $F(3,111)=4.003$, $p=0.010$, and on Self-direction, $F(3,111)=2.949$, $p=0.036$

For Benevolence, as shown in Table A.1., the mean score is 0.405 for Patients, 0.576 for Family Members, 0.534 for Nurses, and 0.838 for Physicians. Post hoc Tukey HSD Tests indicate that Patients and Physicians different significantly in their mean scores ($p<0.05$, $d=-1.05$)

For Conformity, the mean score is 0.680 for Patients, 0.546 for Family Members, 0.262 for Nurses, and 0.082 for Physicians. Post hoc Tukey HSD Tests indicate that the mean score of Physicians has a different significant from that of Patients ($p<0.05$, $d=-1.16$) and Family Member ($p<0.05$, $d=-1.03$)

For Tradition, the mean score is 0.107 for Patients, −0.340 for Family Members, −0.293 for Nurses, and −0.501 for Physicians. Post hoc Tukey HSD Tests indicate that the mean score of Patients has a significant difference from that of Family Members ($p<0.05$, $d=1.63$) and Physicians ($p<0.05$, $d=1.80$)

For Self-direction, the mean score is −0.040 for Patients, −0.058 for Family Members, 0.072 for Nurses, and 0.306 for Physicians. Post hoc Tukey HSD Tests indicate that Family Members and Physicians different significantly in their mean scores ($p<0.05$, $d=-2.023$)

Acknowledgments The study reported in this chapter was funded by a SRG Grant # 7002198 at the City University of Hong Kong. An earlier version was presented at 7th Global Conference—"Making Sense Of: Dying and Death: Care, Dying and the End of Life," organized by Inter-Disciplinary. Net, 8 to 10 November 2010, Prague, Czech Republic. Thanks are due to Germaine Cheung for her help in drafting the grant proposal, and to Ellen Yan for her assistance in designing the questionnaire of the study.

The fieldwork conducted in the two public hospitals and the study was respectively approved by the research ethics committees of the Kowloon Central/Kowloon East and Kowloon West Clusters of the Hospital Authority in Hong Kong.

References

Akabayashi, A., B. T. Slingsby, and I. Kai. 2003. Perspectives on advance directives in Japanese society: A population-based questionnaire survey. *BMC Medical Ethics* 4:5.

Biegler, P., C. Stewart, J. Savulescu, and L. Skene. 2000. Determining the validity of advance directives. *Medical Journal of Australia* 172:545–548.

Bito, S., S. Matsumura, M. K. Singer, L. S. Meredith, S. Fukuhara, and N. S. Wenger. 2007. Acculturation and end-of-life decision making: Comparison of Japanese and Japanese-American focus groups. *Bioethics* 21 (5): 251–262.

Bowman, K. W., and P. A. Singer. 2001. Chinese seniors' perspectives on end-of-life decisions. *Social Science & Medicine* 53 (4): 455–464.

Chan, H. M. 2004a. Informed consent Hong Kong Style: An instance of moderate familism. *Journal of Medicine and Philosophy* 29 (2): 195–206.

Chan, H. M. 2004b. Sharing death and dying: Advance directives, autonomy and the family. *Bioethics* 18 (2): 87–103.

Colclough, Y., and H. M. Young. 2007. Decision making at end of life among Japanese American families. *Journal of Family Nursing* 13 (2): 201–225.

Hospital Authority. 2002. HA Guidelines on Life-Sustaining Treatment in the Terminally Ill. http://www.ha.org.hk/visitor/ha_visitor_index.asp?Content_ID=200776&Lang=ENG&Dimension=100&Parent_ID=10138. Accessed 5 Nov 2014.

Hospital Authority. 2010. Guidance for HA Clinicians on Advance Directives in Adults. http://www.ha.org.hk/visitor/ha_visitor_index.asp?Content_ID=200776&Lang=ENG&Dimension=100&Parent_ID=10138. Accessed 5 Nov 2014.

Kwak J., and W. E. Haley. 2005. Current research findings on end-of-life decision making among racially or ethnically diverse groups. *The Gerontologist* 45 (5): 634–641.

Law Reform Commission of Hong Kong. 2006. Substitute decision-making and advance directives in relation to medical treatment. *The Law reform commission of Hong Kong*. http://www.hkreform.gov.hk/en/docs/rdecision-e.pdf. Accessed 5 Nov 2014.

Ohara, S. 2000. We-consciousness and terminal patients: Some biomedical reflections on Japanese civil religion. In *The moral status of persons: Perspectives on bioethics*, ed. G. K. Becker, 119–127. Amsterdam: Rodopi.

Ohi, G. 1998. Advance directives and the Japanese ethos. In *Advance directives and surrogate decision making in health care*, ed. H-M. Sass, R. M. Veatch, and R. Kimura, 175–186. Baltimore: Johns Hopkins University Press.

Ritchie, J., and L. Spencer. 1994. Qualitative data analysis for applied policy research. In *Analyzing qualitative data*, ed. A. Bryman and R. G. Burgess, 173–194. London: Routledge.

Sahm, S., R. Will, and G. Hommel. 2005. Would they follow what has been laid down? Cancer patients' and healthy controls' views on adherence to advance directives compared to medical staff. *Medicine, Health Care and Philosophy* 8:297–305.

Sanchez-Gonzalez, M. A. 1997. Advance directives outside the USA: Are they the best solution everywhere? *Theoretical Medicine and Bioethics* 18 (3): 283–301.

Schwartz, S. H. 1992. Universals in the content and structure of values: Theoretical advances and empirical tests in 20 countries. In *Advances in experimental psychology.* vol. 25, ed. M. P. Zanna, 60–62. San Diego: Academic.

Thompson, T., R. Barbour, and L. Schwartz. 2003. Adherence to advance directives in critical care decision making: Vignette study. *BMJ* 327:1011–1017

Tse, C. Y., A. Chong, and S. Fok. 2003. Breaking bad news: A Chinese perspective. *Palliative Medicine* 17:339–343.

Tuschida, T. 1998. A differing perspective on advance directives. In *Advance directives and surrogate decision making in health care,* ed. H-M. Sass, R. M. Veatch, and R. Kimura, 209–221. Baltimore: Johns Hopkins University Press.

Veatch, R. M. 1998. Ethical dimensions of advance directives and surrogate decision making in the United States. In *Advance directives and surrogate decision making in health care,* ed. H-M. Sass, R. M. Veatch, and R. Kimura, 66–91. Baltimore: Johns Hopkins University Press.

Vollmann, J. 2001. Advance directives in patients with Alzheimer's disease: Ethical and clinical considerations. *Medicine, Health Care and Philosophy* 4 (2): 161–167

Wicclair, M. R. 1999. Ethics, community and the elderly. In *Ethics and community in the health care professions,* ed. M. Parker, 135–153. New York: Routledge.

Part V
Risk Assuming, Organ Donation, Medical Research and the Family

Chapter 11
Families and Medical Decisions to Assume Risks for the Benefit of Others

Ana S. Iltis

11.1 Individual and Family Interests in Health Care Settings

A sign on a highway features a crash test dummy driving and encourages drivers to slow down and drive safely by reminding them: he doesn't have anyone waiting for him at home—you do.[1]The appeal to the stake family members have in each other's well-being and to the care and concern family members have toward each other is unmistakable and powerful. It should be taken more seriously in the clinical and biomedical research settings. Particularly when individuals are making decisions that will expose them to risks primarily or exclusively for the benefit of others, the decision making process should, when possible, include the family. "Family" refers to people connected by blood or marriage (and perhaps in some settings by alternative bonds) and who constitute what Lainie Friedman Ross describes as an "intimate family." Such a family is one in which "there is significant interdependence of the members and a commitment to the well-being of each other and of the group" (Ross 1998, p. 5).[2] Different cultural traditions and circumstances will yield different understandings of how "far" the family extends and who counts as being part of an intimate family. None of these details are important here. This is not a sociological analysis of the family. This essay offers a philosophical argument for explicitly and proactively engaging families in certain types of decisions made in the clinical and biomedical research settings and could accommodate different understandings of the family. For individuals who exist outside of an intimate fam-

[1] Such a billboard was visible along highways near Doha, Qatar during a February 2012 visit.

[2] Ross is addressing ethical issues in medical decision-making regarding children, and thus speaks explicitly of adults and non-emancipated minors for whom those adults are responsible. I use the term more broadly.

A. S. Iltis (✉)
Center for Bioethics, Health and Society; Department of Philosophy, Wake Forest University, P.O. Box 7332, Winston-Salem, NC 27109, USA
e-mail: iltisas@gmail.com

R. Fan (ed.), *Family-Oriented Informed Consent*, Philosophy and Medicine,
DOI 10.1007/978-3-319-12120-8_11

ily, the arguments here are not relevant. However, clinicians and researchers should not assume that most patients and potential research participants exist outside of an intimate family whose participation in health-related decisions is unnecessary or even inappropriate. Family engagement is justified and important for two main reasons: family members may have a stake in each other's well-being and may also be concerned with advancing and protecting each other's interest because they care about each other's well-being.[3] These two considerations give us important reasons to take families seriously in the health care setting. These considerations are especially important when individuals are asked to assume health risks primarily or exclusively for the benefit of others.

Many decisions individuals make in the clinical setting involve giving informed consent for health care interventions aimed at treating or ameliorating a medical condition they have or preventing or diagnosing disease in them. These decisions involve exposure to varying degrees of risks or chances of benefit, but the primary goal is to advance a patient's health-related interests. Other decisions involve exposure to risks primarily for the benefit of others. In the clinical setting, the paradigm case of such decisions is the living organ donor who assumes risks and burdens associated with living organ donation primarily for the benefit of the recipient. The donor may experience psycho-social benefits, particularly if the recipient is a friend, colleague or relative, but the primary purpose of donation is to benefit the recipient. The recipient experiences all of the anticipated health benefits associated with donation, while the donor assumes many of the health risks. There is no reason to believe that donating an organ will make one healthier.

In the biomedical research setting, research participants bear health risks primarily for the benefit of others. Research refers to "a systematic investigation, including research development, testing and evaluation, *designed to develop or contribute to generalizable knowledge*" (Code of Federal Regulations, Title 45, Sect. 46.102d). Although some research offers the prospect of direct benefit to research participants, the *primary purpose* is to secure generalizable knowledge for the benefit of others and not to make the research subject better off. In the United States, for research subject to federal research regulations, the risks associated with studies generally are justified by the potential benefits to participants or to others.[4] For example, for a study to be approved by an institutional review board (IRB), the IRB must determine that:

> Risks to subjects are reasonable in relation to anticipated benefits, if any, to subjects, and the importance of the knowledge that may reasonably be expected to result. In evaluating risks and benefits, the IRB should consider only those risks and benefits that may result from the research (as distinguished from risks and benefits of therapies subjects would receive even if not participating in the research). The IRB should not consider possible

[3] The concept of duty to one's family, particularly to one's parents, might provide a different justification for family engagement in contexts in which such a duty is recognized. See, for example, Chap. 9.

[4] In some cases, research risks must be justified in part by the potential for direct benefit to participants or other special considerations. See, for example, the regulations governing research on children and prisoners (Code of Federal Regulations, title 45, sec. 46, subpart D, subpart C).

> long-range effects of applying knowledge gained in the research (for example, the possible effects of the research on public policy) as among those research risks that fall within the purview of its responsibility. (Code of Federal Regulations, Title 45, Sect. 46. 111a.2)[5]

Notice that it does not require that participants be expected to benefit. Some research offers participants no prospect of direct benefit or a very modest possibility of benefit and alternatives available outside the study offer the individual a better personal risk-benefit ratio. The focus here is on studies offering little or no prospect of direct benefit.

There has been an interest in some settings of encouraging stakeholder consultation or community consultation for broad scale decisions. For example, the state of Maryland passed a statute requiring counties to establish policies to prohibit bullying, harassment and intimidation in the schools. The statute requires that counties consult with stakeholders such as parents, teachers, administrators, students and other community members in developing the policy (Md. Code Ann., Educ. § 7-424.1(c) (2010). For some smaller scale decisions there also is an expectation of stakeholder consultation. For example, victims (if they are alive) and victims' families may be consulted when a convicted criminal is up for parole. This awareness of consulting with others who may be affected by a decision has not ushered in sustained discussion of recognizing family members as stakeholders who should be included in medical decision making processes in the West.[6] James Lindemann Nelson and John Hardwig have argued for the importance of including the family. Hardwig argues that patients who are part of a family have a duty to consider the implications of their decisions for their families: "To be part of a family is to be morally required to make decisions on the basis of thinking about what is best for all concerned, not simply what is best for yourself" (Hardwig 1990, p. 6). He defends the bold conclusion that all competent family members directly affected by a medical decision, including decisions aimed at advancing a patient's health-related interests, ought to have an explicit role in making the decision:

> [W]e need to consider the autonomy of all members of the family, not just the patient's autonomy. Considerations of fairness and, paradoxically, of autonomy therefore indicate that the family should make the treatment decision, with all competent family members whose lives will be affected participating. (Hardwig 1990, p. 9)

Nelson, whose conclusion is more modest, argues for a change in the culture of health care to acknowledge the present legal authority of patients to give and withhold informed consent but that takes family interests into account in medical decision making: "Family interests are not impertinent to medical decision-making; they ought to be heard and paid due heed, and conflicts met with efforts to mediate, to facilitate consensus, or to forge acceptable compromise" (Nelson 1992, p. 11).

[5] There are additional regulations governing research risks in research on certain populations such as children and pregnant women.

[6] To engage family members in a decision does not mean that they are granted authority over the individual or that they may force the individual to make a particular decision. The extent to which such authority is recognized will depend on a number of factors and will vary among cultures. The relevant point here is that they should be included in a discussion.

While a literature defending the role of the family has developed among individuals working in bioethics in China and Hong Kong (see, e.g., Fan and Tao 2004; Chan 2004; Cong 2004), this has not been the case in the West.

11.2 Informed Consent and Informed Decision Making

In the West, we emphasize the importance of individual informed consent for most health care decisions. Informed consent is, ideally, the culmination of a decision-making process that involves disclosure of information relevant to the decision (such as risks, potential benefits, alternatives). Information must be disclosed such that the decision maker can understand and appreciate it. The person making the decision must have the capacity to make such decisions and must have the legal authority to do so (i.e., the person must be competent). Finally, the decision must be free and voluntary (for further discussion, see Faden and Beauchamp 1986; Berg et al. 2001). Specific requirements for research subject to the U.S. federal regulations governing human research are outlined in the Common Rule and in the FDA regulations (cf. Code of Federal Regulations, Title 45, Sect. 46; Title 21, Sect. 50). Individual states also may specify requirements for informed consent in the clinical or research settings.[7] Although informed consent is the culmination of a decision making process that involves the give and take of information and the evaluation of that information, often the focus is on the documentation of consent using a signed copy of a written form rather than the process. The call to engage the family in the decision making process de-emphasizes the signature on a form and points to the significance of the decision making process.

The reasons for which we have come to see informed consent as an important requirement in the clinical and research settings include a desire to protect the authority of individuals over themselves (autonomy), to protect the welfare/well-being of individuals, and to respect individuals' right to bodily integrity (and avoid assault and battery) (See Berg et al. 2001, pp. 18–24). Some of these justifications for informed consent play an important role in helping us understand the importance of family engagement in health related decisions, as discussed below.

There are some exceptions to the requirement to obtain free and voluntary informed consent of an individual in both the clinical and biomedical research settings. In the clinical setting, these include (1) medical emergencies in which there is a presumption in favor of acting to prevent death or permanent harm (Berg et al. 2001, pp. 76–78). (2) Therapeutic privilege allows clinicians to withhold information that ordinarily would be disclosed to patients if doing so is necessary protect patients from harm (Berg et al. 2001, pp. 79–80). (3) Patients may choose to waive their right to give informed consent, although first they must be told that clinicians are obligated to provide them with information about proposed interventions and

[7] For a summary of state laws regarding informed consent, see King and Moulton (2006) and the Appendix of the decision in *Ketchup v. Howard*, 247 Ga. App. 54 (Ga. Ct. App. 2000).

obtain their informed consent or refusal, and that they have a right to make decisions regarding their treatment (Berg et al. 2001, pp. 85–87). (4) Under some circumstances, treatment is compulsory because the individual poses a serious threat to others and the informed consent of the patient is not required for treatment (Berg et al. 2001, pp. 90–91). Examples of individuals who might be subject to compulsory treatment include persons with tuberculosis who threaten the public's health. (5) The informed consent of individuals who are incompetent cannot be obtained and hence is not required (see Berg et al. 2001, Chap. 5). Individuals may be legally incompetent because of different factors, including their age or their cognitive status. The fact that a person is not legally competent and cannot give informed consent does not necessarily mean that clinicians or researchers may make decisions on the person's behalf. In many cases, a legally authorized representative or surrogate will have the authority to give informed permission (or refusal) for treatment or research participation. For example, parents of a minor ordinarily will have the authority to receive information about proposed interventions and accept or reject them. Moreover, a person who once was competent may have expressed wishes about the kinds of treatments he would or would not want in different circumstances (as in an advance directive) or may have appointed a specific person a durable power of attorney for health care, granting that individual authority to make informed decisions on his behalf. (6) In the research setting, there are specific circumstances under which research may be conducted without the informed consent of subjects (see Code of Federal Regulations, title 45, sec. 46.116c; Department of Health and Human Services 1996; Faden and Beauchamp 1986, pp. 35–36). Nevertheless, typically informed consent (or the informed permission of a surrogate or legally authorized representative) is required prior to commencing research.

11.3 Informed Consent and the Family

Individuals who choose to involve their families in clinical research participation decisions may do so, yet clinicians and biomedical researchers do not always facilitate or encourage such involvement and sometimes such engagement is possible only if patients demand it. For example, patients may be asked to sign forms and "give consent"[8] to treatment or research participation at times when family members are not present. Patients may be asked to enroll in a study during a doctor's appointment when they are alone. Some informed consent documents may state that persons may take a copy of the form home to discuss it with their family, but many people may never see that or appreciate the importance of doing so because they are not explicitly encouraged to do so or asked whom else they'd like to involve in making a decision. Sometimes patients are encouraged to bring someone to an

[8] "Informed consent" is in scare quotes because in at least some cases patients are not in a position at the time they are asked to give consent to fully understand and appreciate information or ask questions.

appointment in case they have difficulty remembering what a clinician tells them, but this is different from encouraging patients to bring or otherwise engage their families in decision making. Personal experience suggest that sometimes clinicians miss easy opportunities to engage families and to encourage patients or potential research participants to involve their families in decisions or even to mention the possibility and importance of engaging their families. In other cases, there may be institution barriers to family engagement (Bishop 2015).

Rather than merely allowing family engagement if a patient seeks it out and finds a way to involve his family, proactive steps to involve families in making decisions in the health care setting should be taken when possible, particularly when individuals are going to assume risks primarily for the benefit of others. There are two main reasons for this: (1) family members have a unique and legitimate stake in each other's well-being (Hardwig 1990; Nelson 1992), and (2) family members typically want to protect or advance each other's interests, so they can help to protect and promote a patient's or potential research participant's well-being.[9] While this is true of all decisions individuals make, they have a special interest when the primary goal of the risk-imposing intervention or interaction under consideration is to benefit others. Such decisions *prima facie* threaten the overall well-being and interests of the family. At best, the outcome in such cases will be neutral, but there is no expectation that the individual or family will be better off (in most cases).[10] The assumption of risk in these cases is not justified by a potential for direct benefit; the risk-benefit ratio for the individual (and generally for the family) is negative. There are special considerations when such a risk-benefit ratio is proposed in health care settings by individuals whom most would not suspect of asking the potential benefactor to do things that might harm them if those things are not also for their own good.[11] This unusual dynamic may render individuals particularly vulnerable to misunderstanding the risks involved and what they are being asked to do, i.e., to expose themselves to risk of harm primarily or exclusively for the benefit of others. Families may help individuals understand the unusual request, though no doubt they too may not fully understand the nature of the request since it is contrary to what one expects from health care professionals. Thus family engagement may help to fulfill the goal of enabling individuals to protect their own interests and well-being, one of the justifications offered for informed consent.

Families have particular interests both because they are stakeholders in family members' well-being, and because they ordinarily want to protect the interests of individuals in the family (and of the family overall). Two types of decisions individuals make in which families have a special interest because they involve a person

[9] Recall that one of the reasons informed consent ordinarily is required is to protect individual well-being (see Berg et al. 2001, p. 11).

[10] In special circumstances, the family might end up better off over all or other individuals in the family might end up better off. This is addressed further below.

[11] This claim is related to concerns about the therapeutic misconception. Therapeutic misconception refers to circumstances in which "clinical research subjects fail to recognize the ways in which research participation may involve the sacrifice of some degree of personal care" (Appelbaum et al. 2004, p. 1).

being exposed to risks and burdens primarily or exclusively for the benefit of others are decisions to be a living organ donor and decisions to participate in research. Each of these raises special concerns about potential conflicts among the interests of family members and of individual versus overall family interests. These are discussed separately below, followed by a discussion of objections that might be raised against family involvement in both types of decisions.

11.3.1 Research Participation

There are different types of research studies, ranging from clinical trials to observational studies, to surveys, chart reviews, genetic marker studies and others. The risks and potential benefits associated with these range from minimal to significant. Some may expose individuals to risk without any prospect of direct benefit, others pose minimal risk, and others involve individual risk and a prospect of significant benefit. Regardless of where along these axes a particular study falls, its primary purpose is to secure generalizable knowledge. In other words, its primary purpose is to benefit others. Except when legally authorized representatives or parents/guardians give permission for children, when consent is waived (see Code of Federal Regulations, Title 45, Sect. 46.116.c), when a study is exempt from the federal regulations governing research (Code of Federal Regulations, Title 45, Sect. 46.101b), or when a study meets emergency research exemption requirements (Department of Health and Human Services 1996), individual research participants give informed consent on their own behalves.

The studies that fit the paradigm under consideration in this essay are those that expose individuals to more than minimal risk primarily or exclusively for the benefit of others. Some studies might pose a modest potential for direct benefit, but the pursuit of treatment outside a study offers a better risk-benefit profile to the individual. Others might pose no prospect of direct benefit or a very, very small (some would say theoretical) possibility of benefit.

Consider two studies in which there is no prospect of direct benefit to participants. First, imagine a phase one study of a new drug designed to treat psychosis enrolling healthy volunteers. Participants will be placed into a cohort and each cohort will receive a slightly higher dose than the previous group. Participants will stay overnight in a research facility for 10 days and be monitored closely. Second, imagine a phase one study of a new potential chemotherapeutic agent to treat cancer.[12]

Now imagine a series of hypothetical studies in which there may be a prospect of direct benefit to research participants, but the studies pose risks to individuals that make the individual risk-benefit ratio for participants negative and alternatives that have better individual risk-benefit profiles are available outside the studies. Imagine

[12] There is evidence that a small percentage of patients who participate in phase one oncology trials do experience some response to the study drug (Horstmann et al. 2005). This does not mean that they are cured as a result of their participation or even necessarily that they live longer or have a better quality of life. A response could be a reduction in tumor size.

a clinical trial testing a new potential drug to treat hypertension. All participants will be people known to respond well to hypertension medication; they will be currently on medication with their hypertension well-controlled. Participants have a one-in-two chance of being assigned to each arm (placebo or study drug). All participants will go through a two week washout period and then will be on study medication for 12 weeks. Second, imagine a cross-over study evaluating a new medication for the treatment of bipolar disorder in which all subjects will be treatment-naïve. One half of the subjects will be on placebo and the other half will take the study drug. In addition to the risks of the study medication, participants forgo the opportunity to begin treatment with a standard, approved medication. The third hypothetical study will compare two different forms of contraception, an approved oral contraceptive and an experimental weekly injectable contraceptive. All subjects will take a pill (either an active pill or a placebo that is made to look and taste like the active drug) and all will take a weekly injection (either a placebo or study medication). Finally, imagine a small pox vaccine study to test different doses of a small pox vaccine in normal healthy volunteers who have not been previously vaccinated against small pox and thus have no protection from small pox.

These studies share several key features. First, they all have a negative risk-benefit ratio for individual participants. In some cases participants have a small prospect of direct benefit, but they could face better odds by seeking treatment outside of the study. In other cases, there is no prospect of direct benefit or an extremely small prospect. Second, they all involve the possibility that not only individual participants will end up worse off but that some possible adverse outcomes could set back the interests of the family or impose additional burdens on the family.

It is not uncommon for research protocols to require participants to avoid becoming pregnant or fathering a child during the study and for a period of time after the study ends. One would think that if a study explicitly requires a third party who is not in the study to do a particular thing, e.g., a woman whose partner is in a study must avoid becoming pregnant, then the prospective subject would consult that other individual. But that might not be the case. Of course, a potential participant might make a conscious decision not to discuss the matter, but the failure to discuss the requirement imposed on the third party by the participant's involvement in the study might be the result of something else. For example, the information might be embedded in an informed consent document and the person may not even notice or realize the significance of a requirement. Or, they might be invited to join a study "on the spot" and thus not realize that they should delay enrollment until they have discussed the study with others. It seems reasonable to do what one can to ensure that they understand the gravity of not only what they are agreeing to do but of what they want their relatives to do. There are many family interests in these trials beyond procreation. For example, if a participant has a bad outcome, the family may incur costs, loss of income, or face additional responsibilities of caring for the injured person. A person with poorly managed hypertension is at increased risk of a cardiovascular event, which surely could be of concern to the family both because they care about each other's well-being and because family members' interests might be set back if a person has a stroke. Family members of a person newly diagnosed with bipolar disorder have an interest in initiating treatment and ensuring that the person receives the best care possible, yet clinical trials do not aim primarily at doing what is best for the participant (e.g., changing medication doses in response to behavioral or physiological changes). There are other ways in which family interests may be jeopardized by these studies as well. Many informed consent documents include a statement such as: "you are encouraged to (or you may) take this form home and discuss it with you family." But merely telling people in the context of a very long form that they may take the document home and discuss it does not mean that they will do so, or even know that they can do so, or should do so. Moreover, when it appears that one is being asked to sign a form on the spot, individuals may not realize that it would be a good idea to wait to talk with family members first. The simple statement that potential participants may discuss the study with others is insufficient. There should be an assumption that the family engages in the process—a much more direct and explicit effort than mild encouragement or merely "allowing" a person to take a document home to share with his family. Potential research participants could, for example, be asked to discuss a study with their family and then be asked later what their family members thought. The assumption should be that if a person is going to assume risks primarily for potential benefit to others, that is, if they are going to agree to participate in research that poses risks and no prospect of direct benefit, then the assumption should be that the family is involved in the decision-making process. A patient who wishes to refuse participation in these types of studies generally should be able to do so without family involvement, since refusal does not threaten his welfare and thus the conditions set forth earlier for family interest generally do not apply.

Although researchers might think that they are asking *individuals* to assume risks for the advancement of general knowledge, in fact they may be imposing risks and burdens on participants' families. As a matter of respect for the integrated nature of the family (Hardwig 1990; Nelson 1992), they must recognize that requests for research participation may in fact involve families in important ways. Moreover, it is in potential participants' interest to have persons who have their best interests in mind helping to make decisions since that is not the perspective or focus of the researcher who seeks consent for participation.[13]

Special concerns might emerge when participants will be paid for their participation in studies. Someone might worry that family members interested in money might encourage or even aggressively push an individual to expose himself to a negative individual risk-benefit ratio for the money. The individual or family members involved might treat this like any other decision in which some people expose themselves to risk to make money or to save money. Such tradeoffs, the family members might argue, are routine and there is no reason to worry about such efforts to influence behavior because of money. Many would object to accepting risks in research for the purpose of making money and would insist that such pressure is deeply problematic (see, e.g., McNeil 1997). Although such pressure could be real,

[13] For an analysis of the role of the family in research decision making in China, see Deng (2015).

we should remember that payments to participants in research that is subject to the federal regulations will be reviewed by an IRB and they are required to ensure that the request for consent does not involve undue influence. While possibility of family pressure exists, there are some limits in place that should decrease likelihood or severity. Moreover, there are important reasons to include the family as outlined here and elsewhere (see Hardwig 1990; Nelson 1992). Other objections to family engagement are explored below.

11.3.2 Living Organ Donation

A second type of decision made in the health care setting in which persons will be exposed to risks primarily for the benefit of others is living organ donation. Living donors currently account for slightly over one third of all kidney donors in the US (34%) and for about 4% of liver donors (Organ Procurement and Transplantation Network/United Network for Organ Sharing). The use of organs from deceased donors also has implications for the family (see, for example, Cai 2015 and Truog 2008). Living donation turns a healthy person (the donor) into a patient who is exposed to numerous risks, including the risks of surgery, and who ends up minus a healthy body part (kidney) or part of a healthy body part (liver). Physical and psycho-social risks have been documented (Jowsey and Schneekloth 2008; Lentine et al. 2012; Friedman et al. 2006; Young et al. 2008; Soneji et al. 2008; Boudville et al. 2006; DuBay et al. 2009; Parolin et al. 2004; Taner et al. 2008; Middleton et al. 2006; Parikh et al. 2010). (For further discussion of the experiences of living donors, see DuBois et al. 2012.)

Given the psycho-social and physical risks of being a living organ donor, and the fact that there is no prospect of direct health benefit to the donor, the same reasons that support inclusion of the family in research decisions apply here. First, family members care about the well-being of the potential donor. Having family members learn more about the risks involved and discussing the procedure with the potential donor may help the potential donor to better understand and appreciate the information relevant to making the decision. It may be good for individuals to have family members engaged and engagement may advance a major goal of informed consent, namely understanding and appreciating information. Second, family members may be adversely affected if risks materialize, e.g., death, disability, lost work time. Given the integrated nature of the family (Hardwig 1990; Nelson 1992), those interests should not be ignored.

The role of the family in living donor decisions often is more complicated than in the research setting because the intended recipient may be another family member or the person may be asked to donate as part of a paired exchange or donation chain to benefit a family member.[14] If the potential donation is an altruistic donation in

[14] Paired exchanges involve two donor-recipient pairs, i.e., pairs in which one person is willing to donate a kidney and one person needs a kidney. However, they are not suitable donors for their intended recipients. Instead, the donor from pair 1 and the recipient from pair 2 are compatible and

which the donor is giving a kidney to a stranger, many of the issues are similar to the research participation scenarios. The individual will assume risks for no prospect of personal health benefit and anticipated benefits will accrue to unknown individuals. Cases in which the intended organ recipient is a friend or colleague also are similar to research participation in that the anticipated benefits are not expected for the individual or his intimate family members despite the risks. The individual's interests and possibly the family's interests are threatened for no prospect of direct benefit; the individual's and family's risk-benefit ratio is negative. No doubt in these cases the fact that the intended beneficiary is an identifiable life changes the emotions and other morally relevant features of the decision. Nevertheless, the family's interests in the potential donor's well-being are significant and the decision making process should explicitly encourage involvement of the family in the decision making process. It is insufficient to mention in a written consent document "you are encouraged to talk with your family or friends…."

If the potential organ recipient is a relative, additional issues emerge. The intended recipient might be an immediate family member such as child or parent, but it also could be an adult sibling, cousin or other relative. Here the risk-benefit ratio for the individual still will be negative, but in some cases the individual and or other family members may think that the potential benefit to another family member (and possibly to the family overall) justifies the risks. In some cases, the donor or the family might be better off overall if the recipient receives a transplant. If they judge that overall the family will be better off with the donation, they may encourage or pressure the individual to expose himself to the risks of donation. We should expect that different family members and different families will think differently about who "counts" as a family member for whom it is worth putting the potential donor at risk. In some families a husband might accept a kidney from his wife but be reluctant to see his wife exposed to the risks of donating a kidney to her sibling or cousin. A wife might accept the risks of her husband being a donor to his brother, but not his cousin and so on. Judgments of where risk is appropriate will vary among families, but the point is that there will be multiple interests at stake, not just the well-being of the donor. The intricacies of intimate family relationships are beyond the scope here. Depending on specific relationships involved in particular living donor situations the structure of family engagement in the decision making process may vary. In some cases, the family may have a stake in a person exposing himself to risk for the benefit of others because potential benefit is internal to the family.

Engaging families in these cases may result in pressure for the potential donor to donate rather than contribute to protection of the donor's interests, which raises questions about whether it is appropriate to encourage family involvement. There are two reasons to encourage family engagement in decision-making even in such

---

the donor from pair 2 and the recipient from pair 1 are compatible. Each donor gives a kidney to the recipient from the other pair. Donation chains involve more people in exchanges that are not simply one-to-one as in the paired exchange. In a chain, one donor may give to a recipient who has a willing but incompatible donor. That willing donor then donates to someone else, who also may have a willing but incompatible donor. That donor then agrees to donate to another recipient an do so. This can be repeated serially to create chains of varying lengths.

cases. The first is that the family is a stakeholder and the goal of engaging the family is not to talk people out of making decisions that involve risk but to engage those who are close stakeholders. The second is that by engaging the family in the decision-making process, they may become more informed and less likely to assume that person must donate to save a relative *no matter what*. What we know about family donation is that donors often experience tremendous pressure—both explicit and implicit—to donate (see DuBois et al. 2012). Presently there is an interest in emphasizing individual autonomy so that the potential donor makes his or her own decision free of such pressure, but we have anecdotal evidence that there is significant family pressure to donate. That pressure can come from people who have not been told first-hand what risks donors face because they are not included formally in the informed consent process. It is difficult to imagine that intentionally sharing information with the family about risks would result in *greater* pressure than what many report feeling now. The family already may be influencing donors but family members may not have relevant information and understand the risks. Informed family members may come to a new appreciation for the risks and interests of potential donors.

11.4 Conclusion

Despite the role the family may play in protecting an individual's interests and the possibility that members of an intimate family may be stakeholders who could be adversely affected by a decision, some may reject any explicit encouragement of family engagement. One reason for this may be that they believe that individuals have the right or authority to make their own decisions and there is no justification for pursuing family involvement unless an individual chooses to do so independently. It simply is not the family's business, unless the individual chooses to make it their business, even if they have interests at stake and even if it might be good for the individual. It is worth asking whether health care professionals should ask individuals to jeopardize family interests without explicitly engaging the family. In other words, maybe it *is* their business. Even one who acknowledges that they have a stake might insist that the legal authority of individuals over themselves must be recognized and that family engagement undermines this authority. Nothing in this essay indicates that the family would assume decisional authority. Rather, the focus of this essay is on what a good informed consent process involves, and I have argued that it should involve the family.[15]

[15] Some have argued that the insistence on individual informed consent grounded in the notion of an autonomous individual in authority over himself fails to reflect the reality of dependence and relationships within families, particularly in the face of illness and impending death. See, for further discussion, Bishop (2015), Cherry (2015), Choi (2015) and Nash (2015).

One reason for supporting family engagement is that it advances the goal of promoting the understanding and appreciation of relevant information prior to making a free and voluntary choice. The involvement of multiple individuals who care about a person's well-being may be especially important in health care settings when individuals are asked to assume risks primarily or exclusively for the benefit of others. Insofar as someone defends family engagement on the grounds that doing so can help protect individual interests, some might argue that it is inappropriately paternalistic to include the family. However, recall that informed consent in its ideal form is paternalistic. For example, informed consent requires health care professionals to disclose information and not simply to wait to see what the patient wants to ask. Health care professionals are expected to offer information and to be sure the person can understand it. It is in the individuals' interest to have others who care for them involved, particularly when what they are being asked to do is not something that is expected to advance their interests and might in fact set them back. This is a way of helping ensure that they are informed when they give consent.

Others may argue that the family does not have as great a stake in some cases as do others, such as employers, or that people outside the family have an equal or greater stake. As such, all parties with comparable stakes should be involved and the family should not get special treatment. There may be some special contractual relationships in which people have agreed to give others a role in making decisions, e.g., a professional athlete who as part of his contract gives team physicians some role in establishing a treatment plan. However, in general, contractual or other non-family relationships are different from family relationships. Special obligations that exist among family members give the family a special role (Hardwig 1990; Nelson 1992). Having a stake alone is not what justifies family engagement. It is, in part, that it is family members who have a stake and that family members are special (Hardwig 1990; Nelson 1992).

Finally, some may not reject the call to engage families for principled reasons but instead hold that to do so would be impractical and costly. Nothing here suggests that family engagement must be achieved in a specific way, yet the criticism that a practice would be too burdensome suggests that a specific proposal—one that is costly and impractical—is required by the arguments here. Creative planning could make family engagement a financially and logistically realistic pursuit. Moreover, the claim that something is costly and hard to do is not an excuse for avoiding moral obligations. If family engagement in the informed consent process is the right thing to do, as I have argued here, then the proper response is to find appropriate ways to fulfill the goal.

To encourage involvement of the family requires more than allowing people to take a form home or making a vague statement in a written form that they are allowed to share it with other people, or allowing family to remain present at the bedside if they happen to be visiting at the time something is being decided. To encourage family involvement requires at least an explicit expression of the value of family and family involvement, and a conscious effort to remove institutional or structural barriers to family engagement. As part of the informed consent process, health care professionals should seek to engage the family, particularly when a person is being asked to assume risks primarily or exclusively for the benefit of others. This essay is not meant as a "how to engage the family guide." The goal of this essay is to demonstrate the importance of doing so. Explicitly encouraging and facilitating the involvement of the family and recognition of the importance of family in the decision making process when individuals are considering assuming risks primarily or exclusively for the benefit of others not only recognizes that intimate families are stakeholders in each other's well-being but also acknowledges the nature of the family and the concern family members typically have for each other. Informed consent is justified in part because it seems overall best to allow individuals to protect their interests by making informed decisions regarding their bodies. Fostering family engagement where the risk-benefit ratio for individuals is negative is a means of advancing this goal of protecting individual interests while still acknowledging the legal authority individuals have over themselves in the West.

References

Appelbaum, P., C. Lidz, and T. Grisso. 2004. Therapeutic misconception in clinical research: Frequency and risk factors. *IRB: Ethics and Human Research* 26 (2): 1–8.

Berg, J., P. Appelbaum, L. Parker, and C. Lidz. 2001. *Informed consent: legal theory and clinical practice*. New York: Oxford University Press.

Bishop, J. (2015). Dependency, decisions and a family of care. In ed. R. Fan.

Boudville, N., G. R. Prasad, G. Knoll, N. Muirhead, H. Thiessen-Philbrook, R. Yang, P. Rosas-Arellano, A. Housawi, and A. Garg. 2006. Meta-analysis: Risk for hypertension in living kidney donors. *Annals of Internal Medicine* 145:185–196.

Cai, Y. (2015). On family informed consent in the legislation of organ donation. In ed. R. Fan. XXXXXXX.

Chan, H. M. 2004. Informed consent Hong Kong style: An instance of moderate familism. *Journal of Medicine and Philosophy* 29 (2): 195–206.

Cherry, M. J. (2015). Individually directed informed consent and the decline of the family in the west. In ed. R. Fan XXXXX.

Choi, K. (2015). The ideal of autonomy and its misimplementation. In ed. R. Fan. XXXXXXX.

Cong, Y. 2004. Doctor-family-patient relationship: The Chinese paradigm of informed consent. *Journal of Medicine and Philosophy* 29 (2): 149–178.

Department of Health and Human Services. 1996. Waiver of informed consent in certain emergency research. *Federal Register* 61:51531–51533.

Deng, R. (2015). The Chinese mainland informed consent of human medical research: A **multi-decision model** of Family-oriented. In ed. R. Fan. XXXXX.

DuBay, D. A., S. Holtzman, L. Adcock, S. Abbey, S. Greenwood, C. Macleod, A. Kashfi, et al. 2009. Adult right-lobe living liver donors: Quality of life, attitudes and predictors of donor outcomes. *American Journal of Transplantation* 9:1169–1178.

DuBois, J. M., A. S. Iltis, and S. G. DuBois, eds. 2012. Narrative symposium: Living organ donation. [Special issue] *Narrative Inquiry in Bioethics* 2 (1):1–37.

Faden, R., and T. Beauchamp. 1986. *A history and theory of informed consent*. New York: Oxford University Press.

Fan, R., and J. Tao. 2004. Consent to medical treatment: The complex interplay of patients, families and physicians. *Journal of Medicine and Philosophy* 29 (2): 139–148.

Friedman, A., T. Peters, K. Jones, E. Boulware, and L. Ratner. 2006. Fatal and nonfatal hemorrhagic complications of living kidney donation. *Annals of Surgery* 243 (1): 126–130.

Hardwig, J. 1990. What about the family? *Hastings Center Report* 20 (March-April): 5–10.

Horstmann, E., M. S. McCabe, L. Grochow, S. Yamamoto, L. Rubinstein, T. Budd, D. Shoemaker, E. Emanuel, and C. Grady. 2005. Risks and benefits of phase 1 oncology trials, 1991 through 2002. *New England Journal of Medicine* 352 (9): 895–904.

Jowsey, S., and T. Schneekloth. 2008. Psychosocial factors in living organ donation: Clinical and ethical challenges. *Transplantation Review* 22:192–195.

King, J., and B. Moulton. 2006. Rethinking informed consent: The case for shared medical decision-making. *American Journal of Law and Medicine* 32:429–501.

Lee, I. (2015) Filial duty: Moral foundation of substitute decision making. In ed. R. Fan. XXXXXXX.

Lentine, K., M. Schnitzler, H. Xiao, D. Axelrod, C. Davis, M. McCabe, D. Brennan, S. Leander, A. Garg, and A. Waterman. 2012. Depression diagnoses after living kidney donation: Linking U.S. registry data and administrative claims. *Transplantation* 94:1–7.

Middleton, P. H., M. Duffield, S. Lynch, R. Padbury, T. House, P. Stanton, D. Verran, and G. Maddern. 2006. Living donor liver transplantation-adult donor outcomes: A systematic review. *Liver Transplantation* 12:24–30.

McNeil, P. 1997. Paying people to participate in research: Why not. *Bioethics* 11 (5): 390–396.

Nash, R. (2015). Toward a shared decision: Against the fiction of the autonomous individual. In ed. R. Fan. XXXXXX.

Nelson, J. L. 1992. Taking families seriously. *Hastings Center Report* 22:6–12.

Parikh, N., D. Ladner, M. Abecassis, and Z. Butt. 2010. Quality of life in donors after living donor liver transplantation: A review of the literature. *Liver Transplantation* 16 (12): 1252–1258.

Parolin, M. B., C. Lazzaretti, J. Lima, A. Freitas, J. Matias, and J. Coelho. 2004. Donor quality of life after living donor liver transplantation. *Transplantation Proceedings* 36:912–913.

Ross, L. F. 1998. *Children, families and health care decision making*. New York: Oxford University Press.

Soneji, N., J. Byas, and V. Papalois. 2008. Long-term donor outcomes after living kidney donation. *Experimental and Clinical Transplantation* 6 (3): 215–223.

Taner, C., M. Dayangac, B. Akin, D. Balci, S. Uraz, C. Duran, R. Killi, O. Ayanoglu, Y. Yuzer, and Y. Tokat. 2008. Donor safety and remnant liver volume in living donor liver transplantation. *Liver Transplantation* 14:1174–1179.

Truog, R. 2008. Consent for organ donation-balancing conflicting ethical obligations. *New England Journal of Medicine* 358:1209–1211.

Young, A., L. Storsley, A. Garg, D. Treleaven, C. Nguan, M. Cuerden, and M. Karpinski. 2008. Health outcomes for living kidney donors with isolated medical abnormalities: A systematic review. *American Journal of Transplantation* 8:1878–1890.

Chapter 12
On Family Informed Consent in the Legislation of Organ Donation in China

Yu Cai

12.1 Introduction

Individual consent is commonly adopted in organ donation legislation worldwide. However, this kind of legislation may cause practice to deviate from law and disparage its authority. Consequently, we have designed an organ donation informed consent system with family informed consent as its basis, whose rationality is not confined to Confucian society. In order to prove its rationality, we need to determine the foundation of the right supporting family-directed organ donation. In addition, such a right needs to be analyzed by reconstructionist Confucian bioethics.

12.2 The Status of the Organ Donation Consent System and the Conflict Between Law and Practice

The first successful organ transplant in a human being was a kidney transplant between identical twins in Boston, Massachusetts in 1954. Subsequently, transplant-related technologies developed rapidly and many more transplants were performed throughout the world. Nowadays, the ability to transplant organs successfully is almost taken for granted, and this kind of technology has saved many lives.

Globally, organ donation legislation requires express informed consent of the donor for any living organ donation. For cadaveric organ donation, however, there is a distinction between express consent and presumed consent. Express consent is more in line with the ethical nature of people's autonomy and is divided into strong and weak systems. In the strong system, an individual's willingness to donate during his lifetime can legitimate harvesting his organs, while in the weak system, the

Y. Cai (✉)
Law School, Yunnan University of Finance and Economics, 237 Longquan Road, Kunming, 650221 Yunnan Province, China
e-mail: yucaicn@vip.sina.com

R. Fan (ed.), *Family-Oriented Informed Consent*, Philosophy and Medicine,
DOI 10.1007/978-3-319-12120-8_12

refusal of one's family can override the sentiments expressed by the deceased prior to death.

Few countries have adopted weak express consent laws like Japan and Lebanon have. However, in countries with strong express consent laws, such as China, doctors still tend to solicit opinions from the family, even if the deceased expressed a wish to donate prior to death. Organ harvesting cannot be implemented if the family members oppose it. In other words, practice conflicts with the law in most countries with strong express consent systems; in effect they are actually practicing in line with a weak express consent system.

As for organ donation systems stipulated by law, the most effective is a registration system. Throughout the world, registration systems are different: in countries like Austria, people can opt in and opt out of organ donation; while in some other countries, for instance, the United Kingdom, only those who opt in are recorded, and in Portugal, only those who opt out are recorded.

China began transplanting organs in the 1960s with the practice developing rapidly. Currently, the technological capabilities and success rates related to organ donation in China have caught up to those in developed counties. At present, the major national law regulating organ donation in China is the "Human Organ Transplant Ordinance," promulgated by the State Council in 2007. According to this law, adult organ donation, whether cadaveric or living, only requires consent from the organ donor.

In March 2010, a pilot project on human organ donation was jointly launched by the Red Cross Society of China and the Ministry of Health in 11 provinces and cities throughout China. In those 11 areas, including Tianjin, Liaoning and Shanghai, recruitment, procurement, and allocation practices were fully explored. Any citizen who wished to be a cadaveric organ donor could register as a volunteer donor in the subordinate organ donation office of the Provincial Red Cross Society. One year later, volunteer donors were recruited nationwide. In 2007, the first batch of organ donation cards (for organ and cornea donation) were designed, which included the signature, contact information, names of immediate family members and emergency contact information of the applicant. In the event of accidental death, this card serves as evidence of the person's expressed willingness to serve as an organ donor. This demonstrates significant progress regarding organ donation in China.

Currently, the legal setup for organ donation has not yet been perfected in China. The National People's Congress and its standing committee's legislation, "Organ Transplant Law," is imminent.

12.3 Constructing a Family-Based Informed Consent System

In order to avoid the above-mentioned conflict between law and practice, which has occurred in the Chinese organ donation system, the legislation of the "Chinese Organ Transplant Law" should adopt a family-based informed consent system. Given that families worldwide can be considered ethical entities linked by the tie of love, the rationality of family-based informed consent can be extended beyond China.

12.3.1 Conceptual Analysis

The concept of "family" invoked in family-based informed consent for organ donation refers to the "family" constituted by the donor and their next of kin only (namely their spouse, parents and children). As the love between the next of kin is the most natural and real, the fundamental basis of the rationality of family-based informed consent is provided.

So-called "family-based informed consent" refers to the joint decision made by the family on a voluntary basis to embody the spirit of the family ethic and to maximize the family's interests. Obviously, to realize the family ethic and its benefit maximization, the decision must be based on information communication and mutual agreement. As a joint decision reached by the family members, it is a unified wish of the patient and his/her family where the possible divergence of wishes between them can be healed over by full discussion and mutual concession, once family-based consent is given. If a unified decision cannot be made by the family, it simply means that the family has been split.

12.3.2 The Design of a Family-Based Consent System and its Advantages

The proposal is to convert individual informed consent into a family-based (common) model where the express consent of the individual citizen during his/her lifetime is a decision jointly made with his/her family and reflected on a consent document signed by the donor and a representative of his/her family. This would allow doctors to remove the donor's organs without having to seek out informed consent from his/her family after the death.

In such a family-based system, the citizen's right to dispose of one's corpse and the family's right to ownership of the body (organs) can both be respected. In addition, a family-based model would also shorten the time between death and organ harvesting, which helps ensure the successful transplantation of the organ(s). Further, doctors can avoid the embarrassment of having to ask the family if the organs can be harvested and one does not risk harvesting organs without the family's consent. This contributes to harmony between the doctors and patients. Lastly, derogation of law's authority is avoided because the doctor is not forced to act against the letter of the law.

It should be noted that in the case where a citizen refuses to register for donation during his/her lifetime, family-based informed consent should not be sought and the individual's refusal should be taken as final. At the same time, if one has not provided express consent or refusal before death, the family is entitled to dispose of the organs as they see fit as they belong to the family. As such, the family can choose to donate the organs by having a representative sign an informed consent letter.

In the case of live organ donation, family-based consent is also necessary and an informed consent letter should also be signed jointly by the donor and his/her family representative.

12.4 Justification of Family-Based Informed Consent: The Family Affairs Component of Health and Family Ownership of Organs

The legal foundation of family-based informed consent for organ donation lies in the family component of health and family ownership of organs.

First, further definition of the concept of "family" is necessary. The so-called "family" generally refers only to a family constituted by the next of kin (only including oneself, one's spouse, one's parents and one's children). In such a family, the parents' kindness to their children, the children's esteem towards the parents, and the affection between the parents are derived from the most natural, tenacious and enthusiastic force of human nature, constituting the strong and abiding cohesion of the family. In addition, between men and women, marriage is an ethical relationship with a legal protection that eliminates some uncertainties.

12.4.1 The Nature of the Family (A Unified Ethical Entity)

12.4.1.1 Love

Undoubtedly, love is the ethical relationship and prescriptive nature of a family. "Love is the desire for union of the separated" (Tillich 1995, p. 38). The "love" among family members means to realize the unity between others and oneself instead of isolating oneself by considering only oneself. On the contrary, only through abandoning one's independent existence and recognizing oneself as the unity of oneself and another, as well as the other and oneself, can we obtain self-consciousness. The love within families, for example, the father loving his son or the wife loving her husband, is the most natural love. The first component of love is that one does not want to be isolated and lonely, otherwise one will be incomplete. The second component is that one finds oneself in another person, namely by obtaining recognition from others and vice versa. As such, love is a most inconceivable contradiction, which by no means can be solved by reason. Love creates contradiction on one hand and resolves contradiction on the other. As the solution of contradiction, love is the unity of ethics (Hegel 2008, pp. 162–163).

12.4.1.2 From "An Independent Person" to "A Member of the Family's Ethical Union"

The essence of marriage is that two parties want to voluntarily form "a person" and both are willing to abandon their own natural and individual personalities for that unity. "This union is a self-restriction, but in fact it is their liberation, because in it they attain their substantial self-consciousness" (Hegel 2008, p. 164).

The prescriptive nature of the family is love, a feeling that one is unified with others. Therefore, it's just because people are aware of this unity that they are no longer an independent person but become a member of the family as an ethical entity. In other words, the independent personality of the individual has already been dispelled in a family and has merged into the direct and natural ethical entity of a family in accordance with the rules of love, as a result of which, affection, kindness, esteem, trust, and the commonality of the individual is experienced. That is to say, the individual and the family (the ethical entity) are directly united as one "person." Here, no individual personality exists because the family unity is the ethical entity.

In this family unity, freedom is not the ability to do whatever one desires or wants, otherwise everyone will lose their freedom because the desires and wants of family members frequently contradict each other. Acting according to the ethical rule of love enables family members to actualize their real freedom within the family, because their desires and wants are mediated and regulated by the rule so that their decisions and actions will not conflict with each other. The ethical rule of love requires everyone to observe communitarianly-established moral virtues and norms to treat one's family members with love. Just as the law exists for realizing freedoms (such as a freedom from being robbed) in the state, the ethical rule of love can be understood to function in the ethical union of a family for realizing freedom within the family. Real freedom exists in ethical entities only. That is to say, only when the individual abandons his/her natural personality to live in accordance with love within the ethical entity of a family can he/she obtain true freedom and ontological self-consciousness in the family. The right to live within the ethical union of a family is the basic right one enjoys in the family. Only when a family is disintegrated can the individual break away from the family and become a person with an independent personality existing outside the family again.

12.4.2 The Service Responsibility of Family Members Towards the Family—The Family Affairs Attribute of "Individual Health"

In the ethical union of a family one is no longer an independent person but a member of the family, a unified ethical entity. Each member of the family must be dedicated to the ethical unity of family. Therefore, physical health and longevity is not an individual matter, but is a common matter for the good of the family unity, that is, personal health is a family matter.

12.4.3 From "the Independent Individual's Right Over One's Organs" to "the Common Family Right over the Organs of its Members"

12.4.3.1 The Independent Individual's Right Over One's Organs

A human being is a subject aware of its subjectivity. As a person, I know I'm free inside myself. In order to be an existence in accord with the idea of freedom (the concept of freedom and its actualization), human beings must be provided with an external field of freedom (Hegel 2008, p. 56). While personality is the thing a human being should have qua human being, it should not stay in the abstract ego but be combined with external things to recognize the ego. That is to carry through the free will in the external things, making them a part of oneself. Therefore, possession is the natural existence of personality as well as freedom. To respect one's possession is to respect his/her personality. In other words, human beings are entitled to embody their will in external things to make them "mine." "I as a free will am an object to myself in what I possess and thereby also for the first time am an actual will, and this is the aspect which constitutes the determination of property, the true and rightful factor in possession" (Hegel 2008, p. 61).

At the same time, as I'm living in the organism of a body, my body is my external existence. But as a person, I own my life and body as I own other things, as long as my will is inside it. Only when I am willing do I have limbs and life. "I possess the members of my body, my life, only so long as I will to possess them. An animal cannot maim or destroy itself, but a human being can" (Hegel 2008, p. 62). The organs removed from my body are also the natural existence of my personality and freedom and therefore people possess their organs. Just as Audi said, we can assume that the body as a whole is not our property, but still insist that part of the body is our property as long as we do not insist that we have all the parts of the body, as the assumption of possessing all parts of a thing would imply our possession of it (1996, pp. 141–143). Indeed, the Law Reform Commission of the United States found that, if people did not have ownership of the component parts of their own bodies, it would be more difficult for us explain why we can donate organs (Price 2002, p. 237).

12.4.3.2 The Common Family Right Over the Organs of Its Members

As we discussed previously, a person who has an independent personality and who does not belong to an ethical entity, enjoys the ownership of his/her own body. Will this remain the case if two such persons form a family? In the unified ethical entity of a family, the results are completely different. The organs of the individuals who have lost their independent personalities gain a common family attribute.

The rule that a family is an ethical entity ("a person") applies at all times in all countries due to its direct and natural nature. As a personality, family has its external reality in possessions, namely family properties. Except for possessions, as a universal and persistent personality, a family also can establish a stable and lasting industry, namely a resource. Here, the selfish needs and desires of the individual are converted into a kind of caring and gain for the family community (Hegel 2008, p. 171). As family property is a common possession, not any member of the family has special possession but only the co-ownership of common possessions. In this regard, the family members' ownership of their individual organs is also converted into a co-ownership by the family unity of all members' organs. Consequently, the individual has lost one's right to dispose of one's organs and, therefore, it is reasonable that family-based informed consent is necessary for organ donation.

To summarize, family-based informed consent for organ donation is grounded in the family affairs component of "health" and family ownership of organs.

12.5 Family-Based Informed Consent for Organ Donation from the Perspective of Reconstructionist Confucian Bioethics

Laws and regulations are only suitable for the society and culture from which they originate. If one considers the reconstructionist Confucian bioethics of Ruiping Fan, there is powerful support for a family-based informed consent model mentioned above. Furthermore, reconstructionist Confucian bioethicists also share the same understanding of the nature of family, that is, the unified ethical entity.

12.5.1 Reconstructionist Confucian Bioethics

The reconstructionist Confucian bioethics established by Ruiping Fan is a construction of contemporary Confucianism that faces social reality while mastering the core of Confucianism. Through analysis and comparison Fan properly interprets Confucianism in such a way that his theory provides direct and concrete Confucian resources for reforming modern policies and institutions. Meanwhile, "the core of the reconstructionist Confucian bioethics is its fundamental nature and its most important promise to stick to principles and solve problems in real life." (Fan 2010, p. 2). As such, the Confucian proposal for feasible and spiritually suitable organ donation regulations in China has sought its foundations in reconstructionist Confucian bioethics.

12.5.2 Family-Based Informed Consent for Organ Donation and Confucian Core Values

Reconstructionist Confucian bioethics is based on close family ties (the love between family members), one of the core values of Confucianism, and is in line with the essential nature of the family's unified ethical entity. Similarly, since family values and the ethical entity of the family in reconstructionist Confucian bioethics share the same nature, they can provide people with guidance for living correctly. Therefore, "the truth that contemporary Western bioethics fails to comprehend," (Fan 2010, p. 9) is reflected in reconstructionist Confucian bioethics.

12.5.2.1 Family-Based Informed Consent for Organ Donation from the Perspective of Family Ties in Confucian Society

One of the core Confucian values is the value of the family. In Confucianism, the core of the family is established through family ties, the source of kindheartedness. In particular, the feature of Confucian kindheartedness is being considerate to those close to you and extending that consideration to others. According to Mencius, one should "revere the elders in one's own family and extend this reverence to all elders;care for the children in one's own family and extend that caring to all children" (*Mencius 1.7*; Chinese Teachers Union Institute of Education Sciences 2005, p. 3). This is the principle of Confucianism, which highlights filial piety and fraternal duty as the basis of kindheartedness. Youruo, disciple of Confucius, said, "The Superior Man concerns himself with the fundamentals. Once the fundamentals are established, the proper way (*tao*) appears. Are not Filial piety and fraternal submission the root of all benevolent actions?" (Confucius 2002, *The Analects 1.2*). According to Mencius, "Every child loves their family members; whey they grow up, they show respect for their elderly brothers. Family ties (the love between family members) means kindheartedness; respect for the elderly means obligation. Selflessness helps to reach afar." (*Mencius13.15*; Xia 2002, p. 417). Therefore, kindheartedness is an extension of family ties while obligation is an extension of respect for the elderly. Furthermore, Confucianism focuses on kindheartedness as the supreme ethic and highlights filial piety and fraternal duty as its basis. Under the filial piety and fraternal duty of Confucianism, the most basic moral ethic is to love your family (Zhang 2009, pp. 80–82).

The family tie is the most natural and stable ethical relation. The components of family ties are to abandon one's independence and to realize the unity between one and one's family and vice versa. For example, the story of cutting off one's flesh to show filial obedience to one's mother in the twenty-four filial exemplars clearly shows the unity between an individual and one's family and the unity within one's family. Through family ties the family becomes a unified ethical entity, "a person," and all assets owned are shared. In this way, the organs of each family member are shared by the family. Consequently, in accord with family ties, an individual will not view their health as their own, but the common affair of the family, as the purpose of health is to add happiness to the family. Naturally, because organ donation harms one's health, family-based informed consent for organ donation is required.

In summary, family-based informed consent rests on ancient Confucian ethical ideas, that is, the theory of "kindheartedness." Confucianism focuses on kindness as the supreme principle, and filial piety and fraternal duty as its basic ethics.

Family-based informed consent for organ donation is in accord with the complete moral essence of Confucianism as filial piety and fraternal duty.

12.5.2.2 Family-Based Informed Consent for Organ Donation from the Perspective of Family Values

Through family ties, or the love between family members, Confucianism has always attached importance to the family, which does not usually view one's life and living as an individual matter, but rather as a link of the family and the life of the family. Therefore, life activities are not only for individual enjoyment, and may not even be enjoyable for the individual, but are rather to increase the happiness of the family and to strengthen the connection between individual's life and that of the family's. As such, Confucianism supports the traditional idea that instructs one to "cultivate the moral self, regulate the family, maintain the state rightly and make all peaceful" (*Book of Rites-Great Learning*). That is, the life of an individual does not only belong to the individual, but to one's family or even one's country.

Such active devotion to and involvement with one's family makes the individual's happiness largely dependent on the family's happiness. Therefore, individual health is not so much a concern of the individual, but a family affair. Additionally, personal belongings do not exist in the traditional Confucian family, only common assets of the family exist. Hidden individual assets (casually called personal funds) are forbidden and, of course, an individual does not have the right to dispose of family assets.

The reconstructionist Confucian bioethics based on the Confucian family values outlined above requires all potential patients and organ donors to be understood as part of a family. Outside of the family it is hard to appropriately understand the individual (Fan 2010, p. 9) because the donor is closely involved with the family and other family members. In Confucianism, the unified family entity makes the concepts of "freedom" and "equality" unnecessary. The individual, whose health and longevity is not an individual matter anymore, has to be dedicated to the ethical entity of the family and better serve the family and the unified family entity. Consequently, as material property due to separation with the body, organs are family property, a common asset, and the individual, therefore, does not have the right to dispose of one's organ and the whole family must decide together.

12.5.2.3 Family-Based Informed Consent for Organ Donation from the Perspective of "Concordance"

Another core Confucian value supported by reconstructionist Confucian bioethics is "concordance." The word "concordance" first appeared in oracle bone inscriptions and inscriptions on ancient bronze objects. "Concordance corresponds to two characters in Chinese, the original meaning of the first one means harmony while the original meaning of the second refers to the upper and lower lips of people's mouth, meaning coordination" (Wu 2005, p. 51).

In China, the traditional idea of concordance is profound, long-standing and well established, consisting of two components: general comprehension and philosophy. General comprehension consists of the relationship between humans and nature, humans and humans, and humans with the nation and society, within which the concordance of humans themselves and that between humans and spirits resides. As for concordance at the philosophical level, Professor Zhang Li-wen defines it as meaning that "the conflicts and harmony of various elements among nature, society, interpersonal relationships, souls and cultures are the aggregation of new structures, new things and new lives formed by quality elements during the conflicts and harmony" (Zhang 1997, p. 51). This indicates that every being consists of conflicts and harmony. Conflicts and harmony are everywhere from astronomical objects to ants and blades of grass. Every phenomenon in the universe has concordance while every concept embodies concordance and every conflict and harmony is the existence of concordance (Yan 1999, pp. 45–47).

"Concordance" is also called "harmony," meaning the unity in a variety of things with different natures, rather than the "collision-free situation" (Zhang 1999, pp. 10–14). "Harmony" refers to the harmonious existence of elements with different natures while "combination" refers to the mastery and comprehension of things with different natures (Cheng 1998, pp. 29–35). The concept of concordance, which includes harmony, peace, neutralization, fusion, association and cooperation, focuses on mastering the relationship among things in a comprehensive manner. Concordance does not mean removing the differences among things, rather, the premise of its existence is the common difference of things. It tries to reach an overall balanced, harmonious and cooperative state through overcoming and settling conflicts. In short, the concept of concordance is a comprehensive idea focused on harmony, which Chinese traditional cultures deem to be the best state (Li and Liu 2005, pp. 38–40). In addition, concordance is also the best state for the relationship of the family.

In Confucianism the concept of concordance emphasizes that the relationship between people should focus on harmony and pursue unity and coordination within its ranks. Confucius took "harmony" to be a worldly ideal. He states, "harmony is the most precious in practicing etiquette (rituals); the principles of former kings from these were the most ideal, where both the great and small matters depend upon it" (Confucius 2002, *The Analects 1.12*). He also said that the middle way is the basis of the world while harmony is its highest standard (*Doctrine of Mean*). He regarded harmony as the standard of "achievement" and "a man of noble character" would reinforce the extreme importance of harmonious interpersonal relationships.

A Confucian family is the unity formed by combining different individuals who gave up their individual personalities. The unified family entity itself reflects the "combination" of different elements. Family-based informed consent for organ donation reflects the will of the person concerned and the harmonious unity of that person's family, which displays the "harmony" of the harmonious coexistence of different elements. The pursuit of harmony is one of the traditional features of Chinese culture. However, for Confucians harmony is the supreme ideal. Therefore, family-based informed consent corresponds to the core value of Confucian "concordance" insofar as it represents "unity" and "harmony".

Concordance is one of the core values espoused by reconstructionist Confucian bioethics. Consequenty, it is never possible for reconstructionist Confucian bioethics to support the autonomous decision of an individual to become an organ donor because it may harm the harmony of the family. As such, only when the decision to become an organ donor is made within the family can it be rationalized by reconstructionist Confucian bioethics.

12.6 Family-Based Informed Consent for Organ Donation from the Perspective of Etiquette (Ritual)

It is without question that ancient Chinese ethics held the theory of "kindheartedness" at its core. Confucius focused on kindness as the supreme principle. However, if "self-denial and a return to propriety means kindness," following the rituals would be the major expression of kindness. In particular, kindness can only be realized in certain interpersonal relationships through loving conduct (Fan 2002, pp. 346–372).

"Etiquette" refers to the family and social rituals commonly practiced in the Confucian community (Fan 2012) "As for Confucianism, the cultivation of virtues, especially the cultivation of virtues through its power, is realized through etiquette (ritual). We should understand that kindheartedness with virtues comes from our following etiquette (ritual). Sound etiquette (ritual) transforms people into real human beings" (Fan 2010, p. 9). Family-based informed consent for organ donation is no doubt the proper behavioral system and norm for practicing the core values of Confucianism.

It should be noted that the reason why the family-based informed consent system is thought of as confirming the family's consent is because it merely reflects the normal state of the family. The point of creating such a system is so that the law defines the normal state and makes it lawful. Consequently, the embarrassment of not following existing laws and the damage done to the unified family entity by unethical laws are avoided.

It is important to note that Fan's reconstructionist Confucian bioethics takes the core value of Confucianism as primary while etiquette is that which needs to be reconsidered to adjust to the current situation. When the world changes, outdated etiquette that fails to adjust to the new practical environment has to be replaced by new etiquette or it will dispel or even distort the core value of Confucianism. "A complete picture of Confucian virtue ethics embodies a sophisticated reflective equilibrium between ritual practices and general principles " (Fan 2012, p. 8).

For example, in Confucianism, the edict stating that the "body is given by one's parents; doing no damage to it is fundamental to filial piety" (Confucius 2007, *Classic of Filial Piety 1*) is one reason to prevent the Chinese people from donating organs. It also distorts the broader spirit of love embodied by "kindness" in Confucianism. The above edict is an old etiquette adapted for a time period when people could not imagine organ transplantation technologies. In the contemporary world, one should not only define the "body as given by the parents" in a broader manner (perhaps so that it represents the whole of individual character), but also establish a new etiquette based on it. We can use "body as given by the parents" as the rationale for family-based informed consent. Given that one's whole being comes from one's parents one does not possess an independent individual personality in the family, therefore, I am a unity with my parents and other family members. My happiness is their happiness while their sadness is my sadness. I can devote my whole being to them and they will do the same. Because the "body is given by the parents," the family tie (the love between family members) among family members is the inheritance and association of flesh and blood. As a result, one's choice to become an organ donor should not be one's own, but should require the informed consent of the family. Upon getting the "approval" of one's family, one has preserved filial piety for one's parents and has discharged one's responsibility to the unified family entity.

Consequently, Confucian family values make family-based informed consent easy to achieve. From their different perspectives, every family member makes the harmony and continuous development of the family their responsibility, making them willing to participate in family-based informed consent. At the same time, the maximal individual interest of family members and the maximal family interest are easier to coordinate if family-based informed consent is present. That is, family members are more willing to respect family-based informed consent even if it goes against one's individual interests.

In summary, family-based informed consent for organ donation is an implication of the family as a unified ethical entity. In addition, family-based informed consent is never only applied to Confucian families, but to any family with "home" and "love".

Acknowledgements I would like to thank Professor Ruiping Fan and an anonymous referee for their comments and suggestions to the earlier versions of this paper. Their comments and suggestion help to improve the paper a lot.

References

Audi, R. 1996. The morality and utility of organ transplantation. *Utilitas* 8:141–143.

Cheng, S. 1998. Development of Chinese concordance thinking through generations. *Forum on Chinese Culture* 1:29–35.

Chinese Teachers Union Institute of Education Sciences, ed. 2005. *Mencius*. Beijing: Academy of Audio Publishing.

Confucius. 2002. The analects. In *The four books and five classics,* ed. Y.-Q. Xia. Hohhot: Inner Mongolia People's Publishing House.

Confucius. 2007. *Classic of filial piety*. Beijing: Chinese Textile Press.

Fan, R. 2002. Reconsidering surrogate decision-making: Aristotelianism and Confucianism on ideal human relations. *Philosophy East and West* 52:346–372.

Fan, R. 2010. *Contemporary Confucian bioethics*. Peking: Peking University Press.

Fan, R. 2012. Confucian reflective equilibrium: Why principlism is misleading for Chinese bioethical decision-making. *Asian Bioethics Review* 4 (1): 4–13.

Hegel, G. W. F. 2008. *Outlines of the philosophy of right.* Trans: T. M. Knox. New York: Oxford University Press.

Li, L., and L. Liu. 2005. Concordance thinking—non-lawsuit—court mediation. *Journal of Nanning Municipal Community School of the Communist Party of China* 4:38–40.
Price, D. 2002. *Legal and ethical aspects of organ transplantation*. Cambridge: Cambridge University Press.
Tillich, P. 1995. *Morality and beyond*. Louisville: Westminster John Knox Press.
Wu, X. 2005. On "He-he" thoughts and setting up the socialist harmonious society. *Qinghai Normal University Journal* 6:51–54.
Xia, Y.-Q., ed. 2002. *The four books and five classics*. Hohhot: Inner Mongolia People's Publishing House.
Yan, X.-C. 1999. Analysis of current value of concordance culture. *Go* 9:45–47.
Zhang, L. 1997. Modern value of Chinese concordance spirit. *Study of Social Science* 5:48–54.
Zhang, W. 1999. On the philosophical basis and practical significance of concordance. *Journal of Qinzhou Normal Junior College* 2:10–14.
Zhang, D. 2009. *Chinese ethical thinking study*. Nanjing: Jiangsu Education Publishing House.

Chapter 13
The Informed Consent of Human Medical Research in Mainland China: A Family-Based Binary Decision Model

Rui Deng

13.1 Introduction

Medical research on human subjects is a necessary link in the development of biomedical technology. "Human subject means an individual who is or becomes a participant in research, either as a recipient of the test article or as a control. A subject may be either a healthy individual or a patient" (U.S. Code of Federal Regulations, title 21, part 56, sec. 102e). Medical research on humans—unlike clinical treatment—aims to test and verify unknown treatments or new drugs, and to gain new scientific knowledge. This process involves unpredictable factors and even serious risks, so the informed consent of subjects is crucial. Together with the Institutional Review Board (IRB), informed consent is considered one of "two pillars" which protect human subjects (Zhai and Qiu 2005, p. 423). The current model of informed consent in mainland China is individual-based and autonomy-oriented, which emerged in America in the 1970s. But the concept of individual and autonomous compared with the west have different meanings in mainland China. With this in mind, the aim of this paper is to find an appropriate model of informed consent. There are, of course, other measures to protect human subjects: e.g. improving the moral virtues of researchers or strengthening the IRB; I will not be concerned with those. With regard to the subjects, my essay will not concern subjects who do not have the proper ability to understand information or make a decision. Rather, my subjects are independent people who are older than 18—not teenagers or children—who are without intellectual disabilities—such as Alzheimer's disease.

The idea of informed consent originated in Western society, history, and culture; its core is personal autonomy, which is based on individualism. When this model

R. Deng (✉)
College of Humanities and Social Sciences, Shanxi Medical University, 56 South Xinjian Road, Taiyuan 030001, Shanxi, China
e-mail: shddengrui@gmail.com

R. Fan (ed.), *Family-Oriented Informed Consent*, Philosophy and Medicine,
DOI 10.1007/978-3-319-12120-8_13

was introduced in China, it was, in turn, influenced by traditional Chinese culture. Because of this influence, there are many differences between Chinese and Western ideas about the nature and practice of informed consent. In human subject research, these differences include doubt about the ability of subjects to make a decision, the motivations behind the decision, and the rights of the subject's family. Further, the process of informed consent itself is complicated. The process relies heavily on the decision-making context, such as the information provided by the researcher, the type of human subject research being carried out, the characteristics of human subjects, as well as their family status. In sum, the process of informed consent is far more complex than just signing one's name on an informed consent form (ICF). Therefore, the assumption that lets human subjects enjoy their right to informed consent unilaterally is defective in many cases. Thus, in order to protect the human subjects' interests, we must find an appropriate model of informed consent and find a way of putting it into practice. The full understand of decision-making information is the most important to an appropriate model of informed consent, but fully understand information depends on many conditions, one of them is to improve the subjects' understanding ability, family members participating the process of decision-making can improve the subjects' understanding ability, which is an appropriate model.

In mainland China there are many discussions about informed consent in clinical practice. But discussions of informed consent in human research are inadequate. So in this essay I will analyze two cases that occurred in mainland China. Firstly, the two cases will indicate that family participation in decision-making is key. Next, I will present a model of informed consent, what I call a "family-based, binary decision model." I want to argue that this model is necessary. For that purpose I will discuss the factors that influence human subjects' decision making. Further, I will make a defense of this model based on Confucian ethics. Finally, I will present some suggestions as to how to implement this model.

13.2 The Importance of Family Participation: Two Case Studies in Mainland China

13.2.1 Case A

Between 1998 and 2001, the Korea Cancer Center Hospital and the Zhejiang University Cancer Institute of China carried out a cooperative research project investigating how ginseng might help prevent colorectal cancer. They recruited more than 500 subjects from the Maqiao district in Haining City, Zhejiang Province, in China (see Wang 2006; *China Business Journal,* May 21, 2005; *Southern Metropolis Daily*, April 6, 2005). The subjects were instructed to take ginseng pills from Korea for 3 years. During the research period, some of the subjects suffered from hypertension, nosebleeds, dizziness and other adverse reactions. Some subjects discontinued the trial, other subjects even died. For example, an old man participated in the trial and died because of brain hemorrhage in June 2000. He was then 68 years old. A woman who participated in the trial often felt dizziness after she took the pill, and her blood pressure became very unstable. Likewise, an old lady who took two ginseng pills every week for 3 years was placed in the hospital due to hypertension, dizziness and headaches. Less than 3 years after the end of the trial, she suffered from uremia and died in February 2004, being 64 years old. There were many other subjects with similar symptoms.

In this case, all subjects were from rural areas, most of them were elderly, and with little education. In fact, some couldn't even write their own names: the old lady had to press her hand print on the informed consent form. Owing to her illiteracy, she did not know the consents, functions and rights involved in the ICF at all. We know that if a subject—such as this old lady—has no reading ability or little education, we should let other people with no conflict of interest participate in the decision-making. In this case, her son noted that he wanted to know more about his mother's participation in the drug trial. But the only ICF with his mother's hand print had been taken back by the doctor or nurse before he could ask for informataion from it. We know the conventional approach is that the subject must possess a copy of the original ICF. So this case indicates that family members—such as adult children, parents, or spouses—want to participate in decision-making. But the implementation of informed consent is not standard, and family decision-making cannot be achieved if the ICF is taken back before he can discuss it with family. But this case presented further complications. Many educated children of subjects worked in other cities; some of the elderly didn't want their children to worry about them; some of the subjects misunderstood the trial as the government's attempt to show concern for the rural elderly.

13.2.2 Case B

In 1991, the governments of China and the U.S. approved a collaborative research project funded by the U.S. with research carried out in mainland China (Wang et al. 2004). The aim of the study was to investigate whether a multivitamin containing folic acid can reduce the risk of fetal neural tube defects. The trial required that female subjects take the multivitamin containing folic acid both before and after pregnancy. In order to ensure the smooth progress of this project, investigators carried out a preliminary experiment for about 1 year, from 1991 to 1992. 503 young women were selected for this preliminary experiment. All of them came from the countryside of northern China, were in their first marriage, and were never pregnant before. They had a low level of education, only half of them had the experience of going out to work, and the work was mainly agricultural. In the process of deciding whether to participate in the trial, 88% of the women's husbands, 36% of their parents-in-law, and 32% of their parents participated in the discussion. No one decision was made independently by the women's parents or parents-in-law.

Research from the preliminary experiment indicated that women do not generally seek the help of their social support system on their own initiative. But in some cases women need their husband or parents-in-laws to share the responsibility, for example when the women have been influenced by rumors, or when their decision suffers opposition of the social support system, or when some abnormal or unexpected things occur during the trial process and the women cannot explain them clearly. There are two different kinds of situations about parents or parents-in-law participating in decision-making. One situation is when parents or parents-in-law and a young couple share the very same viewpoint, or when the parents or parents-in-law are unable to participate in the decision making because of illiteracy. In this case, parents or parents-in-law tend to remain silent or understate their view, showing an attitude of non-interference or non-opposition. Another situation is when parents or parents-in-law have different perspectives, or abnormalities—such as abortion—appear during the process of taking the medicine being tested. In this case, they will tend to express their opposition. The research also indicated that the women's family members, immediate family members, intimate friends and village doctors are the most important supporters. Group psychology also plays a very important role when they choose compliance or non-compliance in the research process. Women have strong interactions with their partners who participate in the same project. By the influence of group psychology, the women are willing to make the same decision as their partners. They usually have the view that the more people participate in the research project, the smaller the risk for each person. The conclusion is that a family-based binary decision model is necessary and effective.

13.3 The Aspects of a Family-Based Binary Decision Model

Through analysis of the above two cases, I claim that a family-based binary decision model is an appropriate model for informed consent decision-making in human research in settings with a Confucian-based culture. The main aspects of this model are:

1. The person who makes a decision is binary; it includes not only the subjects, but also their families. In mainland China, "the nuclear family and stem family are still the main family structure not only in urban but also in rural" areas. (Shen et al. 2009, p. 22). Given this fact, it is common to have three generations living under one roof: parents (or a parent) live with one of their married sons or daughters and their kids. Even if in recent years the stem family appears to be declining, more and more the "temporary stem family"[1] (Yao 2012) carries the same value

[1] The three generations in a traditional stem family live together and have dinner together over a long period of time even till the older generation die. But the situation has changed with the development of the society. Subjectively, Chinese have more autonomy and hope to have more freedom. Objectively, the married children work or live in different city from their parents, so even parents

and function as the traditional stem family. Of course, other people also propose their opinions during decision making, but they are not the decision-makers. Thus, a decision-maker circle comes into being around the subject, including the subject's family members, relatives, friends, colleagues and companions who join the same research trial, etc. This circle is not an interest group, but a support group with mutual dependence and trust.

2. In this binary decision model, the interest of the subject is the first interest, and the interest of family is a secondary interest. The purpose of the family's participation in the decision-making is to consider the risks of the trial from more angles, and to propose their opinions in order to protect the interests of subject. Family interests cannot override the interests of subject. The person who makes the final decision should be the subject himself.
3. In this binary decision model, the family has a right to agree to or refuse the subject's decision to participate in research or not. That is to say, the family does not have the right to require or force a family member to participate in a trial. This is because participating in research (not for treatment) is always an issue which disturbs the normal life or the usual treatment procedures of a family member. Extra familial consent is needed in order to protect individual family members. Accordingly, for any trials, if a person decides not to participate, the family cannot require or force him/her to participate. However, for some significant trials, if a person decides to participate, he still needs to get the consent of his/her family. If the family denies consent or has different opinions, in most cases the person will not be able to participate in the trial, such as Case B.
4. The role of the family in agreeing or diagreeing with the decision of the subject will vary according to the situation of the subject. If the subject is rational, independent, and the trial has less risks, then the family's right is mainly limited to agreeing with the decision of the patient. If the trial is more important and the risks to subject or his/her family are greater, then the family has a greater right to deny the decision. Of course, the final decision should be made by the subject. The fact that the family has different rights in different situations is mainly according to Ren (仁 benevolence)—the loving of the family member.

In essence, the nature of a family-based binary decision model is a kind of transfer and share of the subject's rights. This model is supported by two kinds of considerations. Firstly, because the actual informed consent situations are imperfect, during the process of reaching a contract, subjects inevitably have various limitations.

and their married children want to live separately. But this state is temporary—when the married children have their baby, they often have not enough time to look after a baby, so one or both the older parents live together with their children and their grandchild. Sometimes, if their economic situation is good enough, the older parents live in another house which is very close to their married children and their grandchild—that's "the distance of a bowl of soup." The distance is very appropriate: both of them have independent living space, and they can take care of each other. If the grandchild grows up, the older parents may go back to their hometown. But if the older parents are too old to live separately, they will live together with their married child again. So, living together or not can be temporary in different situation. But eventually most will live together for quite a long time, so this type of family was named"temporary stem family" by Yao Jun.

For example, subjects might be unable to make a decision because of the lack of relevant knowledge; or they might not know what to do because of being in an unfamiliar environment; or they might not know what to do because of too much information. If normal judgment is affected by the above factors, they will probably make a wrong decision. Secondly, there are ethical considerations. Imposing the responsibility of making decisions on subjects unilaterally will weaken the responsibility and obligation of doctors and investigators. If this is so, informed consent will be more and more a kind of legal formalism, and the result will harm the interests of subjects.

13.4 The Necessity of a Family-Based, Binary Decision Model

The subject can only know the written information of the consent form. And yet it is what is *not* stated in the contract that really affects the behavior of investigators and subjects, such as who are the research investors and cooperators, and what the relation among all the parties is. So in such circumstances, the subject's decision whether to participate or not is very complex. Now, family participation is highly necessary, and such participation shows a different scope and depth.

13.4.1 Problems with the Guidance of Investigators

The investigators play a leading role in the course of implementing informed consent. The information provided by them is key in helping subjects to make a decision. Incentives for investigators to enroll subjects into clinical trials are diverse and create conflicts of interest (Alpert et al. 2006). Information from investigators is always deficient due to the impact of the interests behind the project.

Firstly, there is withholding of information. The key pieces of information that affect decision making are not mentioned, such as adverse events, and whether there are other available treatment projects that can be chosen if the subject does not want to participate in the trial. In fact, the things mentioned above have detailed rules in research plan, adverse events will be noticed to IRB within 24 hours, the available treatment projects written in the research plan, but those information is not told to the subjects in detail in order to save time or recruit subjects quickly, which is different from the way in America and other western countries. Secondly, investigators can avoid important points and dwell on trivial ones. They can understate the risks and injuries that may occur during the trial, and emphasize the potential benefits of the trial. Thirdly, investigators may talk vaguely. The investigator may not be very clear about the risk assessment and the treatment after injury. Finally, investigators may oversimplify matters. The presentation of trial procedures, trial methods and risks can be excessively brief.

In this process, the investigators tend to let the subject make the decision alone for the following reasons. The first reason is the impact of the concept of self-determination. The investigators think that self-determination is an international principle and should also be obeyed in mainland China. So a subject can sign his/her name on the informed consent form by himself/herself, and need not consult with his/her family. Therefore not only in the design of the form but also in practice, there is no space for the family to participate. The second reason concerns the needs of the trial. Investigators know that allowing more people to participate in determining the participation of the subject will lower the planned speed of the research project. Further, some people originally intend to participate in the trial, but after they consult with family or friends, change their idea and abandon the trial. The potential subject loss and the necessity to recruit new subjects will extend the time of the research project. The third reason is standard practice. Unlike clinical treatment, family consent is not a necessary condition in human subject research; rather, standard practice only requires the signature of the subject. Investigators hope the decision is made by the subject himself, so this hope will have an impact on the doctor or investigator during the process of informing the patient, such as exaggerating the effectiveness of the trial products, avoiding or evading risks and injuries, and emphasizing the concept of personal autonomy to subjects. It can be seen that the lack of information and the guidance of the investigator deeply affect the rational judgment of the subject. Therefore, we propose that family or friends should participate in the informed consent process, to avoid medical paternalism in human research and to eliminate the risks raised by value guidance from investigator.

13.4.2 Family Considerations in Different Types of Human Research

In different phases and different types of human medical research, the interests of scientific research and those of the human subject conflict. Subjects will consider different factors when they make decisions, so the expectations of the family are also different.

In phase I of a clinical trial, the subjects are healthy people, the trial can't bring direct benefit to the subjects, but can have a lot of uncertain risks. In such a case the interests of the subject and those of the researchers are seriously in tension. To protect subject interest, the family's participation in decision-making is necessary in theory. But quite the contrary, in practice subjects in this phase have to make decisions by themselves. The cultural reason is filial duty. First, an important aspect of filial duty is that subjects should care for themselves and not make their parents worry. Participating in a phase I human trial may bring about risks and damages to the subject's body. This is contrary to the doctrine of the *Xiao Jing* (*The Classic of Filial Piety*), which states that "Our bodies—to every hair and bit of skin—are received by us from our parents, and we must not presume to injure or wound them. This is the beginning of filial piety." (身体发肤, 受之父母, 不敢毁伤, 孝之始也,

Sturgeon 2013b, The Scope and Meaning of the Treatise) As a member of a family, a subject undertakes the obligation of not worrying his/her family; hence, the subject might not want to consult his/her family. Secondly, Chinese culture thinks the motivations of healthy people to participate voluntarily in human subject research are questionable. Especially when the subjects are young adults or university students, their participation mostly aims to obtain high remuneration—given their usually weak economic condition. The *Xiao Jing* also states, "When we have established our character by the practice of the (filial) course, so as to make our name famous in future ages and thereby glorify our parents, this is the end of filial piety." (立身行道, 扬名于后世, 以显父母, 孝之终也, Sturgeon 2013b, The Scope and Meaning of the Treatise) So in Chinese culture, the choice of being human test-subjects is not considered honest labor, and this behavior can't be of spoken to other people. It would even bring disgrace to the family. Human subjects are aware of this, so they don't want their family to know, so that family participation during the informed consent process is missing. Of course, there are many objective reasons for why family advice could not be obtained. Because many subjects study or work in a different city than their family, and because the family's attitude to trial may be ignorance or misunderstanding, the two sides often hinder effective, accurate and timely communication about the trial. It should be noticed that even though the main family members might not be able to participate in decision making effectively, subjects can obtain some advice from other members of the family, such as those similar in age, those who are more liberal, or those with whom they have an especially intimate relationship. So at this phase, subjects prefer to make decisions together with their good friends or other subjects who participate in the same trial.

In Phase II and III clinical trials, subjects are often clinical patients, and participating in the trial can solve some medical problems or treat a disease to some extent. In these phases, the conflict between the subjects and the scientific research are reduced to a certain extent, and the relationship between them is similar to the relationship between doctor and patient in clinical treatment; the research activity is similar to clinical activity. According to a recent survey, "most patients (79.5 %) are still willing to share the right of informed consent with their family in clinical treatment" (Wu and Liu 2012).It looks similar in human research. Moreover, in both phases, the recruitment and selection of subjects is non-independent. Unlike Western countries, in mainland China, a large percentage of the population with all kinds of relationships live together. Hence, recruiting subjects might involve having them introduced by relatives, friends and colleagues; doctors might seek patients to be subjects; or subjects might be assigned directly by leaders.[2] Of course, all of

[2] The 'leaders' I refer to are, in Chinese culture, ordinary people (i.e., not necessarily government officials) who are in charge of some part of an organization—for example, a school or company. Given their leadership role, people tend to trust them. For example, I have seen a head teacher (a teacher in charge of a class) of a university recruit subjects from his class during the break and encourage student leaders to recruit students actively. To give another example, a newspaper reported that since a village official allowed a new drug trial to be carried out in his village, more than a hundred people trusted him and participated in the trial. See, Li Guangming. "Dozens of Farmers in WangJiang County AnHui Province Became Subjects Inexplicably". *Legal Daily*.

them, usually with good intentions, think that providing such information and making subjects join the research process can bring them benefits. So in the two phases, not only the subject, but also the family, sees it as desirable to let their family member participate in the decision of informed consent, which can also include the subject's friends, other subjects who participate in the same trial, or other patients suffering from the same disease, and so on.

Genetics research is another special type of research, because the family's genetic characteristics, disease susceptibility, and even life expectancy could be known from subject's genetic information. The genetic information of the subject is shared with his/her family members. So when a person is making a decision whether to participate in genetics research or not, he can't decide alone, because the decision is connected with the interests of other people. Therefore, it's necessary to make a binary decision when participating in genetics research; the subject must consult with family members and make decision together with them.

13.4.3 Different Individual Characteristics of Human Subjects

The decision is made ultimately by the subject alone or after consultation with his/her family, which also relies on the subject's ability to understand, make decisions and other factors. The individual characteristics of human subjects include the following three aspects.

First, the literacy and knowledge of subjects will affect their reception and understanding of trial information. If subjects lack literacy and medical knowledge, they will not understand the trial. Carrying out informed consent effectively will be extremely difficult, and even if they agree and sign, their consent is mostly passive. In a survey, subjects who signed the very same informed consent form were asked whether the information provided is clear or sufficient. The majority of the subjects with good education answered 'no,' while the majority subjects with lower education said 'yes' (Wang et al. 2011). These results mean that the subjects with lower education do not have a good understanding of the information. They have a lower standard for the information and think that the existing information is enough. But in fact, their understanding of the trial is not sufficient.

Second, due to the long-term influence of traditional culture, subjects lack the willingness to engage in informed consent. On one hand, we have the traditional proverb that speaks of the "doctor with parental heart" (医者父母心) which makes subjects believe that doctors or researchers can make an appropriate decision that most accords with their interests as subjects. On the other hand, under the influence of traditional Chinese culture, which emphasizes family values, subjects have a higher willingness to transfer their right to consent, and their family has a higher willingness to participate in decision making (Wang et al. 2011). For instance, in

---

October 29, 2010, http://www.legaldaily.com.cn/index/content/2010-10/29/content_2332847.htm?node=22668 .

rural China, researchers often need to have the consent of the mayor before obtaining subjects' consent.

Third, the subjects' psychological quality could influence the implementation of informed consent. When a person is recruited to be a subject of a trial, in most cases, he has become physically weak, and maybe has been confused because of the threat of disease. Perhaps he wants to get money from the trial, or believes the research project could bring benefits to his illness or to his social status, or became a subject because he was introduced by an acquaintance, or feels embarrassed to refuse, and so on. In a word, various reasons make it so that the subject can't express his/her willingness completely when making a decision, thus they are unable to protect their own rights.

13.4.4 Influence from Family, Friends, and Peers of the Subject

Due to differences in the structure, size, economic situation, social backgrounds, and the cultural backgrounds of the family, subjects have different degrees of dependence on their family's decision. For example, for a rich family it is difficult to accept that a member of their family participate in a human trial. They all have medical insurance and have enough money to pay for the regular treatment, so the free experimental treatment with uncertain potential risks presents no attraction. On the other hand, the poor family often lacks relevant knowledge about the human trial. They usually think the trial is dangerous for their body, that it will make healthy people fall ill, and the make patient worse. So they would not agree that a member of their family should participate in human trial.

A subject's friends will also have an impact on the decision-making process. Some subjects have many friends in different lines of work and will select one or more of them to consult in the trial. This is especially true when the subject's friends engage in medical or legal work. They can provide significant advice from their professional expertise. Also, friends who are very familiar with the subject's life, values and intentions also can make important suggestions.

Moreover, subjects who participate in the same research can also play important roles in decision making. Presumably they live in the same area, the same community, attend the same hospital, the same ward, and so on. Owing to a conformist mentality, the subject will prefer to make the same decision as other subjects. If other subjects decide to participate, the subject in question will also participate. If most subjects decline, the subject in question will change his original positive idea. Of course, such decisions may not always be reasonable, but we should continue to pay attention to how other people play a role in the decision-making process.

13.5 The Ethical Defense of the Family-Based Binary Decision Model

I must emphasize that I do not want to deny the individual rights of subjects. My aim is to strengthen the subjects' capacity for decision-making, and reach a truly effective informed consent. The Nuremberg Code emphasized in its first proposition that human subjects must consent voluntarily, and human subjects must have the capacity to consent (Vollmann and Winau 1996). This means that subjects might *not* have a proper capacity to consent. Maybe they have a limited or inadequate capacity, in which case they are weak and should be protected. To address this I propose a family-based binary decision model in the decision-making process in mainland China. And this model, with its particular theoretical basis, can be defended by traditional Chinese Confucian ethics.

13.5.1 Emphasis on Family Value

From the structure of Chinese society, we know that the value of family is very important. In mainland China, everyone's identity and social role is confirmed in the family. An individual is a part of the family, so the most important decisions in one's life, such as education, health, hospitalization, employment and marriage, are all made by discussing them in the family. We believe that a person does not necessarily have the "reason, experience or capacity to make contract with others equally" (Fan 2010) when he/she is faced with deciding about a significant problem, such as whether or not to participate in a trial. When a person must make a decision whether to participate in a trial or not, because of their vulnerable status, they are not always able to make a reasonable judgment which fits their best interests. So, mutual consultation is always more proper than arbitrary decisions by oneself. In fact, too much emphasis on individual autonomy can be harmful to individuals (Fan 2011).Confucian virtue ethics asks people to make decisions based on family—as opposed to emphasizing individual decisions—so they can care for and help each other, especially about important events.

13.5.2 Emphasis on Filial Piety and Family Harmony

According to traditional Chinese culture, the family is the most basic political, economic, cultural and social life unit. The *Li Ji* (*The Classic of Rites*) taught people that the proper way of life was to first cultivate oneself in order to regulate one's family; this would, in turn, order one's states, and finally spread virtue throughout the kingdom.[3] This indicates that family harmony is a necessary stage of development

[3] The theory above is summarized from the chapter of DaXue ("The Great Learing") in *Liji* (*The Classic of Rites*). The original text is "the ancients who wished to illustrate illustrious virtue

for both personal development and national prosperity. Family harmony means the mutual obligations and responsibilities of family members.

Firstly, there are mutual obligations and responsibilities between parents and children. The *Mao Shi ZhengYi* states: "if we can be friendly to our family members, not abandon our friends, and not leave behind old friends, people will naturally be led to become more honest" (Mao and Zheng 1980, p. 410).[4] Further, Confucius affirms in the *Zhong Yong* (*The State of Equilibrium and Harmony*) that "benevolence is the characteristic element of humanity, and the great exercise of it is in loving relatives"(仁者人也, 亲亲为大。Sturgeon 2013d, p. 20). It means that the first and most important thing is to love parents. Filial piety is the most important of all virtues. Filial piety is love from the heart. Filial piety means making parents happy, comfortable; sons and daughters must not act contrary to the wishes of their parents. Filial piety also means that sons and daughters must strive to do everything to minimize their parents' concerns for them; the kindness and care is mutual. Secondly, there are mutual obligations and responsibilities between husband and wife. A harmonious marital relationship is the primary factor in family harmony. Spouses who discuss the research trial embody charity. Thirdly, there are mutual obligations and responsibilities between siblings. Brothers and sisters who live in harmony show respect to parents and elders. Hence, individuals and families are in a close relationship; the concern of one member often requires the family to make a decision. Family autonomy is often associated with personal autonomy. In human research, whether a family member participates in research is not regarded as a purely personal problem, but as a family problem. So informed consent decisions tend to have family involvement; such decisions are considered the responsibility of the family.

13.5.3 Emphasis on Friendship

The Confucian attaches great importance to friendship. According to Confucianism, friends should follow Li (礼etiquette), Ren (仁benevolence), Yi (义righteousness), Zhong (忠loyalty), Shu (恕forgiveness), and Xin (信integrity) with each other. Friendship has a special status among the five cardinal kinds of relationships because of its particularity. Its particularities are selectivity, equality, integrity and mutual encouragement (Wang 2007). The *Analects* states: "friendship with the upright, friendship with the sincere, and friendship with the man of much

throughout the kingdom, first ordered well their own states. Wishing to order well their states, they first regulated their families. Wishing to regulate their families, they first cultivated their persons". See, Sturgeon 2013c.

[4] The *Mao Shi ZhengYi*, also named *Mao Shi Xu*, is a book annotating the *ShiJing* (*The Book of Poetry*, a famous classics in Western Zhou). The authorship of Mao has not been confirmed, but it is generally thought that it is MaoHeng in the Han Dynasty. This book has no English version, so I translate the sentence myself. The Chinese text is: 亲亲以睦友, 友贤不弃, 不遗故旧, 则民德归厚矣.

observation—these are advantageous "(Sturgeon 2013a, JiShi, p. 4). There is horizontal equality between friends—friends both fulfill obligations and share rights. Friends can help each other because the relationship is based on uprightness and is not for personal gain. So mere personal interest can be avoided. The help of a good friend can provide a more objective view and meaningful advice about for informed consent. Another particularity of friendship is integrity. *The Analects* states: "in his intercourse with his friends, his words are sincere." (Sturgeon 2013a, XueEr, p. 7) The relationship between friends is not hierarchical. Their relationship is maintained by good faith. These features provide the possibility for the subjects to get effective help from their friends.

13.6 Some Suggestions for Implementing the Family-Based Binary Decision Model

The implementation of a family-based binary decision model needs to be supported by policy or ethics. But only a few special Chinese guidelines on clinical trial have the regulation that requires family to participate in decision-making. This is because the trials are special, such as the clinical gene trial and the clinical vaccine trial etc. For example in 1999, the policy of "The Guidelines on Application of Clinical Trials in Genes" published by China Food and Drug Administration (CFDA) stated: "The gene treatment can start only when the patient *and* family understand the information adequately and the family sign the ICF" (1999, Appendix 9; II Content of Application Material, 6 Ethics). In 2003, it was revised to state: "The gene treatment can start when the patient *and* family understand the information adequately and not only family but also patient sign the ICF"(CFDA 2003, II Content and Quality Control. 8 Ethics). Both items highlight the importance of family members during decision making. These policies display the concept of a family-based binary decision model. The practice that patients themselves and family members sign the ICF together, which respects the right of informed consent of patients while following traditional Chinese culture, might be a better solution at present. But this policy is only for the gene research, but not for other medical research. Most guidelines just need the subject's agreement and signature. I will now present some suggestions to implement a family-based binary decision model.

Firstly, a family-based binary decision model needs to be written into the national laws and regulations, eapecially some important guidelines published by the National Health and Family Planning Commission of the People's Republic of China or China Food and Drug Administration, such as "Ethical Guidelines for Biomedical Research Involving Human Subjects", (NHFPC 2007) and "Guidelines of Ethical Review in Drug Clinical Trials"(CFDA 2010). I suggest that these guidelines should put forward some clear demands or provisions, such as: "in research projects involving human subjects, the investigator or doctor must provide information to the subject and his/her family, and have both sets of signatures. If they do not procure the signature of the family, investigators should explain why this research does not need the family's advice." Accordingly, IRBs should pay more attention to trials with more or uncertain risks. If the research project is very complex, the duration of the research project is very long, or the research involves family interests, the IRB should require that both the subject and the family to sign the informed consent form.

Secondly, the current informed consent form in mainland China has no unified format, and the contents are too simple. Thus the ICF do not have enough information for subjects. Therefore, a review of the process of informed consent is more important than a review of a signature. The IRB should play a supervising function, and should make the family really participate in the decision-making process. The process could be supplemented with a survey and individual interview to inquire into the process of decision-making, and find some appropriate ways to implement a truly binary decision.

Thirdly, the IRB should serve the function of education and training, announcing the advantage and necessity of binary decision process. It should demand that subjects read the ICF with their family or friends, consult each other, and then make a decision together. Of course, the result of such requirements would increase the workload of the investigator, and even make it difficult to recruit subjects. But I believe that, in the past, the recruitment of subjects has been simple and cheap in mainland China. This is due to the subjects' inability to protect themselves and due to the lack of social support systems. It is impossible to improve the ability and autonomy of subjects in a short period, and the social support systems will not be set up quickly. Relatively speaking, family support is more readily available in all social support systems, and binary decisions are easy to implement through informed consent.

Acknowledgments This essay was funded by the "Humanities and Social Science Research Fund Plan of the Ministry of Education of the People's Republic of China" (12YJA720003) and the "Soft Science Project in ShanXi Province" (2013041084-01).

References

Alpert J. E., and M. M. Biggs, L. Davis, K. Shores-Wilson, W. R. Harlan, G. W. Schneider, A. L. Ford, et al. 2006. Enrolling research subjects from clinical practice: Ethical and procedural issues in the sequenced treatment alternatives to relieve depression (STAR*D) trial. *Psychiatry Research* 141 (2): 193–200.

China Food and Drug Administration. 1999. Regulations of application in new biological products license. *China Food and Drug Administration*. http://www.sfda.gov.cn/WS01/CL0053/25884.html. Accessed 10 Oct 2012.

China Food and Drug Administration. 2003. The quality control methods guidelines on gene treatment and preparation. *China Food and Drug Administration*. http://www.sfda.gov.cn/WS01/CL0237/15708.html. Accessed 11 Oct 2012.

China Food and Drug Administration. 2010. Guidelines of ethical review in drug clinical trials. http://www.sfda.gov.cn/WS01/CL0058/55613.html. Accessed 11 Oct 2012.

Fan, R. 2010. Building Chinese bioethics—pursuing the excellence and permanence of Chinese culture. *Chinese Medical Ethics* 5:6–8. [In Chinese.]

Fan, R. 2011. A better society from the perspective of a Confucian. *Beijing Cultural Review* 12:70–3. [In Chinese.]
Mao G., and X. Zheng. 1980. *Mao Shi ZhengYi, Ruanyuan Proof Zhushu of Thirteen Classic.* (阮元校刻十三经注疏本). Beijing: ZhongHua book company.
Vollmann, J., and Winau, R. 1996. Informed consent in human experimentation before the Nuremberg code. BMJ 313(7070):1445–1449.
National Health and Family Planning Commission of the People's Republic of China. 2007. Ethcal guidelines for biomedical research involving human subjects. *National health and family planning commission of the people's republic of China*. http://www.nhfpc.gov.cn/qjjys/s3581/200804/b9f1bfee4ab344ec892e68097296e2a8.shtml. Accessed 15 Oct 2012.
Shen, C., D. Li, and F. Zhao. 2009. *Changes in the urban and rural family*. Chongqing: Chongqing University Press. [In Chinese.]
Sturgeon, D. ed. 2013a. *The analects of confucius*. In *Chinese text project*. http://ctext.org/analects/xue-er/ens. Accessed 12 Jan 2013.
Sturgeon, D. ed. 2013b *The classic of filial piety*. In *Chinese text project*. http://ctext.org/xiao-jing/ens. Accessed 12 Jan 2013.
Sturgeon, D. ed. 2013c. *The classic of rites* (*Liji*). In *Chinese text project*. http://ctext.org/liji/da-xue/ens. Accessed 14 Jan 2013.
Sturgeon, D. ed. 2013d. *The state of equilibrium and harmony*. In *Chinese text project*. http://ctext.org/liji/zhong-yong/ens. Accessed 21 Jan 2013.
Wang, Y. 2006. Reflections on legal issues in a clinical drug trail: The "Korean Ginseng Pill Incident." *Hospital Management Forum* 12:60–4. [In Chinese.]
Wang, W. 2007. Discussing the ethics of friendship in ancient China. *Jianghan Tribune* 12:106–9. [In Chinese.]
Wang, D., Z. Du, M. Zhao, Y. Yang, F. Gao, and Y. Qin. 2011.Study on issues of informed consent in the perspective of patients. *Medicine and Philosophy* 32 (5): 38–42. [In Chinese.]
Wang, H., J. D. Erickson, Z. Li, and R. J. Berry. 2004. Evaluation of the informed consent process in a randomized controlled clinical trial in China: The Sino-U.S. NTD project. *The Journal of Clinical Ethics* 15:61–75.
Wu B., and J. Liu. 2012. A survey of cognitive effects of family factors on informed consent in patients. *Medicine and Philosophy* 33 (6): 28–30. [In Chinese.]
Yao J. 2012. "Temporary stem family": The transformation and strategy of family structure in cities—based on the case study in N City. [In Chinese.] *Youth Studies* 3:85–93, 96.
Zhai X., and R. Qiu. 2005. *An introduction to bioethics*. Beijing: Tsinghua University Press. [In Chinese.]

Part VI
Family Shared Decision Making, Truth Telling, and Advance Directives

Chapter 14
Toward a Shared Decision: Against the Fiction of the Autonomous Individual

Ryan R. Nash

14.1 Introduction

Western, specifically American, medical ethicists, in a reaction against a particular narrative of medical paternalism, have called for the use of a shared decision making process. Increasingly, this process is primarily aimed at honoring the autonomy of the patient. However, the concept of autonomy invoked often assumes an isolated, idealistic, individual deciding amongst neutral options from an objective position free from all biases. However, such an individual original position[1] is fictional. American physicians, complicit in this false assumption, have been reduced to providers of options instead of givers of guidance. This paper describes a better path to achieve a shared decision that respects and serves patients. It urges physicians to employ an appropriate degree of directiveness. Further, it argues that one may enhance the true autonomy of patients by engaging their rich contextual influences and biases. This will include communal decisions, decisions that include the family or other significant community. The common experience of communal decision making amongst certain families and religious communities in the United States are positive examples. Using the frame of end-of-life care decisions, including considerations of decision burden and bereavement, the superiority of family or community oriented decision-making is explored with arguments of inherent value and positive potential consequences.

[1] The similarity of Rawlsian theory and the current prevailing bioethics in the United States and Europe cannot be over emphasized. For consideration of the original position see Rawls 1971 and 2001.

R. R. Nash (✉)
Center for Bioethics, Wexner College of Medicine, The Ohio State University, 2194 Graves Hall, 333 West 10th Ave, Columbus, OH 43210, USA
e-mail: Ryan.Nash@osumc.edu

R. Fan (ed.), *Family-Oriented Informed Consent*, Philosophy and Medicine,
DOI 10.1007/978-3-319-12120-8_14

14.2 The Challenge of Family and Community in the West

The rise of technology and increase in worldwide wealth, or at least the appearance of an increase in worldwide wealth [2], has brought radical changes to numerous facets of life, not least of which is how we receive medical care. Peoples in the West faced enormous change over the last century with an almost doubling in life expectancy, the effective prevention and treatment of many diseases, and unprecedented access to food, shelter, and clothing. Often these latter three are found beyond just necessity: food that is on demand and consisting of representative culinary tastes from around the world, shelter in homes with comforts our forefathers would scarce imagine, and clothing that is shipped around the world with the click of a button and too often discarded with equal ease. Historically, man struggled, toiled, and had the shadow of death ever before him. Today in the US, for the first time in human history obesity is a sign of poverty, cell phone service, cable TV, and a computer with the obligate Wi-Fi access are heralded as rights in some intellectual and political circles. And death is institutionalized, medicinalized, and hidden.[3,4] Expansion and advancement of technology, industry, and wealth have brought luxury. This luxury is now becoming an expectation or even a right. With this change the West has realized a shift from the old humanist trinity of liberty, equality, and fraternity to pleasure, health and security with pleasure being the chief pursuit[5]. The pursuit of pleasure[6] is evident in many facets of life but perhaps most vividly in the radical shift in sexual ethics. A consideration of sexual ethics is appropriate in setting the stage for consideration of family oriented decision making not only in that it further elaborates the goals of pleasure as opposed to the more traditional pursuits, it emphasizes a shift that directly threatens family. It is by sex that families are formed and all too often it is by aberrant sex that families are destroyed. Regarding sexual ethics, Western culture has moved from venerating the virtuous virgin to validating the vanguard of vulgarity. In only a few short decades, we have witnessed the move from sex within lasting marriage with a positive expectation of offspring, to sex without the burden of pregnancy, to sex without the shackles of wedlock, to sex with the gender, object, or visual stimulation of choice. The individualistic

---

[2] Wealth can have many different measures such as money, goods, property, and even social considerations. Wealth is often relative to the given time of consideration. Thus many poor of today have necessities of life that poor of previous generations lacked. However, beyond the material it is arguable that in many ways modern cultures are impoverished socially, culturally, and religiously.

[3] See: Illich 1974, 1976, p. 96.

[4] This theme is repeated throughout Bishop 2011.

[5] See: Juvin 2010, p. 158. It should be pointed out that pleasure in my treatment, consistent with Juvin's, is referring to an immanent self-serving gratification that focuses on immediacy.

[6] Pleasure is potentially a complex topic that this paper will not explore in great detail. For the purposes of this paper pleasure is to be defined as temporal pleasure that serves the self in tangible ways. In contrast, it is not referencing the gratification of an altruistic act, the joy of deep intimacy, or wonder of eschatological joy.

pornographic age that is upon us seeks pleasure in ways previously unimaginable, or at least in ways that were not socially acceptable.

To further emphasize the move to the pursuit of pleasure I offer an educational anecdote. I recently asked a class of 160 medical students engaged in studying neuroethics a hypothetical question. I asked for their sentiments on the following proposition: If I could guarantee that they would be basically safe and basically healthy (the subordinate two goals of the new humanism met) would they want to be plugged into a machine that would bring them continuous simulated pleasure through direct brain stimulation. The scenario was met with much attention—some shouts of "yes" or "why not" immediately came, especially from the men. They followed with their most articulate efforts to rationalize their ethic, passions fully awakened, with all of the theories they could remember from earlier foundational lectures or perhaps undergraduate philosophy, political science, or psychology class.

Pleasure being the chief of contemporary goals means that the ultimate enemy is no longer sin, death, and the devil as the Christian founders imagined, or the loss of liberty as our humanist founders imagined, but suffering. Thus, life becomes increasingly mechanized toward this new goal.[7,8] So long as pleasure is possible and suffering abated then liberty may be forfeited. Liberty is collapsed into an individualistic autonomy that serves to enforce the will to flow with the *zeitgeist* and pursue pleasure. Trapped in a mechanized ethic of autonomy there is little true liberty. Divorced from family and rich culture the prevailing secular morality dominates.

The goals of the new humanism have increasingly become rights in the minds of the growing masses. Subsequently, governments increasingly are turned to provide such claim or positive rights. The culture increasingly looks to government to provide mechanized rights including medically provided erections for men, contraception and abortion for women, divorce on demand, provisions for children born out of wedlock, and medical care and institutionalization of the aging, sick, and dying. The vast mobility of persons in the modern era in addition to the loss of being tied to a region or land and the resulting challenges this brings in relating and knowing kin (and kith) gives excuse for the latter. With the above cultural shifts in view it is a wonder that a culture that pursues individualistic pleasure with merely a prerequisite amount of health and security to achieve this goal keeps the word "family" in its lexicon. In fact, in the prevailing Western culture one must define family as many competing definitions have arisen. This paper refers to the traditional family defined as married parents with children (immediate and expanded kin beyond the nuclear family may apply in many circumstances as well). This was the consensus definition of the West's Christian past but is now under challenge. It is with this backdrop that I stage a consideration of shared decision making in family and community.

[7] See: Juvin 2010, p. 38, 175.

[8] See: Vlachos 1999, p. 16.

14.3 Shared Decision Making as Aspiration

Unfortunately, many Americans focus on a misdiagnosis or at least an incomplete or partial diagnosis. Frequently we hear the diagnosis of a fractured health care system or political system when the true diagnosis is fractured family and community. The treatment prescribed is often a simulacrum of true community—chaplain instead of priest, social worker instead of aunt, counselor instead of old friend or sibling, health navigator instead of adult son or daughter, improved documentation on transitions instead of a physician that actually knows the patient. Or the promised cure with improved health policy or payment structure is given as essential treatment. Yet the cancer of the destruction of family and community is far more challenging to treat and the above are merely focused treatments on metastatic tumors. To add iatrogenic injury, physicians have cast off the traditional role of guides. Thus patients are increasingly isolated in decision making. With the above considerations it should be clear that a truly shared decision making that includes family and community, if desired, must be aspirational in the West. Nonetheless, it is an aspiration that is in reach. Family and community, despite their rapid deterioration are the pillars of traditional Western culture and a decision making that respects them may be possible. At least three correctives are required to reclaim a family or communal shared decision making: (1) the concept of the autonomous individual need be challenged, (2) physicians need to accept the burden and responsibility of giving guidance, and (3) subcultures within Western culture that hold to family and community need be turned to as guiding examples. The remainder of this paper will attempt to sketch a path for these correctives.

14.4 The Dogma of Informed Consent

The rise of the dogma of informed consent in the West is best understood not in consideration of the bedside but of the laboratory. Research ethics flourished in the twentieth century. After the Nazi research atrocities, many of which were in line with practices elsewhere, including the US, codes such as Nuremburg and Helsinki were formed. The codes further developed in particularity in the Belmont report. In this narrative, it is clear that the goal of informed consent is to protect the subject from the researcher. Take for instance a subject in the Tuskegee syphilis experiment. If the proper assumption of permission is given informed consent, that is, its purpose is to ensure we do no harm to unconsenting innocents,[9] then most would reasonably suspect that if a researcher asked a poor black man in Alabama if it would be 'OK' to not treat him for a treatable disease in order that the natural progression of disease (his coming needless illness and suffering) would be better articulated, such a researcher would be refused if not harmed for his lack of basic

[9] See: Engelhardt 1996, p. 123.

respect. It is clear that in research informed consent is designed to give an opportunity for permission or a right to refuse. When applied to clinical care that is not part of a formal research trial the goal of informed consent perhaps becomes a bit more robust. It has been argued that patients who are informed of their treatment plan and partner in it have better outcomes. The potential reasons for this are many, but perhaps it is the sharing of expertise. The physician shares medical information, such as you will need to take the antibiotic four times daily, while the patient shares their expertise of themselves, which may include the low likelihood that a drug will be taken four times daily and asking for a different option. This making of 'good' or effective plans is surely a goal for informed consent, but the constitutive goal remains to protect patients from physicians or the health care provider, or the health care system. It remains a process of permission giving patients the opportunity to exercise a right of refusal.

The right to refuse is a formal, forbearance, or negative right, not a human, claim, or positive right. The only positive or claim right the patient has in the US is for the expectation that the standard of care is met, and even this is arguably a forbearance right related to not being harmed. The exercising of the right to refuse can be based on any internally rational reason including what seems to be external foolishness. It can be based on relationship, authority, etc. But this assertion is increasingly challenged. There is a trend to redefine autonomy to be more than a right of refusal or self-determination but an ability to reflect on the content transmitted by the physician in a "meaningful way." Some of the more militant emphasizing this shift in autonomy redefine the role of the physician as a partner and educator that may help enable the patient to escape the confines of their culture, family, or religion to find enlightenment in the more reasonable approach of the prevailing secular moral sentiment. One group has suggested that the long practiced respect given to Jehovah's Witnesses in their refusal of blood products should be refused and replaced with an assumption that the patient is a vulnerable, deluded victim of misinformation.[10] They go so far as to question their capacity to make an informed decision given their presumed delusion. Their motive is likely one of compassion to help a group that they believe is wrong-headed but this vastly oversteps the conventional use of delusion in decisional capacity evaluation and represents a radical secular fundamentalism. Regarding the former, it is recognized that there is a diversity of approaches to decisional capacity evaluation. Some hold to an internal rationality standard, that the choice accomplishes the goal of the patient. Others employ an external rationality standard, which too often means, "I do or do not agree with your choice." If the evaluator disagrees then the patient is deemed to potentially lack the reasoning ability to make an informed choice. Were it practically easier to take liberties away in the US then many physicians, yielding to the wisdom of psychiatrists or clinical ethicists[11], would indeed circumvent their patient's directives. Moreover, these shifts change the role of consent. Instead of a right of refusal, informed con-

[10] See: Louderback-Wood 2005; Guichon and Mitchell 2006.

[11] Decisional capacity evaluations are done by a host of other healthcare team members as well, such as psychologists, counselors, social workers, nurses, chaplains, etc.

sent's goal becomes protecting the patient from oneself or protecting the patient from their family, culture, or society. Thus, the physician or healthcare worker can become a sort of secular missionary to free people to secular humanist ideals. This violates the essential aspect of informed consent discussed earlier, that it provides an opportunity for informed refusal in order to protect the patient from physicians or health care and from unwanted medical evaluation or treatment. This movement that insists on an anthropology of a decapitated Cartesian mind combined with an overt anti-religious and anti-cultural stance is a setting for a Rawlsian dystopia. The basis of this remains an individualistic and rationalistic anthropology.

14.5 Individualism—Not So Rugged

The roots of rugged individualism in the West run deep. Theologically, we could speculate that it is as old as mankind, for the two sins of the Garden of Eden are turning to the things immanent for salvation and for each person to become a god unto themselves. But as we are want to trace the narrative of ideas, we could begin the tale with Plato or Aristotle, or Francis of Assisi or Duns Scotus as Louis Dupre[12] has done, or Luther and Calvin as Jacques Barzun[13] did, or Anselm and Thomas Aquinas as Orthodox Christians are usually inclined to do. Regardless, the philosophical culmination of individualism as an anthropologic assumption is reached in Renee Descartes and Immanuel Kant. A full refutation of individualism is beyond this paper. However, it remains clear, as Fergus Kerr suggests in his book on Wittgenstein, that we remain prisoners, due to a failure to consider metaphysics, to the ascendant philosophical school of the past 350 years represented by Descartes and Kant.[14] Kerr notes that a thorough refutation has occurred,

> the picture of the self-conscious and self-reliant, self-transparent and all-responsible individual which Descartes and Kant between them imposed upon modern philosophy may easily be identified, in various guises, in the work of many modern theologians [and I add physicians and bioethicists]. It is a picture of the self that many modern philosophers, Wittgenstein certainly among them, have striven to revise, incorporate into a larger design, or simply obliterate (1997, p. 5).

Kerr in considering Wittgenstein's critiques rightly reveals the deifying attempt of individualism,

> What a hydra-headed creature it is! The self who is free to survey the world from no point of view within the world often turns out to be the self who is totally impenetrable to anyone else—in this being once again rather like the hidden God of classical theism (1997, p. 18).

This theological end Wittgenstein seems to suggest had a theological beginning in Blessed Augustine. Wittgenstein strongly challenges Augustine's description in the Confessions of his acquisition of skills as an infant. Augustine thinks and knows

---

12 See: Dupre 1993.

13 See: Barzun 2000.

14 See: Kerr 1997, p. 3.

prior to his action and relatedness. He observes, reflects and wills himself to new accomplishments. Wittgenstein in the words of Kerr rebuts,

> we are agents in practical intercourse with one another—not solitary observers gazing upwards to the celestial realm of the eternal forms, or inwards at the show in the mental theatre. What constitutes us as human beings is the regular and patterned reactions that we have to one another. It is our dealings with each other—in how we act—that human life is founded (1997, p. 65).

Kerr also engages Wittgenstein's consideration of the limits of discursive reason in summarizing

> Wittgenstein wants us to acknowledge that the stability there is, such as it is, is already given in the customs and practices of everyday human intercourse. The given cannot be explored or explained any more deeply because it is the foundation of every kind of exploration and explanation. If you like: the given cannot be discovered except by showing how it makes possible all that we do and suffer (1997, p. 69).

But considering our present task of challenging individualism Wittgenstein's

> general strategy, certainly, is to remind us that we learn to think and feel in the context of a legacy of extremely complicated social interactions with one another—not by introspecting our own inner lives and making inferences from them about those others (Kerr 1997, p. 209).

Yet despite this and other rebuttals the individualistic assumptions in American bioethics persist. Combining this individualism with the above mentioned pursuit of pleasure our present condition is aptly described in the conclusion of Herve Juvin's *The Coming of the Body*.

> [This] is where we are. At the end of a secular evolution, the body is now real only to be transfigured: to escape from reality through processes of fabrication. Every individual's own body is his primary object of desire, most valuable possession and sole inheritance, the accumulation of experiences his only true expense. While medical, psychiatric and aesthetic artifice act as insurance against a setback that might restore to the body its share of reality, of hazard or risk. Even psychoanalysis itself, in its dive into the abysses, comes up against that dense wish, in whose formation it has played no small part: the wish to produce oneself, understood as the wish to be detached from one's environment, heritage and origins. After the intoxicating excitement of the ownership of the body and the production of bodies, we remain suspended between two eternities, of the species and of the soul. And we are discovering, to our confusion, that the much-vaunted advances meant to liberate mankind from its connections also release deeper and more harmful impulses (Juvin 2010, p. 175).

14.6 Family or Communal Decision Making

Given the fracture of the family in the West[15], it is as one might expect that the current individualistic system exists as an expedient way to deal with clinical dilemmas. The real dysfunctions and fractures observed in families offer the anecdotes that drive hospital policy and professional sentiments. This is similar to the

[15] See: Murray 2012, pp. 149–167.

development of healthcare law in the US, where case law is built upon extreme cases. Thus, too often, the reaction in both is an overreaction against giving the family respect. Bioethics, healthcare systems, and law increasingly assume that a family is not—and by extension questions whether it can be—in a state of health and is not helpful. This mistrust of family and community becomes a presupposition or at least a practiced assumption. Instead, family is treated as something to be wary of, a problem to prevent or a disease to attempt to cure. Family comes under the gaze of the clinician as a disease. The common disease model of efficient causation[16] can then be applied and the proper techniques[17] can be employed to bring a perceived health, which usually looks strikingly like the clinical team or hospital or clinical bioethicist getting their way. But this mechanized individual decision making perpetuates the problem further by driving a wedge of mistrust between family members. The current system reinforces the overly simplistic view that decision making is merely between the individual and the medical establishment/team/or doctor. Thus, the current system encourages a practice of familial and cultural shunning.

Contrary to the mechanized, individualistic ethic, the patient's decision may be enhanced by family, friends, authorities, and relationships. Instead of following the dictates of the healthcare providers, a balance of power to bring opportunity for informed refusal is needed. This balance is provided by such contextually rich relations as family. To have a process to protect a patient from us, that is, those in healthcare, to only strip the patient naked of all cultural clothing makes them not a respected person receiving care, but a demoralized individual.

Currently, in the name of freedom of choice, the choice not available is desired constraint. Many traditional cultures and religions adhere to a belief that true freedom comes paradoxically from constraint. This sentiment, or confession, is expressed in a plea to God by poet John Donne, "Take me to you, imprison me, for I,

Except you enthrall me, never shall be free,..." (Donne 1971 [1610], p. 314). *Pace* William James[18] and John Dewey,[19] pursuit of the good does not require an open future of unbridled choice. Instead the right, virtuous, or holy may be discovered by following a particular truth and path—even if in opposition to individual preference or proclivity. Constraint can allow attempts to reflect revealed or evident truths of how creation ought to be, thus recognizing oneself as a creation and not the creator and master. Inclusion of family and other intermediate systems allows the choice for constraint. This constraint should not be mandated. It should merely be an option for particular moral communities and families. Practically, this likely means that if a particular familial or communal decision making process ends in disorder then the basic right of refusal in an individualistic mode may become the

[16] See: Bishop 2011, p. 60.

[17] These techniques may include various methods of subtle persuasion from healthcare team members such as bioethicists, psychologists, chaplains, physicians, social workers, case managers, or business officers.

[18] See: James 1911, pp. 127–152.

[19] See: Dewey 1935.

fail-safe. But this individualistic model should not become the enforced norm but an option or lowest common denominator.[20]

14.7 The Role of the Physician

The physician plays a key role as medical expert and often facilitator of the decision-making process. In an overreaction against a prevailing narrative (potentially a false narrative) of paternalism, many physicians have become reluctant to share their recommendations with patients for fear of overly influencing them and diminishing patient autonomy; however, this reluctance deprives patients and families of the physician's expertise. We can consider the role of the physician by considering how directive he ought to be in influencing the decision. The pejorative of paternalism, besides being offensive to fathers, has varied definitions often reflecting the moral leanings of the given author. I offer the precise but not exhaustive definition of a physician deciding for the patient and not discussing the decision with the patient before acting. This is generally only acceptable when certainty and risk are at levels that a physician can presume consent. Far more common is the other extreme shade in the spectrum of directiveness, abdication. To abdicate is to renounce or cast off a duty, function, or responsibility. Many physicians in the name of respect for autonomy, but usually in the spirit of fear or laziness, fail to give the needed information for a patient and their allies to make an informed decision. This often occurs when options are presented as clinically or morally neutral when they are not. The most common being "if you get really sick do you want us to do everything?" Such overarching statements, especially when concerning technologies that may or may not be effective, but whose efficacy is predictable, should be foreign to medicine—yet it pervades. The minimum acceptable level of directiveness should be informational. The physician should be able to share with patient and/or family the disease, prognosis, expectations, potential treatments, and burdens of treatments. This information must avoid false dilemmas or false forced choices. Better yet, in most circumstances the addition of the physician's guidance is needed to really share in the decision. The physician can call upon expertise and experience to offer advice and recommendation. Most laymen would say that is why one would seek a physician. This guidance need not be given behind a façade of objectivity. Whether the physician and patient are moral friends or strangers[21] should be apparent and known. Hiding behind the fanciful veil of ignorance[22] does not encourage sharing

---

[20] Contemporary bioethics often recognizes that some cultures and religions desire family oriented or community decision making yet no real concession is made to allow such. Beauchamp and Childress (2009, pp. 106–107) suggest that the practical way to deal with those requesting family-oriented consent is to ask the individual if this is his preference. This, of course, is the application of an individualistic system and already does violence to any family-oriented consent.

[21] See: Engelhardt 1996, pp. 6–9.

[22] Again, a reference to Rawls. See: Rawls 1971 and 2001.

in a relational process. Physicians need not stop being who they are but can dare to relate. At times this relatedness can include the physician persuading the patient. To move by argument, entreaty, or expostulation to a course of action; in other words, to try to convince is not infringing on a patients right to refuse so long as such opportunity is given. Darker shades of physician directiveness such as manipulation and coercion[23] are generally not thought of as ethically appropriate with perhaps a few exceptions. Physicians accepting a more active role and sharing in the burden of decision making can aid in moving away from the individualistic, autonomous consent process to bring a more shared decision.

Above I have described two seemingly disparate narratives. First, that of the secular fundamentalist using the mechanized ethic of autonomy to isolate the individual patient from family and community then converting the patient to their way. Second, I tell of the pervasive abdication by physicians in giving guidance. These two narratives can both be true in that the abdicating physicians, in their failure to inform and guide patients, bring numerous false dilemmas and "difficult" decisions. To aid in the difficult decisions the experts[24] can be called in. These difficult decision experts are increasingly the secular fundamentalists mentioned. Further, the zealous secular fundamentalist is more likely to work to shape law and hospital policy. Thus, according to the prevailing ethic of an autonomous, individualistic decision the patient is at risk of being encouraged to act against their values and beliefs or they are left alone to introspect on options falsely presented as neutral. Both of these poles miss the mark.

14.8 Lessons from Western Subcultures

My attempt thus far has been to describe the prevailing decision making ethic in Western culture. However, despite the individualistic, mechanized ethic of autonomous decision making being de jure in law and hospital policies, some subcultures in the US force greater de facto liberty. Some examples include a refusal to designate a single surrogate decision maker or proxy in Southern African-American culture. Generally, this culture, even though family and community infrastructure is often lacking, insists on a consensus decision of all family members, often lead by a select few matriarchal figures. If consensus is not found then often a decision is not made and the mechanized process comes to a frustrated halt. Similarly, Orthodox Jews and some Muslims will even challenge legal definitions of death and insist on considerations beyond the policy algorithms. One of the more common exceptions

[23] These terms, particularly persuasion, manipulation, and coercion, have been considered by various authors, including Ruth Faden and Tom L. Beauchamp (1986). These sources often focus on the perspective of the patient and their mind. The current treatment is focusing on the intent of and style employed by the physician.

[24] As previously mentioned these may include clinical ethicists, psychiatrists, counselors, social workers, nurses, or chaplains that have the duty in 'decision support.'

is the before mentioned Jehovah's Witness community with their focused refusal of blood products. They strive to protect the patient from isolation and offer organized defense against any measures to sway the patient away from their accepted canonical position. In the pluralistic West the community (with religious, cultural, or familial basis) examples of exceptions to the prevailing decision making ethic likely abound but may not be the norm.

14.9 Traditional Christianity as a Positive Example

Often Confucian or Asian cultures are offered as the paradigmatic example of a truly shared decision making process, one of family decision making.[25] However, others do exist. As another positive example of a truly shared decision process, we consider how medical decisions should be made in Traditional Christianity, that is the Christianity still represented in the Orthodox Church. Orthodoxy's conception of the 'individual' is only as a self-imposed state of isolation and torment as one gives into the passions. But this is not how things are to be. First, instead of the individual, Orthodoxy confesses the person.[26] This person is a particular hypostasis but also is in unity, by grace and energy, with God. Thus, this paradoxical tension protects from the poles of individualism and collectivism. Second, Orthodoxy appreciates the freedom of choice but warns that such choices are not to be guided by self-deceiving vain reasoning. Operationally this may include a physician, preferably one that knows the patient, sharing the essential information and giving valuable guidance. This is then considered by the sick person, their family, other loved-ones among the faithful, and finally with the priest, the one who has heard the sins of the one that is ill. Together they prayerfully seek true Orthodox wisdom and knowledge, they seek the mind of the Church, the will of Christ. This should resolve in consensus, that all may say, "It seemed good and right to us and to the Holy Spirit." The practical wisdom of this approach is clear. The ill do not bear the burden of decisions alone. They are also told of the disease, for in Orthodoxy knowing of an illness and even approaching death is a blessing, that we may prepare for the birth pangs of death. However, the ill are to follow the Tradition that does not allow them to be identified by their illness. It also encourages hope, a hope that is placed in the surety of Christ not necessarily in the healing powers of medicine, thus they are protected from the chief dangers that many non-disclosure, family-oriented decision making advocates fear. The Orthodox way also provides peace for family members as they share in the process and have the cognitive protections from moral distress that they are loving and honoring the patient, they are hearing the guidance of a trusted physician, they receive the blessing of their priest and thus the comfort that they were seeking the mind of Christ and not their isolated desires. Most importantly, this is the path of seeking Truth. This is not the truth of philosophy or

[25] See Chen and Fan 2010.

[26] Vlachos 1999, p. 117.

bioethics, for in Orthodoxy Truth is the person of Christ and this is the way in which he instructs to follow Him.

A decision making process such as that in Traditional Christianity, though not the norm in the West, surely should be allowed and perhaps encouraged. Thus, the mechanized individualistic ethic that pervades should not be mandated by law or policy, though it may be the default when those able and willing to care and share in a decision are lacking.

References

Barzun, J. 2000. *From dawn to decadence: 500 years of western cultural life 1500 to the present.* New York: HarperCollins.

Beauchamp, T. L., and J. F. Childress. 2009. *Principles of biomedical ethics.* 6th ed. Oxford: Oxford University Press.

Bishop, J. 2011. *The anticipatory corpse.* Notre Dame: University of Notre Dame Press.

Chen, X., and R. Fan. 2010. The family and harmonious medical decision making: Cherishing an appropriate Confucian moral balance. *The Journal of Medicine and Philosophy* 35 (5): 573–586.

Dewey, J. 1935. The future of liberalism. *Journal of Philosophy* 32 (9): 225–230.

Donne, J. 1971 [1610]. *John Donne: The complete English poems. (Holy Sonnet 14).* London: Penguin.

Dupre, L. 1993. *Passage to modernity.* New Haven: Yale University Press.

Engelhardt, Jr., H. T. 1996. *Foundations of bioethics.* 2nd ed. London: Oxford University Press.

Faden, R., and T. L. Beauchamp. 1986. *A history and theory of informed consent.* Oxford: Oxford University Press.

Guichon, J., and I. Mitchell. 2006. Medical emergencies in children of Orthodox Jehovah's Witness families: Three recent legal cases, ethical issues and proposals for management. *Paediatric Child Health* 11 (10): 655–658.

Illich, I. 1974. Medical nemesis. *Lancet* 303:918–921.

Illich, I. 1976. *Medical nemesis: The expropriation of health.* New York: Pantheon Books.

James, W. 1911. *Some problems of philosophy: A beginning of an introduction to philosphy.* New York: Longmans, Green.

Juvin, H. 2010. *The coming of the body.* New York: Verso.

Kerr, F. 1997. *Theology after Wittgenstein.* London: SPCK.

Louderback-Wood, K. 2005. Jehovah's witnesses, blood transfusions, and the tort of misrepresentation. *Journal of Church and State* 47:783–822.

Murray, C. 2012. *Coming apart: The state of white America, 1960–2010.* New York: Crown Forum.

Rawls, J. 1971. *A theory of justice.* Cambridge: Belknap Press.

Rawls, J. 2001. *Justice as fairness: A restatement.* Cambridge: Belknap Press.

Vlachos, M. H. 1999. *The person in the orthodox tradition.* Levadia-Hellas: Birth of the Theotokos Monastery.

Chapter 15
A Confucian Worldview and Family-Based Informed Consent: A Case of Concealing Illness from the Patient in China

Wenqing Zhao

15.1 Introduction

The practice of obtaining informed consent has become routine in clinical medicine. The general rule is that a physician may not perform medical procedures on a patient without her voluntary agreement after having been given proper and adequate information (concerning the nature of the treatment, expected outcomes, reasonable alternatives, material risks, etc.). Of course, informed consent requires that the patient is competent. Failure to obtain informed consent may expose the medical practitioner to charges of assault or negligence. However, this has not always been the case. The basic ideas that animate the popularization of informed consent as a standard procedure developed in the aftermath of the Second World War and the atrocities of the experiments conducted by Nazi Germany. The first official code of informed consent, the Belmont Report, was generated by the National Commission for the Protection of Human Subjects of Biomedical and Behavioral Research in 1978 with no particular concern regarding clinical medicine. In the past 30 years, we have witnessed the continuous inflation of the concept of informed consent, which has gone beyond protecting the rights of human participants in biomedical research and gradually taken over clinical decision making.

I see no cultural objection to the practice of informed consent in biomedical research. It is widely agreed that when a government or any other institutional authority performs a procedure on individuals without their consent it is a form of assault and unacceptable behavior that should be prohibited by law. In clinical medicine, some form of informed consent is necessary to enable the patient and his family to express their views and choices. However, whether the exact same practice should be implemented in clinical medicine across cultures is subject to dispute. Different cultural traditions have different practices for making medical decisions, and

W. Zhao (✉)
Department of Public Policy, City University of Hong Kong, Tat Chee Avenue, Kowloon, Hong Kong, China
e-mail: zhaowenqing711@gmail.com

R. Fan (ed.), *Family-Oriented Informed Consent*, Philosophy and Medicine,
DOI 10.1007/978-3-319-12120-8_15

the moral reasoning behind them needs to be taken seriously before any global judgment, if one is in fact possible, can be made.

The standard Western understanding of informed consent is promoted by the prevalence of liberal individualism. The principle of respect for autonomy requires the medical professional to respect the individual's right to hold views, to make choices, and to take actions based on their personal values and beliefs, as long as they are independent from controlling influences and capable of intentional action (Beauchamp and Childress 2009, pp. 102–103). The problems of autonomy considered in the Western literature often result from conflicts between a patient's dependent condition and the physician's authoritative position, whereas the patient's relationships with her family and community tend to be overlooked. Although Beauchamp and Childress (2009) have discussed the role of community in medical decision making, using the case of a Jehovah's Witness refusing a blood transfusion, their emphasis is placed on respecting an individual's identification with particular cultural tradition or institutional authority (102). The correct source and responsibility for making important medical decisions is attributed solely to autonomous agents. As made clear by Fan, "Although family members often participate in such discussions, serious medical dialogues are undertaken between patients and physicians. The patient is the final authority when it comes to accepting or refusing treatment, and the family plays a secondary role" (2000, p. 89).

Another important concept that endorses individual-based informed consent practices is what Ronald Dworkin (2002) sees as the sovereign virtue of modern democracy, the concept of equality. The value of equality requires health practitioners to treat patients as equals regardless of their individual needs. In reality, we all know that people are shaped by their past experiences and relationships with the world. Their capacity and will to make choices for themselves under extreme conditions differ significantly from one another. An increasing body of psychological and medical evidence demonstrates that having too many choices does not necessarily lead to satisfaction or happiness. On the contrary, the obligation to choose in a vital situation may very likely increase the level of stress and depression for the subject.

In light of the problems mentioned above, we must not take for granted the presumed universal practice of informed consent marked by Western liberal individualism. As Ruiping Fan and Julia Tao have suggested, the so-called standard American bioethics is not really universal (2004a, p. 146). The scope, subject, and manner of implementing informed consent in medical contexts varies across cultural traditions. In Confucian-influenced Chinese society, informed self-determination has never been a social norm for making important medical decisions. The principle of respect for autonomy and equality are not prime considerations when it comes to making a critical medical decision. Traditionally, Chinese people consent to important medical treatments as one family. This means that the fundamental unit for making medical decisions in Asian societies is not the individual patient but the patient's entire family. Such communal decisions are further embedded in Confucian moral philosophy and especially in its understanding of virtuous behavior.

Family-based informed consent and the Confucian worldview are most clearly reflected in the common practice of concealing conditions from patients with terminal illnesses. In Confucian-influenced Chinese societies, it is often seen as unsympathetic to disclose the truth to a patient who is terminally ill, especially if the patient is elderly. Instead, it is thought that the family should shoulder the pain and fear and bear the responsibility of deciding what is best for the patient. That is not to say that the family should never include the patient in making such decisions, regardless of the patient's will or mental capacity. Rather, there is no general algorithm regarding whether to involve the patient in the decision making process. A rather complex equilibrium is involved for the family members reflecting upon the medical advice given by the physician, the will and capacity of the patient, and their shared vision of life, which may or may not lead to concealing certain information or even lying to the patient.

The beauty of Confucianism is that it sees the complexity of life and human psychology because there is no Confucian moral algorithm that tells one something along the line of "if X then Y." But this does not mean that anything goes. Confucian virtues, such as care and benevolence, are the governing characteristics of making medical decisions. The family members, whether they are lying or not, must act on their best understanding of caring for the patient. It is thus morally impermissible, in most cases, for family members to abandon the patient, leave the patient by herself, or treat the patient with cold and indifferent manners. In this sense, it is better to understand the core question of Confucian medical decision-making not so much as a "what" question, "what should we do?," but as a "how" question, "how should we care for the patient?"

This is in contrast with the principle of respect for autonomy which requires the support of many more specific moral rules. Some prime examples given by Beauchamp and Childress include the following:

1. Tell the truth.
2. Respect the privacy of others.
3. Protect confidential information.
4. Obtain consent for interventions with patients.
5. When asked, help others make important decisions (2009, p. 104).

The cluster of rules directed towards respecting individual autonomy requires the physician to tell the truth to the patient and to the patient only. Respect for the privacy of others prevents the physician from disclosing the patient's condition to her family without her consent to do so. The idea that in certain cases the family should make critical decisions on the patient's behalf is seldom considered. However, this fundamentally conflicts with the traditional understanding of informed consent in Chinese society as mentioned above.

Drawing upon a case of concealing illness from a patient in China, I intend to explore the following questions in this essay: Is there a distinctive Confucian way of making important medical decisions? If so, how is it different from a Western individual-oriented approach to informed consent? And moreover, what values and justifications exist in Confucianism for such a practice?

15.2 A Case of Concealing Illness from the Patient in China

First and foremost, the question that needs to be asked is whether there is a way of making medical decisions that is distinctively Confucian? I think the following story can help us better understand the Confucian way of dealing with informed consent:

A "touching story of true love" was recently reported by the Chinese news media; in the story, a beloved husband was diagnosed with advanced lung cancer. He had a wife of over 17 years and a young son. He faced the horrible news of his diagnosis with amazing courage and strong determination to keep living for his loved ones. Unfortunately, the treatments did not turn out well, and his condition worsened every day. His wife was afraid he would feel hopeless and defeated upon seeing his negative test results, so she took the liberty of taking her husband's lab reports to a local copy shop, where she changed the lab results to make the numbers look better. Over 15 months of treatment, she repeatedly changed the test results and presented them to her husband so that he would remain in good spirits. She even stored fake lab reports at the doctor's office in case her husband asked the doctor directly for the results.

This story was widely carried by local newspapers with such headlines as "Fifteen Months Weeding a Lie of Love" (Qianjiang Evening News Staff 2012), and has achieved an even greater presence on the internet. Many Chinese people have commented on this story, saying that it makes them "believe in true love again" (Zhou 2012).

The commentary surrounding this story demonstrates that Confucian-influenced Chinese accept the practice of concealing diagnoses from patients with understanding, approval, and even praise. Indeed, concealing a diagnosis from a patient is often seen as a common practice and can even be regarded as the ethical norm in China. It is seen as acceptable especially under the following two conditions:

1. The subject is a senile patient. In this case the children and grandchildren are considered responsible for making critical decisions to spare the elderly patient, who is generally viewed as psychologically vulnerable and dependent on her family, from the pain of making decisions herself.
2. The case of severe illness that may result in death. In this case, it is also seen as unsympathetic to let the patient face the fear of death. Traditionally, family members take on the responsibility of caring for the patient, which often times leads to concealing the truth from the patient in an effort to shield her from pain and worry.

It is important to note that these two conditions are generic. Confucian-influenced Asian communities find it acceptable or even commendable under such conditions for the family members to exclude the patient from the decision making process. However, to be clear, I am not arguing that all families will or ought to lie to the patient under these circumstances, neither am I arguing that the family must not lie in other cases. In the story of "Fifteen Months Weeding a Lie of Love," the husband was severely ill whilst showing a strong determination to live. The wife took into account the husband's mental capacity for facing negative test results and decided not to let him know about his true condition. This demonstrates that the patient's family needs to act in light of their knowledge of the particular patient when deciding what is in her best interest. There are no hard and fast rules that must be followed, only general patterns for family members to refer to when making critical decisions.

Another fact that needs to be noted here is that Confucian-influenced Chinese society sees no harm in depending on one's family in situations of critical illness. It is seen as a good thing, a virtuous thing, to have elderly parents living with or close to their adult children. It is seen as common, if not ideal, for elderly parents to help raise the grandchildren whilst the adult children take care of the elderly parents when they become sick and weak. Every family member has their role in the family and together they form a strong web of mutual dependency. In this sense, Chinese culture values support, care and harmony among family members over individual preferences, personalities or even privacy.

However, if this story were examined in light of the liberal principlism proposed by Beauchamp and Childress, the perspective will be entirely different. On this account, the wife obviously violated her husband's right to be informed about his medical condition, and thus violated her husband's right to autonomous decision making in matters of his own health. The doctor, who gave the test results to the wife instead of the husband, also violated the fundamental rules and principles of informed consent. It is wrong for the doctor to treat the husband without him being fully informed of his condition and it is also illegal, leaving the physician open to an assault charge. A liberal principlist may argue that the husband's fundamental human rights were violated, and both the wife and doctor are ethically or even legally blameworthy for breaking the first and foremost rule of bioethics: respect for patient autonomy. From the principlist's perspective, this story would not be interpreted as a "touching story of true love," but rather as a "horrible violation of human rights."

I believe that the case of concealing illness from the patient partly reflects the Confucian way of making medical decisions, which is virtue-based and family-oriented. On the one hand, the wife treated her husband according to her knowledge of what was best for him, which shows a Confucian understanding of caring. Her shouldering of all of the fear, stress, and suffering arising from her husband's worsening condition is, according to Chinese norms, regarded as fulfilling her duty as a wife with great courage and love. From the Confucian point of view, a virtuous person need not obey the rule of truth telling regardless of the mental and physical condition of the patient. It is a much more complicated process of weighing the consequences of different decisions, carefully evaluating the patients' mental and physical condition, and most importantly listening to the will of the patient and other family members. In the story above, the husband showed a strong will to live, and the wife was afraid that knowing the bad news would diminish her husband's courage to fight the cancer. Thus, one could hardly say that informing the husband of his condition would really be doing what was best for his care. There are no

Confucian rules or requirements stating that the family must or must not conceal information from the patient. The right decision is always dependent and contextual. In a sense, taking a Confucian approach is often much harder than simply following the principles of beneficence, justice, and, most importantly, respect for autonomy. Confucianism requires people to be more pro-active in doing what they believe to be right; they cannot shy away from their moral responsibility to act by deferring to the rules.

Of course, there is the problem of whether the family can accurately assess the patient's mental capacity to face her illness. Normally, people do not think they know better than the person concerned, what she thinks or how she will feel. However, I think there are good reasons to believe that the family members, especially those who are living with the patient, are capable of making good judgments about the patient's needs and feelings. In fact, we make decisions for our close family members and even some friends all the time. Imagine if you were to pick a movie to watch with a friend knowing that they do not appreciate thrillers. Normally, you would just avoid picking a scary movie without obtaining your friend's advance permission. If by chance your friend wants to watch something unusual, she will probably let you know.

The other question we need to ask is whether the patient is really deprived of her right to choose. I think it is hard to say that the patient is absolutely clueless about their own condition. From the physician's behavior, the medicine they are taking, etc. they probably might guess that their diagnosis is grave. Not to mention that the patient will usually feel very ill in a serious condition. Just like in the case of picking movies, if my friend has grown to like horror movies, she will let me know. If the patient really wants to find out the truth about their own condition, they will take the initiative to ask the people around them and repeatedly demonstrate their will to know. In this case, I think most family members would and should respect the patient's wishes and try to face the difficult situation with the patient.

In the case of "Fifteen Months Weeding a Lie of Love", the wife stopped faking test results when it came to the very final stage. She mentioned in the interview that her husband was simply too ill and it became pointless to keep anything from him (Zhou 2012). But why didn't the husband say anything? It would be hard to imagine that he was really fooled throughout his illness given the fact that he suffered terminal stage lung cancer. But, like most Chinese people, he said nothing about his dying. He let his family believe that he was fooled by their lies of good will. He did it to relieve the stress and anxiety his family was facing upon his departure, so as to maintain its harmony.

So, is this really so bad?

From the relational perspective of Confucianism, the husband is not viewed as an autonomous individual who can and should take responsibility for himself. The family members, the wife and child in this case, are often considered essential parts of not just one's life, but also one's being, and thus should be involved in the decision making process. In many critical situations, the decision does not just affect the patient's well-being; it often has long-term consequences for the entire family. The family-oriented structure of medical decision making differs from a Western individualist model as mentioned above.

Although the wife was well-intended, one may still ask whether her choice was truly in the best interest of the patient, as her behavior was clearly deceptive. In this particular case, it is clear that the husband had shown a strong will to live, and in the beginning stage of the treatment, the positive test result may have had placebo effects that helped him to fight the disease. The more critical question is whether the deception was still in his best interest when his condition worsened and the treatment showed no effect. In order to answer this question, we have to look at the incident in its cultural context. As mentioned above, it is a very common practice in China for the family to make decisions for the patient in situations of critical illness. The patients will do just the same for their family members in similar situations. This means that most Chinese patients have a general understanding of what is really going on; i.e., that there is a high possibility that the family members are embellishing the condition. The practice has been passed on as a common sense or even ritual practice among Chinese people throughout time. It is very unlikely that the Chinese patient has no knowledge of this practice and is totally fooled by the family members. In most cases, the patients know what is going on, but they just choose to not say it out loud and let the family make the decisions for them. In this particular case, the husband could have chosen to go to his doctor directly and ask for the results. There were many ways to know the truth if he really wanted to know, but he chose not to know and rely on his wife. It may be hard for someone who is outside of the tradition to understand this practice. Why go through all these troubles? But for people who have been familiar with the practice their whole lives (they may have witnessed their parents doing so to their grandparents, their friends doing so to family members and so forth), it is like a smooth dance that they do together with their family members in their final stage of life.

Is there any beauty in this Chinese way of dancing with death? I think there is. Chinese people see great value in the idea of being part of a family. Influenced by liberal individualist tradition, Western societies emphasize the value of personal independence and courage to face harsh conditions by oneself. In Confucian Chinese society, however, people think it is one of the greatest personal achievements to have family members to take care of you when you are old and facing the end of your life. A person is evaluated not only based on his willpower as an individual but also on whether he can be weak and have people to rely on when he wants to. The delicate sense of harmony and mutual support among family members are what truly underlie the practice of concealing illness.

What is the professional role of the physician in disclosing information according to the two different approaches? According to Beauchamp and Childress (2009), there are seven elements of informed consent; three of them concern the role of a physician, namely, disclosure of material information, recommendation of a plan, and help in understanding (p. 120). However, the dialogue is limited to the physician and the patient. It is not necessary for the physician to involve the patient's family in the process of disclosure, recommendation, and understanding, so long as the patient is an autonomous agent. Of course, the patient can choose to engage her family in the decision making process, but that remains the patient's personal decision. The rule of respect for privacy requires the physician to maintain the confidentiality of a patient's medical information, unless the patient is no longer competent to make decisions, or the patient approves the disclosure of such information to a third party.

Family-oriented informed consent, on the other hand, requires the physician to recognize the family as a fundamental unit for making medical decisions. The physician would first discuss the patient's condition with some of her key family members, for instance, her parents, husband, and adult children. Then the physician would leave the family members to decide whether to inform the patient about her condition. Some of the early Chinese medical texts about serious disease diagnosis, dating back to the Han dynasty, insist that they should only be disclosed to the family, and then the family should decide whether to inform the patient (Fan 2000, p. 94).

What if the patient asks the doctor directly for information? In this case I think a Chinese doctor may try her best to engage the patient's family and once again shift the communication from physician-patient to family-patient. One could imagine a Chinese doctor saying, "Your family already has all the information. Maybe you can talk to them about the situation." However, if the patient really insists on hearing the truth about her illness, the physician may be put in a rather difficult position. I think the best thing for a Chinese physician to do in this circumstance is still to have the family present when disclosing related information, and to prepare the family in advance for handling this situation. Family members usually will come up with some ways to help deal with such problems.

In the above mentioned case of the husband with lung cancer, the doctor acts in accordance with this Chinese social norm, by disclosing the bad news first to the wife, and then letting the wife decide whether the husband should know. Moreover, when interviewed, the wife said she had stored a fake copy of the test results in the doctor's office just in case her husband asked the doctor directly for them. In this case, the doctor is purposely overlooking the fact that the wife is deceiving the husband, which makes him part of the conspiracy. This is not to say that in the Chinese tradition medical practitioners are encouraged to lie to their patients. There are many ways to "get around the problem" that do not require lying to the patient. With the presumption that the family is acting in the best interest of the patient, the physician will often turn a blind eye to its deceptive behaviors, as long as they are not in conflict with her medical knowledge or professional conduct.

I am not suggesting that this anecdote alone proves the existence of a Confucian ethic that is fundamentally different from Western liberalism. In the next section, I will further elaborate a Confucian view of the world that justifies the practice of concealing illness, and compare and contrast it with an individualistic view of the world. However, we must at least acknowledge how often Chinese families choose to conceal illnesses from family members. The practice of making medical decisions as a family is deeply rooted in the 2000 year history of Chinese society as part of the Confucian understanding of the self and the good life. It is widely believed in Confucian-influenced Asian societies that family members can make the best decisions for their loved ones without them knowing their condition, in which case the patient can be spared from the fear of death and allowed to maintain a positive spirit that may help them get better or at least allow them to live more happily in the time they have left. According to Ruiping Fan and Benfu Li (2004a), the right to medical truth only serves as a fallback right when family members do not care for the best interest of the patient. In this case, the doctor should take the initiative to intervene and tell the patient the truth (Fan and Li 2004a, p. 179). This of course puts more pressure on the physician. The physician must actively participate in the decision-making process by communicating with the family members.

By using the case of concealing illness from a patient, I have sketched out a rough picture of the Confucian way of making important medical decisions, and have explained how it is different from a liberal individualistic approach. This leads to the last but not least important question of the essay: What values and justifications exist in Confucianism for such a practice?

15.3 A Confucian Worldview and Family-Based Informed Consent

I was once at a gathering with a number of Chinese college students, all receiving their education in a Western or semi-Western (Hong Kong) environment. In a sense, they were well exposed to Western liberal vocabularies and no stranger to individual rights. I talked about the case above of concealing illness from the dying husband and asked them whether they thought important medical decisions could be delegated entirely to their families, mostly their parents. Most of them acknowledged the prevalence of the practice in Chinese society and viewed it as a more or less natural way for making medical decisions. Talking with them made me realize that there was something significant about accepting the family's deceptive conduct. Here is a conversation between my friend, Alex, and I, who had just returned from studying in the United Kingdom for several years:

> Alex: Why would you ask such a question?Me: It is for my project. I am trying to see if there is a Chinese perspective on informed consent.Alex: I mean that I don't see why it is a problem.Me: What do you mean?Alex: Personally, it doesn't matter whether my family decides for me or not. I don't see why there is any problem if my parents are the ones to decide what treatment I should have.Me: So you think they will always act in your best interest?Alex: Of course. And I just don't see what your problem is.Me: So you don't see them violating your right to know about your own health condition and to choose your own treatment?Alex: No. We are one family. It doesn't matter. I don't know how to explain it to you. But I just don't see any problems with my parents making health decisions for me. Maybe the Westerners see a problem, but I don't.

At first, I thought Alex was just re-affirming a Confucian family-oriented way of making medical decisions, which is very common among Chinese people. Only later did I realize that Alex was actually making an important point by repeatedly saying "I don't see a problem in my family making decisions for me." Many Chinese people, like Alex, may have learned about the individual way of practicing informed consent, but it is not part of their own way of life. As made clear by Fan, "Generally, the Chinese take for granted that the entire family makes medical decisions for a patient, whether the patient is competent or not" (2000, p. 90). Family-oriented medical decision making is not only a fact of life, but it is also part of a way of life that is deeply embedded in the Confucian view of the world.

What is the difference between a fact of life and a way of life? For instance, Chinese people are accustomed to using chopsticks. It is a fact that Chinese people traditionally use chopsticks instead of forks to pick up food and bowls instead of plates to hold the rice. However, a Chinese way of life does not depend on using chopsticks. Nowadays, more and more young people use forks, but that does not mean they are less Chinese or that they view the world differently. Using forks is not something that would threaten the peacefulness of a Chinese person's worldview. Differences in the facts of life are real and we must take these differences seriously when engaging with other cultural traditions. However, changes in such facts have little impact on who we think we are in relation to the rest of the world.

Unlike a fact of life, a way of life is itself part of how we view the world. Oftentimes, we take the traditional way of life for granted. In the process of practicing a traditional way of life, our understanding of the world is translated into a deeper conviction and belief about how the world really is. This persuasion then again feeds our understanding of the world and re-affirms the practice of ethics. Health is an important contributor to a good life. The way we make such decisions is part of how we understand the world, what we think of human relationships, and most importantly, what the natural way of living is. The reason why my friend, Alex, thinks a family-oriented informed consent process is unproblematic is because it is already part of who he is, and to challenge it is to challenge the fundamental way he understands the world.

The individual-oriented approach to informed consent comes out as part of a liberal individualist picture of the world. Liberal individualists see autonomous agents as the fundamental units of the world. The Western liberal conceptual schema puts much weight on the tension between individuals and the rest of the world. It is often suspected that "others" (society, government, institutional authority, family and etc.) are always seeking to control and restrain individual freedom and to prevent individuals from exercising their wills, whereas the individual is always struggling to break free. This is not to say that Chinese people do not face tensions between themselves and their government, however, the government is never assumed to be evil in the first place. In a Confucian conceptual schema the idea of individual cannot be separated from one's family and community. The sense of oneness is much stronger than it is in the Western liberal tradition. The liberal picture of the world fosters an awareness of protecting individual liberties from the state by constructing a set of rules, such as respect for autonomy and respect for privacy in the practice of ethics. Most basic for an autonomous individual is being able to decide which treatments are performed on one's body. Individual-based informed consent is then one of the key practices that construct the individual-society picture of the world.

On the other hand, Chinese patients are not unaware when they make critical decisions that influence their quality of life in the long run. They simply have no intention to break their usual life pattern composed of interdependent relationships among family members. This ethical practice is deeply rooted in the way Confucian-influenced Chinese people view the world. Confucianism views families, instead of individuals, as the fundamental units of the society. From a Confucian perspective, an individual and her close family members are fundamentally "one" bonded by the intimate emotional ties of caring and love. Of course, the oneness shared between family members is a dynamic reflection of past life experiences, an understanding of the current situation of the family, and also ritual practices shared among them. Family members will still disagree with each other from time to time, and there may even be feelings of anger, frustration, and disappointment. But this does not change the fact that close family members are a community that makes important decisions together in almost every important aspect of life.

Confucian ethics encourages family members to depend on each other, especially in difficult life moments. There is nothing shameful in depending on your family members to take care of you when you are sick. Confucianism recognizes the fact that all humans are vulnerable when faced with the threat of death and it is only natural to rely on trusting family members during this time. It is not only natural, but also a result of life-long moral cultivation. Confucianism believes that humans are not isolated individuals and that moral perfection can only be reached in a relational setting. Generically speaking, a man is most intimately connected to his own household because he is born to them in the very first place. Loving one's family members, especially one's parents, is the starting point of one's moral development. Chinese people think that the most desirable thing is not to have good fortune when you are young, but to be surrounded by good family members when you are old. To live among one's children is the reward of a life of good deeds.

We must take into account cultural diversity when making serious medical decisions. First, we need to acknowledge the existence of family-oriented approaches to informed consent within Eastern traditions that are dramatically different from standard Western practice. Second, we should not just regard these as differences in facts of life, but rather differences in ways of life. In this sense, it is not just an exotic practice that could be "improved" over time. Family-oriented medical decision making reflects how Chinese people view the world. Moreover, as a way of life, it marks the moral identity of Chinese people. Hopefully, this distinction can help us to truly understand and respect different ways of making medical decisions.

It is important to note that I am not arguing as long as the decision is made by family members it is necessarily in the best interest of the patient, or good. Neither am I saying that only family members can make good decisions for the patient. I am, however, trying to articulate a Confucian worldview in which the mutual dependence among family members is seen as a great value of mankind. Due to the limit of this paper, I cannot give a full-blown account of Confucian theory of good life, but I have highlighted the essential features of the theory in previous passages.

The question remains: Is it truly in the best interest of the patient when the family members make critical decisions for them? Let me try to answer the question of "best interest" in reply to an objectivist theory of good life. I think that in order to say "the family members are (or are not) deciding in the best interest of the patient", one first has to define what the concept of "best interest" means, in other words, what is good for the patient? Is there one and only one good way of life? I think there is no good evidence to support an absolute, unifying and objective theory of good life. The idea of "best interest" could be misleading as it indicates that there is a "best" thing for the patient. What we can see, more obviously, is that humans are all shaped by a certain cultural tradition. This means all conceptions of the good life are rooted in specific cultural traditions, and what is good for a person needs to be understood in relation to a particular culture.

It seems like we can ask another question instead of the question of best interest: Which way makes one happier when facing death and terminal illness: facing it independently by one's own will and courage, or leaving it to one's family members? Confucian value supports the latter. Confucianism thinks that at the end of the day one can be vulnerable in the face of death and pain. To be able to rely on one's family members in the final stage of life is the good thing to do, and it makes the patient happy. As mentioned, Chinese people think that family is the way towards many essential goods of life, and that only through a happy family life can one become a virtuous person. The case of family-oriented informed consent is a reflection of a full-blown theory of good life that is developed and tested by Confucian-influenced Chinese people throughout centuries.

I am not arguing that people from all cultural traditions need to follow this pattern of practice. The understanding of good practice is diverse: Most American people who are deeply influenced by liberal individualism would think for a grown child to move back with their parents and rely on them to take care of the grandchildren is a sign of failure. On the other hand, many Chinese people think that to live in the same space with their parents and children is the ultimate happiness of life. This concludes my answer to the question of "best interest": the conception of "interest" or "good" has to be understood in its cultural context; there is no unifying theory that defines the ultimate good way of life. Confucianism, like many other cultural traditions, offers a unique way of helping people face death and terminal illness. This way of life has been practiced by Confucian-influenced Chinese people throughout centuries and greatly promoted the happiness and interest of Chinese people. We have to take seriously the value of a Confucian way of family-oriented informed consent when forming policies and laws.

15.4 Conclusion

As shown in the case of concealing illness from patients in China, there are significant differences between the Confucian and the liberal individualistic approaches to informed consent. The liberal ethic locates the authority of informed consent in autonomous agents, whereas for Confucianism the decision making authority often rests with the patient's family. Traditionally, in the case of severe illness, Chinese people believe that the patient should depend on her family to shield her and to carry the responsibility of making critical decisions for her. Thus, it is common for Chinese family members to conceal information from other members who are ill, especially when the person is elderly. As for medical practitioners, they will often provide the information to family members first, and then let them decide what is in the best interest of the patient. However, in the Western liberal context, a patient with freedom and competence has the ultimate right to consent to a procedure. Moreover, in the interest of protecting one's privacy, a patient's medical information cannot be accessed by even her closest family members without permission. Communication regarding diagnosis and treatment is restricted to the patient and physician unless the patient has taken the initiative to involve family members.

In this paper I have argued that the differences regarding informed consent are differences not only in facts of life but also in ways of life. The case of concealing illness from a patient in China demonstrates that there are different practices across cultural traditions. Additionally, this practice has deep roots in the Confucian view of the world and human relations that need to be taken seriously. As mentioned above, liberal individualists see the world as made up of individuals, whereas Confucian-influenced Chinese people see families as the basic unit of any moral analysis. Nevertheless, these are both worldviews that people are so accustomed to in particular traditions that they become parts of their different ways of life. In a sense, they are like a pair of inbuilt glasses we inherit from our tradition. We look at the world and ourselves through them without realizing their existence. I hope that introducing the practice of Chinese medical decision making can help us take a step back to reexamine the popular ways of implementing informed consent.

References

Beauchamp, T. L., and J. F. Childress. 2009. *Principles of biomedical ethics*. 6th ed. Oxford: Oxford University Press.

Dworkin, R. 2002. Sovereign virtue: The theory and practice of equality. Cambridge: Harvard University Press

Fan, R. 2000. International perspective: Informed consent and truth telling: The Chinese Confucian moral perspective. *HEC Forum* 12 (1): 87–95.

Fan, R., and B. Li. 2004a. Truth telling in medicine: The Confucian view. *Journal of Medicine and Philosophy* 29 (2): 179–193.

Fan, R., and J. Tao. 2004b. Consent to medical treatment: The complex interplay of patients, families, and physicians. *Journal of Medicine and Philosophy* 29 (2): 139–148.

Qianjiang Evening News Staff. 2012, July 17. "Fifteen Months Weeding a Lie of Love" 十五個月愛的謊言. *Qianjiang Evening News*. http://qjwb.zjol.com.cn/html/2012-07/17/content_1632206.htm?div=-1. Accessed 25 Nov 2013.

Zhou, J. 2012, July 18."Hangzhou wife weeding a lie of love for fifteen months" 杭州妻子十五個月編織愛的謊言. Xinhua News 新華新聞. http://news.xinhuanet.com/local/2012-07/18/c_112469427.htm. Accessed 25 Nov 2013.

Chapter 16
Towards a Good Practice of Family-Oriented Consent: Reflections on Medical Practice in Taiwan

Hon Chung Wong

16.1 Introduction

In Western countries, respect for autonomy always acts as a guiding principle in medical practice and in tackling many thorny issues in Bioethics. For instance, physicians will routinely first inform their patients alone about their illness if they are competent, and only then ask whether they want to discuss with others, such as their families or other acquaintances, to make their own clinical decisions. If the patients do not want persons other than the medical professionals to participate in the discussion, the medical professionals are required to keep it confidential. Though the patients' autonomy seems to be respected in these practices, such respect for autonomy has been challenged by many studies. In particular, the practices seem not to be suitable for non-Western countries, such as Taiwan, in which the patient's family plays an important role in clinical decision making. Hence, with reference to philosophical traditions other than Western liberalism—such as Confucianism—a family-oriented consent approach has been developed as an alternative to the practice of autonomy in place in the Western health care system in order to address to problems arising from it (Lee 2007; Fan 2007).

However, though the family-oriented consent approach seems to be quite fit for the medical practices of Taiwan, since the physicians often tell the medical situation of their patients to their families first, and then ask them whether to tell the truth to the patients—especially in the case when the patients are terminally ill—there are other problems arising from these practices. For instance, if the interests of the patients are in conflict with those of their families, as when families do not care much about the patients' illness, or there are some quarrels about patients' inheritance within the families, the patients' best interest may not be the foremost concern of their families. Moreover, if the understanding about the best option for the patient

H. C. Wong (✉)
Central Regional Research Ethics Center, China Medical University,
No. 91, Hsueh-Shih Road, Taichung, 40402 Taiwan
e-mail: honcwong1970@gmail.com

R. Fan (ed.), *Family-Oriented Informed Consent*, Philosophy and Medicine,
DOI 10.1007/978-3-319-12120-8_16

differs between the patients, their families, and the medical professionals, and if the medical decision making is manipulated by the families, the physicians often follow the families' suggestions, as they fear subsequent recrimination and litigation. Hence, the patients may not be able to receive the best treatment which would otherwise be available to them.

In what follows, I shall criticize the practices of patients' autonomy in Taiwan through an analysis of some cases, in order to articulate the problems that the family-oriented consent approach has to face. I shall then clarify the conception of family in Confucianism, which is regarded as one of the possible bases for this approach. For a Confucian, though a person's family is her main source of personal identity, it does not mean that her views should always be outweighed by the views of her family as a whole. Hence, in the end, I shall make some suggestions for the family consent model based on a Confucian conception of family and my reflections on practices in Taiwan. These suggestions are meant to strengthen rather than compromise patients' autonomy, though the central role of family in patients' clinical decision making is still recognized.

16.2 Challenges to Family Consent Model: Case Studies in Taiwan

In Taiwan, it is a commonplace that healthcare professionals share patients' medical information with their families, as the Physicians Act requires that "a physician shall inform the patient or the patient's family of the status of the disease, treatment principles, treatment, medication, prognosis and possible unfavorable reactions when diagnosing and treating patients" (Ministry of Health and Welfare, 2009), by which patients and their families can make medical decisions accordingly. Though this kind of family consent model can relieve the patients' burdens, it will be a difficult choice for the families if the patient is in a terminal condition. Further, positive medical intervention may be futile to restore the patient's health if palliative care is what he or she really needs. However, the families may think that if they decide to withdraw treatment from the patient, it will just let their seriously ill parent die, and hence violate their duty of filial piety. They may also just not want to acknowledge that the patient's death is coming. Hence, they may insist that positive treatment should be continued, though it will only increase the suffering of the patient and violate his or her best interests, as it is shown in the following case.

> An old woman was sent to a hospital in Taipei City owing to cardiorespiratory failure. Though she had already signed a DNR, her nephew insisted that she should receive extracorporeal membrane oxygenation therapy; otherwise they would sue the attending healthcare professionals. When she woke up, she found that she was intubated. She felt very angry and beat against the sickbed repeatedly. Later she died with great resentment. (J. F. Jan, "Woman died with resentment as her nephew insists CPR be done to her," *United Daily News*, September 12, 2011)

According to the Hospice Care Law, terminally ill patients in Taiwan can sign a DNR to withhold or withdraw any CPR, and they should sign it personally unless they are unable to express their willingness explicitly (Ministry of Health and Welfare, 2013). However, families remain the main barrier for the patient to choose not to resuscitate. Even if a patient has already signed a DNR, his or her family may insist that the physician should try to save the patient's life, whether it is possible or not, no matter what the patient's quality of life would be after the treatment, as is illustrated in the above case. Indeed, some physicians will follow the families' requirement due to fear of subsequent recrimination and litigation by them (Sheu 2007, p. 63). Moreover, the families themselves may also have no consensus about which treatment will be appropriate to the parent:

> Mrs. Kim was an 83 year-old patient with Alzheimer's disease. She lived with her two daughters and a son and they took care of her very considerately. However, her condition had deteriorated so much that they were no longer able to take care of her by themselves. Hence, they sent her to a nursing home. A year later she could not feed herself. Her physician discussed with her children whether a feeding tube should be adopted. They told him their mother had said to them that she would not like to stay alive like this, so they asked the physician not to use the feeding tube and use palliative care instead. Mrs. Kim's eldest daughter, who was living aboard, heard about this later and came back to Taiwan immediately. She was greatly opposed to her siblings' decision and even accused them of intending to kill their mother. She warned the physician that she would take legal action against him if the feeding tube was not used for her mother. (Tsai et al. 2006, p. 517)

In this case, though the children living with the patient said that their mother had told them she would not like to remain alive by tube feeding, there was no written document, such as an advance directive signed by the patient, to prove that this was in fact the patient's will. Hence, others—such as the sister living abroad—might doubt whether the patient actually gave these directives, or whether it is just an excuse to relieve the burden to taking care of the patient. Moreover, as in the previous case, the physician would probably follow the opinion of the patient's eldest daughter owing to his worry about possible legal actions against him, even if not providing a feeding tube for the patient would in the patient's best interests.

On the other hand, healthcare professionals in Taiwan routinely tell the patients' families the diagnosis of the patient first, though the families will usually not tell the truth to the patients, especially in the case where the patient is terminally ill:

> Mrs. B is a retired civil servant. Her husband died many years ago and her children have their own families and do not live with her. However, she used to enjoy her retired life very much and her children did not need to worry about her. A year ago her children arranged a health check for her at a hospital in Taipei City. Her son received the result first and found out that she was in the end stage of adenocarcinoma lung cancer. When he came back home, Mrs. B asked him the result and told him that she would not like to receive any treatment if she had cancer. He decided not to tell her the truth. However, when Mrs. B and her children visited her doctor again, he could not discuss the treatment frankly with them and, as a result, Mrs. B only received treatment for relieving her pain. Mrs. B gradually doubted whether her children had told the truth to her, but they all claimed that nothing bad had happened. One day, she said to them calmly, "I know very well what kind of disease I have. As I have enjoyed many things in my life, it doesn't matter if I am going to die. I would like to go to my hometown once again to see my relatives." Finally, Mrs. B's children told her the truth and she received chemotherapy later as her doctor suggested. They hoped that their mother's situation would remain stable and that her last wish would be realized (Chen and Jeng 1999, p. 40).

Many families in Taiwan would like to keep the patients ignorant of their incoming death. However, one study surveyed and interviewed 617 groups composed of terminally ill cancer patients and their family caregivers in 21 hospitals. These hospitals provide most of the care for cancer patients in Taiwan. The results showed that patients strongly claimed their own right to be informed about their disease to be superior over their family's wishes to keep them uninformed (Tang et al. 2006). If telling the truth to the patient causes them great depression and even leads them to try to hurt themselves, it will be appropriate to keep them ignorant. However, some patients would still like to know whether they are going to die. They may come to have plans surrounding their own death if they know the truth, such as visiting someone or somewhere, or making arrangements about their inheritance, or other related matters. A study in Taiwan showed that patients who were better informed about the prognosis or about hospice care were most likely to play an active role in their own end-of-life care (Huang et al. 2008). If families make them falsely believe that they may have a chance to be cured, or prevent the doctors or nurses from telling them the truth, the patients will think that they have plenty of time to do these things after treatment. Indeed, it is very difficult to conceal the truth from patients, as they are more or less aware of their own health condition. It was fortunate that in the above case the patient talked actively and frankly with her children about her feelings and wishes, and that she finally received proper treatment. However, in some other cases the patients would not know about their conditions until death was near approaching. In such cases, they could be very angry with their families for lying to them, not giving them a chance to receive proper treatment, or realize their last wishes. As a result, they would pass away with deep regret.

Moreover, there were also some cases in which the families did tell the truth to the patients and asked them whether they would like to sign a DNR. However, some may still doubt that whether such decisions really reflected the patients' wishes:

> Mr. Chan was an 83 years old coma patient living in a nursing home. At night, he suddenly had breathing difficulty and was sent to a hospital in Taipei City. He had already signed a DNR but the attending physician in the emergency department had not noticed this when he read his medical record. He performed CPR on him for 40 min. After his situation became stable, the hospital informed his family. However, when they arrived, they condemned the physician for his violation of Mr. Chan's will. The event was reported by the next day's newspaper and many people criticized the family as lacking filial piety toward Mr. Chan. However, Mr. Chan's granddaughter defended her family's actions. Mr. Chan, she affirmed, told them he did not want to be intubated anymore and signed the DNR himself after he had received CPR previously. Further, the families could not bear to see his persistent suffering. (J. C. Lin, "Families condemn physician after saving a dying old man's life," *China Times*, November 27, 2008)

To be sure, the physician in this case should confirm whether the patient signed a DNR before performing CPR on him. However, the physician's mistake would have been avoided if the families had told the staff of the nursing home about the patient's decision, and had required them to inform healthcare professionals of this when the patient was sent to the hospital. Moreover, many people in Taiwan still think that patients' families should require healthcare professionals to save the patient's life if possible, no matter what the consequence might be, as is reflected by the criticism to the family's reactions in this case. Hence, there would be great pressure towards the patient's family if they decide to withhold or withdraw treatment for the patient.

In sum, though the familial relationship in Chinese societies such as Taiwan is generally more intimate than that in the West, we should recognize that the former is subject to change. More and more families are not living together under the same roof. This is due to family members working in different places, the younger generation's preference for their own personal space, or other reasons. Divorce rates are climbing, and more and more adult children have to take care of their separated, aged parents, or only choose to take care of the ones who are living with them. Moreover, economic burdens lead to an increase of dual-earner households, and the growing lifespan leads to a longer duration of long term care for the elderly, with many difficulties in care which are almost never handled by their families. As a result, many elderly people have to stay at nursing homes and some of them have not been visited by their families for a long time. In some cases, the elderly have more than one child but their children are not living together. The responsibility of caring for them is shared by their children. They may stay at one of their children's homes for a few months, and then have to move to another's for another few months, only to move again. Their children may just regard taking care of them as an additional burden, and would like to transfer their responsibilities to their brothers or sisters as soon as possible. It is doubtful whether they really know the wishes or care about the interests of their parents in making medical decisions, as is also the case if children just leave their parents at a nursing home and rarely visit them afterward.

Furthermore, the best interests of patients themselves and those of their families as a whole are not the same thing. As the duration of long-term care extends, many people have to spend more time taking care of the aged members of their families. It not only means that they have to sacrifice their own leisure time, but it also means that they have to abandon their own work, hire a caregiver for the patients, or send them to a nursing home if necessary, which often greatly aggravates their economic burden. Even worse is when a couple has no siblings, in which case they may have to take care of their four parents or even their grandparents at the same time. Though National Health Insurance was instituted in 1995 in Taiwan, not all of the medication patients need is covered by it. Some kinds of medication paid out of pocket are so costly that they aren't affordable for many families, or a large portion of family income has to be spent in order to provide the best medications, which may make it very difficult for families to maintain normal lives for other members of the family. Hence, whether or not to buy these medications is a hard choice, even if they may be the best ones for the patient. Moreover, many people have to take care of their own children and their parents. If all of them have urgent medical or caring needs at the same time, families will often decide to satisfy their children's needs first, or will pay less attention to their parents' needs. Hence, a family always has to think about how to distribute its limited resources properly in order to satisfy the needs of all of its members, otherwise the interests of some members, especially those of the aged members, may be sacrificed in the name of the best interests of the family as a whole.

As a matter of fact, the members of many families, no matter whether they are in a Chinese society or not, may not have fully communicated to each other about how to make medical decision for their ill members. They may not be fully aware of all wishes of the patients and they may not have a consensus about how to distribute their resources to satisfy all needs or even all urgent needs of their members. Hence, if the family consent model is adopted, some procedures are needed in order to protect the rights and welfare of the patient. Moreover, if we carefully examine the Confucian conception of the family, we shall find that the familial collectivism prevalent in Chinese society, in which individuals in a family should have no voice, is not really derived from Confucian thought, as is shown in the next section.

16.3 Confucian Conception of the Family

For Confucianism, all human beings have a mind that cannot bear to see the sufferings of others, which is a moral capacity common to all humans and is essential to the human being qua human being. The virtue of *ren* (humane concern) in its broadest sense is equal to this capacity. If *ren* is understood in this way, it is the highest virtue and is the foundation of all other virtues. The first and foremost duty according to the virtue of *ren* is to maintain good relationships with one's family and adopt proper attitudes towards them, as Yu Tzu, one of the disciples of Confucius, said,

> Are not filial piety and respect for elder siblings fundamental to the practice of *ren*? (*Analects*, 1:2)[1]

The virtue of *ren* can be manifested in various kinds of moral conduct, including filial piety and respect for elder siblings; it is in principle the foundation of such conduct. On the other hand, to practice *ren* everyone should first perform these two duties, since both of them are the two most natural and direct ways displaying this virtue. If someone claims that they act according to this virtue but they have not shown any respect to their parents and elder siblings, the Confucian would regard them as hypocritical. As Mencius said,

> The virtuous man cares about (non-human) living things, but does not love them as if they were people. He loves people as people, but not in the intimate way he loves his family. He loves his family intimately and loves people as people. He loves people as people and cares about (non-human) living things. (Mencius, 7A: 45)

Hence, for the Confucian, our love towards our family should be different from that towards other people, and our attitudes towards human beings should be different from those towards non-human beings. If a man loves other people more than he loves his family, or if he loves non-human beings more than he loves his fellow humans, his behavior would be regarded as violating the virtue of *ren*. However, the above passage also shows that the duties derived from the virtue of *ren* are not limited only to one's family, but also includes those towards other humans and even non-human beings. As Mencius said,

[1] All my citations of Confucian classics are based on those by James Legge (1972) but adapted by me.

> Reverence the elders in your own family and extend reverence to other elders. Show loving care to the young in your own family and extend it to those in other families—and you can rule the world easily in the palm of your hand...Hence one who extends his bounty can bring peace to the world; one who does not cannot bring peace even to his own family... (Mencius, 1A: 7)

Since a man with the virtue of *ren* cannot bear to see the sufferings of others and views them as his own sufferings, he will try his best to relieve their pain and contribute to their proper development as much as possible. The first thing he can do is to extend his respect for the aged and his love for the young in his own family to those in other families. They are also duties derived from the virtue of *ren*. If all of our actions are performed in similar ways and our sense of compassion extends to every human as well as non-human beings, we will become sages in the Confucian sense. Moreover, Confucians think that every human being has the ability to perform these duties. Hence, to act in accordance with the virtue of *ren* persistently and thus to be a sage is just a matter of choice and is not conditioned by any circumstance. Mencius noted the story of King Shun as an example,

> When Shun was living amid the remote mountains, dwelling with the trees and rocks, and wandering among the deer and swine, the difference between him and the rude inhabitants of those remote hills appeared very little. But when he heard a single good word, or saw a single good action, he was like a stream or a river bursting its banks, and flowing out in an irresistible flood. (Mencius, 7A: 16)

Mencius wants to tell us through this story about the moral capacity common to all humans, and all of us are fully able to realize this capacity, no matter the circumstances of our lives, as in the case of King Shun. Given that he lived highly isolated from civilization, he may not have been well educated. Moreover, according to the legend about him, his family treated him extremely badly and even wanted to kill him. However, when he heard a good word or saw a good action, he did the same thing immediately, just like a stream or a river flowing out in an irresistible flood. Though Mencius also recognized that human behaviors would be greatly affected by the circumstances surrounding the person, he believed that all humans can behave like Shun, owing to their own moral capacities.

On the other hand, though Confucians think that to take care of one's family and to adopt proper attitudes towards them is a person's primary duty, and that one's family is one's main source of personal identity, it does not mean that he should always conform to his family's values. Normally, the relationships between various members of family should be reciprocal, as the parents should love their children and the children should respect their parents. Though in some cases the parents have not shown any love to their children, as we saw in King Shun's example, the children are still required to perform their duties towards their parents. Hence, Mencius said that though the proper relationship between father and son is determined by destiny, it is also a duty derived from human nature (*hsing*), so the virtuous man does not regard it as destiny (Mencius, 7B: 24). However, a person's action should not conform to his parents' opinions, if he finds that it is contrary to his own moral judgment. Mencius, once again, gives us the story of King Shun as an example:

> There are three forms of unfilial conduct, and the worst one is to have no posterity. Shun married without informing his parents because of this, lest he should have no posterity. The virtuous man considers that his doing so was the same as if he had informed them. (Mencius, 4B: 26)

Zhao Qi, a commentator of *Mencius* explained that the other two forms of unfilial conduct are compliance with everything the parents want to do that lead the parents to act wrongly, and not to seek an office in the state though the family becomes poor and the parents are aged (Ruan 1978, p. 137). Normally, a person should ask his parents' approval if he wants to get married. However, since Shun's parents greatly hated him, they would disagree with his marriage, so that he would have no posterity. Shun's major concern was not his own welfare but the continuation of the family line. Since not only to inform his parents about his marriage, but also to have children in order to continue his family line are duties of filial piety, Shun had to balance the relative weight of these norms. Clearly, to conform to the latter is more important than the former, so Shun decided to get married without informing his parents. What is shown in this case is that for Confucian thought we should always make a judgment about what we ought to do according to our conscience; though in some cases our decisions may be inconsistent with general ethical norms. Asking parents' approval before deciding to get married may be regarded as a general ethical norm which should be conformed to in most cases, as we believe that most of the parents will provide good advice to their children about their marriage. However, we should not conform to this norm if we encounter a situation similar to Shun's. Similarly, in the context of medical decision making, patients and healthcare professionals should generally ask for the opinions of families and follow their suggestions, since we also believe that most families care about their members' welfare, and know well about their own values and preferences. However, patients and healthcare professionals should do otherwise if the families are alienated from the patients and do not show concern about the patients' welfare.

16.4 Towards a Sound Model of Family-Oriented Consent: Some Suggestions

Though the structure of the family and the intimate relationships among family members are changing, we believe that most of the families in Taiwan are still deeply concerned about the welfare of their members, and usually provide good advice on medical decision making if they are ill. Moreover, since medical decisions for patients greatly affect the welfare of their family members—as is expressed in the saying that when a family member is sick it is as though the whole family were sick—family members should participate in the decision process to choose appropriate treatments for patients. However, family members should not conceal diagnoses from patients if it is not necessary, even if they are in a terminal situation. Rather, they should assist physicians and nurses in telling the truth to patients, and should make medical decision with patients according to patients' best interests.

Though many patients will feel hopeless after knowing that the end of their life is coming, their families should try their best to relieve patients' emotional responses, and to let the rest of their lives be meaningful. Unless patients will hurt themselves after knowing the bad news or insist on waiving their rights to know about their conditions, concealing the truth from them should not be regarded as appropriate.

In order to ensure that family members are playing their proper role in medical decision making, physicians and nurses should know about the wishes and preferences of their patients, the conditions of their families, as well as their relationships with other family members by building trust with them. This trust-building should start with the first visit of patients, and should be gradually developed during the diagnosis and treatment processes. According to Chinese medical ethics, physician-patient relationships should not be regarded as contractual. Rather, physicians should treat their patients as if they were their family and should count patients' sufferings as their own. Hence, physicians should at least show sincere concern for their patients' health, and give patients and their family members enough time to talk about the patients' conditions. Normally, physicians may not have much time to gain knowledge about all the wishes and preferences of their patients, and the situation of patients' families in detail. But these things can be done by nurses through their daily discourse with patients and their families, if patients are hospitalized (Lee 2008, p. 14).

Among the various kinds of information about the patients' situations that health care professionals should come to know, the first one should be the competency of patients for medical decision making. The necessary abilities for patients to decide completely about their health care include the capacity for understanding and communication, the capacity for reasoning and deliberation, and the capacity to have a set of values which is consistent, stable and affirmed as their own (Buchanan and Brock 1989, pp. 23–25). If patients are proved to be incompetent, their family members or proxies have to make medical decisions for them. However, health care professionals should help them to think about how the patients would choose if they were able to decide by themselves, and make decisions for them according to the patients' best interests and their possible preferences. On the other hand, if patients are proved to have the necessary abilities to make decisions, and if family members know first about their condition, they should be encouraged or even be required to discuss with patients about possible treatments, or whether to withhold or withdraw treatment if it has proved futile.

Certainly, health care professionals should consider carefully the severity of the sadness patients will experience and what they will do after hearing the bad news about their health. This is why they should ask family members of their patients for help to understand the patients' possible emotional reactions and for help to tell the truth to patients (Hu et al. 2002). However, family members of the patient usually suffer from emotional pain and they expect that the patient's reaction will be more negative than it would actually be (Sheu 2007, p. 62). Hence, health care professionals should persuade them persistently and let them understand that concealing diagnoses from patients may lead to harmful consequences for their loved one. This can be done through describing previous positive and negative cases to families as a reference. As we noted before, patients may have their own views about their subsequent treatments which are different from those of their families. If they know that their death is coming, they may have plans of their own which may not be known to their families. They may regret it greatly if these plans have not been realized by the end of their lives. Moreover, family members of terminally ill patients may insist that positive treatment to patients should not be withdrawn, which may only increase harm to the patients. Patients have their own right to know what will happen to their bodies. However, if their family insists that patients should not have been told about these, and their suggestions are in conflict with the best interests of the patients according to professional judgments, then health-care professionals should tell them about their present condition, and obtain informed consent directly from them for subsequent treatments.

On the other hand, though we think that most families care about the welfare of their members, we should also recognize that exceptional cases exist. If a patient's children have not seen their parent for a long time after leaving him or her at a nursing home, it is reasonable to doubt whether they know the patient's relevant opinions or even care about the patient's situation. Families may also select from the patient's life history those values which are consistent with theirs and make a medical decision for the patient accordingly (Beauchamp and Childress 2001, p. 101). Moreover, though some families are willing to discuss treatment options with patients, the patients' opinions may be overridden without good reason. Hence, health care professionals or others such as social workers and ethical consultants in hospitals should be well aware of the relationships between patients and their families. They should not only encourage families to discuss with patients before making medical decisions, but also help them to make sure patients' opinions have been fully expressed and respected.

Besides, patients who have become incompetent may have previously written advance directives to instruct their families about how to make medical decisions for them, or they may have procured a durable power of attorney to appoint others to be their proxies instead of their families. Rather than being regarded as undermining the value and intimate family relationships (Fan 2002, p. 351, 2011, p. 60), advance directives should be counted as an effective way to help the families of the patients know about the patients' wishes and preferences when they cannot communicate with them, and the durable power of attorney will also provide an alternative for patients to protect their own rights if their relationships with their families have become estranged or if there are conflicts of interests among them. Indeed, if the relationship between patients and their families remains intimate, they will not want to ask people other than their families as their proxy. Hence, how families get along with patients is affected by other factors, regardless of whether a mechanism of durable power of attorney exists. However, in order to fulfill their duties to obtain patients' informed consent, healthcare professionals should not just ask the patients or their families to sign a DNR or a durable power of attorney form. Rather, they should communicate with them continually and provide advice to them based on an adequate understanding about the patients' values and preferences and their familial relationships. Moreover, they should let their patients fully understand that something unexpected may happen after they determine their advance directives; for example, new medications for their diseases may be developed. Hence, they should suggest to their patients to let their families or their proxies make medical decisions for them according to the patients' best interests, if such unexpected things do happen and the patients cannot express their own will at that time.

16.5 Concluding Remarks

There is something invaluable in traditional Chinese families that should be maintained, such as the emphases on the duty of filial piety and familial intimacy. Yet we should recognize that traditional Chinese families also involve issues, such as patriarchism, which were criticized by many scholars in the New Culture Movement taking place in the early twentieth century, though these problems should not be totally attributed to Confucianism. As we reconstruct the family-oriented consent model for medical decision making instead of the Western individualistic one, we should try to avoid these problems. Moreover, the family structure of contemporary Chinese society such as today's Taiwan is different from the traditional one. The mechanism of family-oriented consent should be modified in order to tackle the problems we now face.

References

Beauchamp, T. L., and J. F. Childress. 2001. *Principle of biomedical ethics*. 5th ed. Oxford: Oxford University Press.

Buchanan, A. E., and D. W. Brock. 1989. *Deciding for others: The ethics of surrogate decision making*. Cambridge: Cambridge University Press.

Chen, S. F., and K. Y Jeng. 1999. Let her know if you love her. *Quarterly Journal of Cancer Care* 4:40–42. [In Chinese.]

Fan, R. 2002. Reconsidering surrogate decision-making: Aristotelianism and Confucianism on ideal human relations. *Philosophy East & West* 52:346–372.

Fan, R. 2007. Confucian familism and its bioethical implications. In *The family, medical decision-making, and biotechnology*, ed. S. C. Lee, 15–26. Dordrecht: Springer.

Fan, R. 2011. *Contemporary Confucian bioethics*. Beijing: Beijing University Press. [In Chinese.]

Hu, W. Y., T. Y. Chiu, R. B. Chuang, and C. Y. Chen. 2002. Solving family-related barriers to truthfulness in cases of terminal cancer in Taiwan. *Cancer Nursing* 25 (6): 486–492.

Huang, C. H., W. Y. Hu, T. Y. Chiu, and C. Y. Chen. 2008. The practicalities of terminally ill patients signing their own DNR orders—a study in Taiwan. *Journal of Medical Ethics* 34:336–340.

Lee, S. C. 2007. On relational autonomy: From feminist critique to confucian model for clinical practice. In *The family, medical decision-making, and biotechnology*, ed. S. C. Lee, 83–93. Dordrecht: Springer.

Lee, S. C. 2008. Theories and models of clinical ethical consultation. In *Clinical ethics consultation: Theory and practice*, ed. S. C. Lee and D. J. Tsai, 3–19. Taipei: Wunan. [In Chinese.]

Legge, J. 1972. *The Chinese classics: With a translation, critical and exegetical notes, prolegomena, and copious indexes*. 2nd ed. Taipei: Wen Shih Che Publishing Co.

Ministry of Health and Welfare. 2009. Physicians act. *Laws and regulations database of the Republic of China*. http://law.moj.gov.tw/Eng/LawClass/LawAll.aspx?PCode=L0020001.

Ministry of Health and Welfare. 2013. Hospice care law. *Laws and regulations database of the Republic of China*. [In Chinese.] http://law.moj.gov.tw/LawClass/LawAll.aspx?PCode=L0020066.

Ruan, Y. 1978. *The thirteen classics with notes and commentaries*. vol. 8. [In Chinese.] Taipei: Shin Wen Feng Print Co.

Sheu, S.-J. 2007. The moral ground of truth telling guideline development: The choice between autonomy and paternalism. In *The family, medical decision-making, and biotechnology*, ed. S. C. Lee, 59–70. Dordrecht: Springer.

Tang, S. T., T. W. Liu, M. S. Lai, L. N. Liu, C. H. Chen, and S. L. Koong. 2006. Congruence of knowledge, experiences, and preferences for disclosure of diagnosis and prognosis between terminally-ill cancer patients and their family caregivers in Taiwan. *Cancer Investigation* 24:360–366.

Tsai, F. C., H. C. Pan, T. M. Wu, T. Y. Chiu, and T. S. Huang. 2006. The ethical-legal issues in advance care planning. *Formosan Journal of Medicine* 10 (4): 517–536. [In Chinese.]

Chapter 17
A Family-Oriented Confucian Approach to Advance Directives in End-of-Life Decision Making for Incompetent Elderly Patients

Yaning Yang

17.1 Introduction

Advance directives (ADs) are planning tools, usually in form of written statements, that enable individuals to preserve control over the decision making process in circumstances where they are no longer capable of making decisions due to serious illness. It can either be a living will, which clearly expresses an individual's refusal or acceptance of certain treatments, such resuscitation, mechanical ventilation, tube feeding, etc., or the appointment of a durable power of attorney, who is empowered to make health care decisions on the patient's behalf when the patient is no longer competent.

ADs are regarded as a quintessential topic in bioethics (Capron 1998). The development of critical care medical technologies over the past few decades has improved our ability to lengthen and, in some instances, save the lives of seriously ill patients. However, for terminally ill patients, advances in medical interventions may only prolong the dying process and their experiences of ill health. Patients, families, and physicians inevitably become involved in the dilemma of end-of-life decision making (e.g., whether to withhold life sustaining treatment), which has not been a problem in the past. In Western countries, especially the United States, ADs give patients the opportunity to express their preferences about the extent and type of medical treatments they are comfortable with, in the event they are incapable of making their own decisions in the future. The underlying assumption of ADs is that

Y. Yang (✉)
Institute for Hospital Management, Tsinghua University China,
Rm 102, 1/F, Building A, Tsinghua Campus, The University Town, Shenzen, China
e-mail: yang_yaning@hotmail.com

R. Fan (ed.), *Family-Oriented Informed Consent*, Philosophy and Medicine,
DOI 10.1007/978-3-319-12120-8_17

a living will or a durable power of attorney will extend an individual's autonomy into the future if one cannot protect his/her own interests. In theory, ADs are meant to relieve the burdens of making decisions for incompetent patients, resolving conflicts between family members and/or physicians, and, to a large extent, accurately reflecting the will of patients for the family and the physician (van Wijmen et al. 2010).

Despite legal advancements in the United States and several Western countries,[1] ADs are still in a "nascent" stage in most Asian and African countries (Kim et al. 2010). In Asia, only some developed areas legally acknowledge ADs[2]. In China, ADs have not been formally recognized through law.

With advances in medical technology and the ease of communicating medical knowledge across national borders, a requirement for policy convergence is emerging. Scholars advocating ADs claim we need a global policy to facilitate the execution of ADs from the West in other cultural contexts. Simultaneously, scholars have found support for a global policy in the documents of international organizations.[3] The principles of autonomy, dignity, and informed consent endorsed by international declarations, can be used to establish a universal basis for the global use of ADs. In the same sense, Kim et al. (2010) argue that respect for patient autonomy must be universally understood and it should serve as the underlying principle for the global promotion of ADs.

Scholars have identified cultural differences as barriers to a global policy for ADs. One serious obstacle is that respect for patient autonomy is not necessarily well-suited for other (non-Western) cultural settings. For instance, the family-oriented Confucian tradition in East Asia has been acknowledged as the guiding influence for end-of-life decision making (Ho et al. 2010; Chen and Fan 2010). In a family-centered society, ADs may conflict with family values (Kim et al. 2010; Blank 2011). In spite of these obstacles, some scholars still believe that ADs should be universally promoted and that international organizations, such as the World Health Organization, should play an important role in doing so.

In the light of the rapid development of advance care planning around the world and the good purpose (i.e., relieving the family and physician of the burden of

[1] In the United States ADs were initially legally acknowledged by the "Natural Death Act" in California and the Patient Self-Determination Act of 1990. Following the United States, many North American and European countries, including Canada, the UK, the Netherlands, Belgium, France, Spain, a number of Australian states, and New Zealand, have adopted ADs, or some forms of ADs (Beširevic, 2010).

[2] For example, ADs were recognized by the Advance Medical Directive Act (AMDA) in Singapore in 1996. According to the AMDA, a person 21-years of age or older, not suffering from a mental disorder, may choose not to be subjected to extraordinary life-sustaining treatment for a terminal illness. Such a person may establish an advance directive at any time so long as there are two witnesses. In Taiwan, advance directives were legislated into being through the 2000 Law of Hospice Palliative Treatment. The law stipulates that terminally ill patients who have been approved by two doctors have the right to craft a living will or appoint a durable power of attorney.

[3] Beširevic (2010) has identified two international documents, the Council of Europe's 1997 Convention on Human Rights and Biomedicine and UNESCO's Universal Declaration on Bioethics and Human Rights, that address the underlying guiding principles of ADs.

surrogate decision making for incompetent patients) it is meant to serve, the push for a global AD policy seems reasonable. However, before tackling the pragmatic problem of global use, there are important theoretical questions that need to be elaborated on in depth. For example, it has not be confirmed that the Western justification of ADs by the principle of autonomy is the only one that matters. In other words, is it the case that Western norms and values, in particular patient autonomy and human rights, are the only justifications for the global acceptance of ADs? Are other normative frameworks, such as Confucian ethics, necessarily obstacles for the global realization of ADs, or is it possible that they might be alternative frameworks capable of supporting ADs?

This essay will argue that individual-oriented approaches to ADs underestimate the influence of cultural diversity on end-of-life decision making and minimize the crucial role of the family in both Western and Eastern countries. These considerations make clear the insufficiency of the individual-oriented approach as the only normative basis for a universal AD policy. Taking into account cultural differences and employing them as alternative normative bases might ease the process of gaining global acceptance for ADs. In particular, incorporating family values into the promotion of ADs in East Asia would ease its implementation. This essay will focus specifically on elderly patients[4] and will employ a Confucian family-oriented approach to argue that some values in the Confucian tradition, for example, *xiao* (filial piety), might be of crucial value to the discourse framing the implementation of ADs for elderly patients in China or other countries in which there is a Confucian tradition.

In sect. 16.2 the important question of whether respect for autonomy is the only normative basis for the global use of ADs will be addressed. In sect. 16.3, through the exploration of Confucian ethics and, in particular, the concept of *xiao*, a family-oriented approach to ADs for elderly patients will be considered. In sect. 16.4, practical suggestions are made from a Confucian perspective for developing an AD policy in China.

[4] The reasons for focusing on elderly patients are as follows. First, with an aging population in China, elderly patients may become the largest patient group considering the use of ADs for their end-of-life decision making. The data from the Sixth National Census shows that, in 2010, 8.87 % of the population was at least 65 years of age (National Bureau of Statistics of China 2011). Second, this essay employs the Confucian concept of *Xiao*, which is a fundamental virtue in Confucian ethics and generally refers to the filial duty of adult children to their elderly parents. For this reason, this essay will not address other family relationships. In China, the elderly population includes those age 60 and above (The Elderly Security Law of P.R. China, 2012). In this essay, the focus is on the filial piety of adult children for their parents, so it is not necessary that the patient's age be above 60. As such, in this essay, elderly patients will be defined as those patients who have adult children (18-years-old and above).

17.2 Is Individual Autonomy the Only Normative Basis for the Global Use of Advance Directives?

There is no doubt that the individual-oriented approach of a global AD policy is based upon traditional Western theories of bioethics. In fact, interest in ADs originated in the United States in parallel with the bioethics movement of the 1970s and 1980s, which emphasized individual autonomy and patient authorization. During this period, medical decision making in the United States witnessed a transformation from a paternalistic model of health care to a patient-centered individualistic model (Fan and Tao 2004). Considerable emphasis has been given to patient empowerment, or patient autonomy, in medical decision making. The emergence of ADs seems to follow the same ethical standard of "return[ing] to the individual the ability to control their dying process, primarily by refusing life-extending interventions" (Blank 2011, p. 202). The Patient Self-Determination Act (PSDA) is the landmark legislation for ADs and is representative of the standard individual autonomy approach to end-of-life decision making. The PSDA stipulates that healthcare providers in hospitals, nursing facilities, home health agencies, and hospice programs, should inform patients of their individual rights under State law to

> make decisions concerning…medical care, including the right to formulate advance directives recognized under State law relating to the provision of care when such individuals are incapacitated, such as through (I) the appointment of an agent or surrogate to make health care decisions on behalf of such an individual and (II) the provision of written instructions concerning the individual's health care (PSDA 1990).

It is very apparent that patient autonomy and self-determination have been highlighted as fundamental normative principles in end-of-life decision making and, more specifically, advance health care planning. This individual-oriented approach has little to do with the family or family values. Some scholars even argue that the family should be entirely excluded from end-of-life decision making because the aim of ADs is to realize self-determination and to refuse the involvement of the family and community in end-of-life decision making. Beširevic claims, "the major impetus for recognizing the need for advance directives was the invention of life-prolonging machines and the idea that death should be a subject of an inviolable private sphere and self-determination and not a community or family matter" (2010, p. 112).

However, this essay argues that cultural diversity is an important reason why the Western ethical framework of patient autonomy is not the only normative basis for ADs. The issues surrounding end-of-life decision making are, to a large extent, shaped by culture. Numerous studies have shown that cultural differences play a profound role in the variation of strategies, options, and ethical judgments regarding similar end-of-life problems (e.g., Cox et al. 2006; Chattopadhyay and Simon 2008; Richter and Eisemann 2001; Werth et al. 2002). In different countries, different policies are adopted to deal with similar problems (Blank and Merrick 2005). Consequently, "there is a danger of making the assumption that ethical frameworks will converge to a western model despite continued evidence of a wide divergence across countries in bioethical perspectives" (Blank 2011, p. 204).

It should be noted that family values that are insignificant in individual-oriented approaches to medical decision making are found to be crucial in other cultural contexts, especially in Asian countries.[5] In China, the health care decision making model is defined as a "family-based and harmony-oriented" model embedded in the "Confucian way of life" (Chen and Fan 2010, p. 573). In this model, the locus of moral responsibility for medical decision making is the family. Patients are treated not as an isolated individual but an inseparable member of the family. For example, studies show that Chinese doctors first inform the family about the patient's diagnosis and whether or not the patient should be informed is the family's decision (Cong 2004; Fan and Li 2004). This model advocates for a harmonious medical decision making process between three parties: the patient, the family and the physician. In contrast to the individual-oriented model, the family is regarded as an autonomous unit, which, as a whole, has the authority to make medical decisions on behalf of its members (Fan 1997). Family cohesion and involvement in medical decision making has been witnessed in many other East Asian countries like Taiwan, Korea, Hong Kong, and Singapore.[6]

Furthermore, it is misleading to regard the family role in medical decision making as inessential in the countries where individual autonomy is the standard for informed consent. Kwon et al. (2009) have indicated that a majority of studies show that even in Western countries, ADs and end-of-life care conversations occur within a family context (Haley et al. 2002), end-of-life decisions are often made by the family (Gochman and Bonham 1990) or the physician (Slomka 1992), and the role of patient autonomy at the end-of-life has been questioned (Winzelberg et al. 2005). In the United States, studies have indicated that patients' family members are involved with end-of-life decisions 60 to 80% of the time and patients often prefer to make end-of-life decisions within the family context (Haley et al. 2002). Kelley et al. (2010) found that a majority of elderly Latinos with serious illness prefer family-centered decision making and limited patient autonomy. Parks et al.' 2010 study found that an advance care planning process with "a family-centered approach" is important and "new tools for advance directives should be developed for decision-making by families rather than the current legal format of naming a single durable POA" (Parks et al. 2010, p. 182).

17.3 A Confucian Family-Oriented Approach to Advance Directives: The Principle of Xiao (Filial Piety)

Before focusing on the concept of *xiao*, a clear definition of the family will be provided from the Confucian perspective. According to Confucian ethics, "it is the arrangement of Heaven (*tian*) that every individual is born to a family, possessing

[5] See: Chattopadhyay and Simon 2008; Ho et al. 2010; Lee 2013; Wong 2014; Deng 2013; Cai 2013; [THIS VOLUME]).

[6] For examples see: Lee 2013; Kim et al. 2010; Tse and Tao 2004; Ho et al. 2010).

special relations to other family members and living one's life inseparably from the family" (Fan 1997, p. 317). The family not only can be defined as a socio-biological unit, namely a group of persons united by the ties of marriage, blood, or adoption, but more importantly, the family is "an eternal social category" (Fan 2007, p. 512) and "a metaphysical reality" (Chen and Fan 2010, p. 576) that "reflects the profound mechanism of the universe, the Dao of Heaven and Earth" (Fan 2007, p. 512). The familial relation is a basic human relation that is not founded on mutual individual consent, but the universal Dao. "The familial relation can never be intentionally abandoned" (Fan 2002, p. 353). As an autonomous social unit, the family has sovereignty over decision making and family members can "make a clear distinction between intra-familial and extra-familial authority" (Fan 1997, p. 317). Family members do not have to cohabitate but they should take part in the decision making process for family issues. Usually, the spouse, parents and adult children of the patient are the main family members involved in medical decision making, however, that is not necessarily to the exclusion of other family members if they play a role in medical decision making. Moreover, it should be noted that Confucian account of the family does not make a sharp distinction between patients and other family members. All family members share in the decisions regarding each other's medical care (Chen and Fan 2010).

As previously noted, arguments have been made that the Confucian emphasis on family values is a cultural obstacle for the use of ADs in East Asian countries. It is assumed, if individual autonomy is not regarded as important as family cohesion, ADs might become unnecessary because "it is taken for granted in Chinese culture that medical decisions should be properly made by the whole family rather than the patient alone" (Fan 2011, p. 306). Kwon et al. (2009) argue that the value of *xiao* (filial piety) has important influence on end-of-life decision making for Asian families. For example, on the one hand elderly patients sometimes wish to forgo life-sustaining treatment because they strongly feel they should not be a burden to their families. On the other hand, in accordance with *xiao*, families may not agree to withdraw life sustaining treatment for an elderly incompetent family member even if it may result in great suffering. "To agree to stop life support may be considered a shameful "face-losing" decision and must be avoided at all cost" (Hui 1999, p. 156). As such, it appears that the value of *xiao* deems the family's interest as more important than the patient's autonomy. In such situations, it seems that ADs are neither necessary nor practicable.

However, it is hard to defend this argument because the meaning of the concept of *xiao* does not imply that it is appropriate for adult children to ignore the will of their elderly parents. A true story reported in China's *Health Newspaper* illustrates why the proper execution of ADs in China is necessary and how it is that it would not diminish the value of *xiao*. On the contrary, *xiao* might be an important normative principle underpinning the implementation of ADs in a Confucian cultural context. In the following citation, a woman named Luo Diandian recalls how her family made the decision to forgo her mother-in-law's life-sustaining treatment when she was terminally ill without a chance of recovery.

> My mother-in-law had been put on mechanical ventilation after a sudden heart attack. Although her heart was still beating, she had no autonomous respiration. Her condition got worse in the following days. Finally, she sank into a deep coma. I remembered that my mother-in-law had mentioned many times that she did not want to be put on a bunch of tubes if she had a grave disease. I understood that we, as her children, should not let her be in such situation that she would not like to be in. However, it was hard to make the decision to remove the machine because I could still feel the warmness of her body. At last, my brothers insisted on withdrawing the life sustaining treatment because they believed it was the will of my mother-in-law. After she passed away, I found a clearly written note in my mother-in-law's diary that she did not want life prolonging treatment to be overused at the end of her life. It was a big relief for me and I knew that we had made the right decision (Yang 2009, p. 003).

In the above case, Luo Diandian's mother-in-law was in a situation where mechanical ventilation would not revive her, but would prolong her period of suffering caused by the disease. The children of elderly patients like Luo Diandian and her brothers encounter the controversial ethical issues involved in surrogate decision making including what principles the decision should be based on, who has legal authority to make decision, and what kind of process should be followed. In the above case, the mother-in-law had something of a living will in her diary stating her refusal of certain treatments at the end of her life. Having to make the decision to forgo life sustaining treatment for their mother without knowledge of her AD put an emotional and ethical burden on the adult children. Once her living will was found, her adult children agreed they had realized her wish and their decision was morally right.

Although *xiao* is not directly mentioned in the story, it played an important role in the children's medical decision making for their mother. For most Chinese people, autonomy is not a conventional concept. The children thought they made a morally right decision because they did not violate the virtue of *xiao,* not because they had realized their mother's self determination.

In the Confucian tradition the family is considered a fundamental and unavoidable unit of society. The cardinal virtue of *xiao,* expressed between a parent and child, is the root of morality. The concept of *xiao* is difficult to translate because it is composed of "a range of attitudes and emotions that attend deep familial relations" (Chan 2010, p. 337) and it is also the "foundational virtue" in Confucian family ethics that "constitutes the source of perfect virtue" and the "main manifestation of humanity (*ren*)" (Fan 2006, p. 3–4). In the classic Confucian books, it is said that "filial piety (*xiao*) and brotherly love (*ti*) are the roots of humanity (*ren*)" (*The Analects* 1:2; cf. Fan 2006, p. 3). "The content of benevolence (*ren*) is the serving of one's parent" (*Mencius* 4A:27; cf. Lau 1970, p. 127)[7]. *Xiao* primarily refers to the moral duty of children to their parents. In particular, it is the filial obligation of children to respectfully and comprehensively take care of their elderly parents.

Although *xiao* is a concept with a two thousand year history, it still has a profound impact on Chinese daily life. Usually, the Chinese word "*xiao*" is accompanied

[7] All quotations from *Mencius* are from Lau 1983.

by the word "*jing*" (reverence), i.e., "*xiaojing,*" or accompanied by the word "*shun*" (obedience), i.e., "*xiaoshun,*" by ordinary Chinese people. Used in conjunction with "*jing,*" *xiao* does not only mean that children should take care of their aged parents' living, such as providing food, clothing, and shelter, but they also should show genuine respect and love for their aged parents and satisfy their parents' mental needs. Confucius said, "nowadays for a man to be filial means no more than that he is able to provide his parents with food. Even hounds and horse are, in some way, provided with food. If a man shows no reverence, where is the difference" (*The Analects* 2: 7; cf. Lau 1983, p. 13)? Used in conjunction with "*shun*" (obedience), *xiao* incorporates the meaning of obedience with the *li* (rites). When Meng Yi Tzu asked about being filial, Confucius' answer was "never fail to comply," which means "when your parents are alive, comply with the rites in serving them; when they die, comply with the rites in burying them; comply with the rites in sacrificing to them" (*The Analects* 2:5; cf. Lau 1984, p. 11).

There are three parts to the concept of *xiao* that are integral for making an argument for a Confucian family-oriented approach to ADs. First, *xiao* is the root of *ren* (humanity), which is the overarching virtue of the Confucian ethics tradition. In a family, *ren* naturally grows between parents and children and then gradually extends to others in general. Family love is not the limit of, but rather the foundation to an ethical life. Without the virtue of *xiao*, that is, without taking responsibility for respectfully caring for one's aged parents, one can neither love others nor cultivate the other virtues for human life. Mencius believes that *ren* is established through our natural compassion or sympathy, which everyone possesses and is expressed when one witnesses the suffering and hardship of others. One experiences a feeling of alarm and distress when witnessing a child about to fall into a well. Mencius considers that natural feeling of sympathy as the seed of *ren*. Compassion is generated from the relation between the caregiver and a child within a family and extended to other kin and, finally, to all others.

Because *xiao* is the root of *ren*, children's moral duty to their aged parents must involve love and sympathy. "Compassion is a moral virtue, not a mere instinct or intuitive feeling" and "our sympathy and moral capacity to feel for others define our humanity" (Tao 1998, p. 607). In this sense, children who possess the virtue of *xiao* would be made unhappy to see their parents suffering and would not do something to aggravate the suffering for their own interest. If continued medical treatment for a terminally ill parent would only result in greater suffering, a child's insistence that everything be done to keep the parent alive would not be considered an exercise of the virtue of *xiao*. As Hui has pointed out,

> Many of these practices based on filial piety are self-serving, benefiting the children of the patient rather than parent himself. At best, it can be seen as an expression of *ren* in order to establish the virtuous character of the self, and at its worst, it is merely an attempt to avoid a strong sense of guilt and (1999, p. 156).

Rather, it is a hijacking of the concept of *xiao*, not an expression of it. By the same token, if a terminally ill parent has a living will, the principle of *xiao* would not require the child to disregard their parent's wishes for the sake of their own interests.

In the story above, Luo Diandian had true compassion for her dying mother-in-law's suffering, which can be understood as an initial expression of the virtue of *xiao* or *ren*. It is because of this compassion that she and her brothers made the decision to forgo life sustaining treatment. If her mother-in-law's living will is taken into account, their decision was in accordance with the principle of *xiao*.

Second, *xiao* embodies the meaning of *jing* (reverence). In the Confucian *Classic of Filial Piety* (*Xiaojing*), Confucius said,

> Of all human conduct none is greater than filial piety (*xiao*). In filial piety (*xiao*) nothing is greater than to revere one's father… Affection is fostered by parents during childhood, and from there springs the child's reverence, which grow daily, while sustaining his parents. The sage was to follow this innate development by teaching reverence and to follow this innate feeling of affection by teaching love (*Xiaojing* IX; cf. Chai and Chai 1965, p. 330)[8].

The concept of *jing* is so important that Confucius contended there is nothing more important than venerating one's father. Like compassion, such reverence originates from the affectionate relationship between parents and children and expands into a respect for others. This respect for others is an important component of the virtue of *ren*. Mencius said, "to feed a man without showing him love is to treat him like a pig; to love him without showing him respect is to keep like as a domestic animal" (*Mencius* 7A:37; cf. Lau 1970, p. 190). It is easily to see that, for Confucians, the elderly are always at the center of the family (Fan 2007). The dignity of the elderly is preserved through the respectful care children provide their parents. In other words, through the virtue of *xiao*, which is practiced within the bonds of the family, the dignity of the elderly is realized. This has been supported by a study of long-term care provided to the elderly in Beijing (Zhai and Qiu 2007). Many respondents in the study expressed the opinion that the dignity of elderly parents embodies the respect and reverence shown by their children and is ultimately manifested by being cared for by their children. Thus, the term *jing* contains a crucial understanding of the Confucian family-oriented concept of dignity for the elderly, which is far different from the individual-oriented dignity developed by Western countries.

The term *jing* implies that children should respectfully care for their elderly parents, not only nourishing their physical bodies, but also preserving their human dignity. If children only care about their parent's physical condition and ignore their dignity and spiritual satisfaction, it is not a real expression of *xiao*. Thus, the implication for end-of-life decision making is that prolonging the life of an incompetent terminally ill parent without caring for their mental needs and dignity cannot be considered *xiao*. In a case where the continuation of life sustaining treatment would only cause the parent more suffering with no hope of recovery, insistence by the children to continue such treatment, despite the parent's spiritual wishes, would not demonstrate *jing* and therefore could not be regarded as *xiao*. Moreover, if a healthy parent establishes an AD, children with the virtue of *xiao* should respect their parent's will. Fung yu-lan, a contemporary Chinese philosopher, indicates that *xiao* includes "conforming ourselves to [our parents'] wishes, and giving them not only

---

[8] All quotations from *Xiaojing* are adapted from Chai and Chai 1965.

physical care and nourishment, but also nourishing their wills" (Fung 1952, p. 359; cf. Tse and Tao 2004, p. 211).

It should be remembered that ADs "are actually a very recent response to a modern problem, namely the worry of many people that they will become victims trapped by medicine's ever-expanding ability to sustain life indefinitely after they lose the ability to voice their wishes about treatment at the end of life" (Capron 1998, p. 262). When one becomes terminally ill, it appears that one loses his/her capacity for self-determination. In this situation one is essentially afraid of losing his/her dignity. As the author of the California Natural Death Act has pointed out, "The image of Karen Quinlan haunts our dreams. For many, the ultimate horror is not death, but the possibility of being maintained in limbo in a sterile room, by machines that are controlled by strangers" (Capron 1998, p. 264). That is to say, the aim of ADs is to a large extent about protecting one's dignity. In the Western liberal tradition, "individual autonomy" is employed as the crucial theoretical basis for protecting the dignity of terminally ill patients. In the Confucian tradition, dignity is developed and protected through the familial concepts of *ren* and *xiao*. In this sense, children who possess the virtue of *xiao* should respect their parents' dignity not only by caring for their parent's living, but also by respecting their end of life wishes.

Third, *xiao* includes the meaning of *shun* (obedience). However, it would be misleading to assert that *shun* means children should always be obedient to their parents. When Meng Yi Tzu asked about being filial, Confucius answers, "Never fail to comply" (*The Analects* 2:5; cf. Lau 1983, p. 11). What Confucius means is that one should not fail to comply with the rites (*li*), not that they should never fail to comply with their parents' wishes (Fan 2006). Confucius answers, "When your parents are alive, comply with the rites in serving them; when they die, comply with the rites in burying them and in offering sacrifices to them" (*The Analects* 2:5; cf. Lau 1983, p. 11). Confucius clearly points out that when parents do something wrong, children should persuade them with reverence. Confucius states,

> In serving your father and mother you ought to dissuade them from doing wrong in the gentlest way. If you see your advice being ignored, you should not become disobedient but should remain reverent. You should not complain even if in so doing you wear yourself out (*The Analects* 4:18; cf. Lau 1983, p. 33).

By the same token, when Tseng Tzu asks whether a son, by obeying every command of his father, can be called a filial son, Confucius states, "In the case of gross wrong, the son should never fail to admonish his father against it" (*Xiaojing*, XV; cf. Chai and Chai 1965, p. 333). In other words, filial children should respectfully persuade their parents from doing wrong and "manage to improve their parents' virtue by communicating with them" (Fan 2006, p. 8). The concept of *shun* is very important for end-of-life decision making because it implies that the process should be in accord with some rites through which the virtues of both the parent and the child can be well cultivated. Communication between the parent and children is far more important than the parent's self-determination. For example, when a parent decides against worthwhile life-sustaining treatment in order not to over-burden the family, through the virtue of *xiao*, children should communicate with the parent and dissuade him or her from forgoing the treatment. The successful implementation of

ADs requires the family to be aware of the document and to insist on its implementation when the patient faces a health crisis. In this manner, the Confucian concept of *xiao* provides a promising framework for establishing effective communication and a trusting relationship between the elderly parent and his/her children. A Confucian family-oriented model of ADs invites the family as a whole to participate in the advance planning of the patient's end of life decisions. However, it does not mean that the patient's own will should be subject to the interests of the family. On the contrary, this model encourages the elderly patient to express his/her wishes and to participate in his/her end of life decision making. At the same time, adult children should respect their parent's will and the final decision should be the result of a discussion between the parent and his/her children, in accordance with the virtues of *xiao*, *ren*, and *li*.

17.4 A Confucian Family-Oriented Policy for Advance Directives in China

This essay has argued for the possibility of a Confucian family-oriented approach to constructing an AD policy in China. There have been some preliminary studies in China exploring the attitudes of patients and their families regarding ADs. Wang et al. (2012) interviewed 75 advanced cancer patients and 112 family members of advanced cancer patients at a hospital in Shanghai regarding their attitudes toward ADs. Their research found that 58.7% of patients and 43.8% of family members were accepting of ADs. Ninety percent of patients indicated that they wanted family members and physicians to understand and respect their end of life wishes. Eighty percent of patients indicated that they wanted their family to be involved in their medical decisions and that the final decision should be based on discussions between the patient, the family, and the physician. This preliminary study seems to indicate that advanced cancer patients in China do not have a negative impression of ADs and support family involvement in end-of-life decision making.

A Confucian family-oriented approach to ADs for the elderly will inevitably be different from an individual-oriented approach. The aim of ADs is to protect the dignity of elderly parents and to cultivate and realize the virtues of *xiao* and *ren*. First, the content of ADs must be in accordance with the requirements of *xiao* and *ren*. In other words, the wishes expressed in a living will should not violate the compassion, love and respect of children towards their parents. The acceptance and refusal of certain kinds of end-of-life treatments should not only reflect the decision of the elderly parent, but should also be acknowledged and accepted by his/her children. It is only with the reorganization of the family that ADs can be implemented. This is congruent with the laws and regulations governing current medical practice[9] and it is acceptable to the Chinese people.

[9] In China, the family already has a recognized position in medical decision making according to current laws and regulations pertaining to medical practice. For example, in China, a family member of the patient is the one to sign the consent forms for surgery (See Cong 2004).

Furthermore, all laws and regulations should legally recognize the whole family as a surrogate for elderly family members. On the basis of the principle of *xiao*, the family, usually the children of the elderly patient, have a moral duty to respect and realize a parent's will and to take part in the appointment of a durable power of attorney. In the United States, appointing a durable power of attorney is one form of ADs recognized by law, which is based on the assumption that a durable power of attorney can accurately represent the true wishes of a patient. In the absence of a durable power of attorney, family members are prioritized in acting as surrogate decision makers for a patient who has lost decisional capacity. Emphasizing the values of *xiao* and *shun*, a Confucian family-oriented approach to ADs is different from the U.S. model. For Confucians, appointing a durable power of attorney is considered one part of the decision making process that requires the involvement of both the elderly patient and their children, in accordance with the virtue of *li*.[10] The surrogate decision maker should be recognized by both the patient and his/her family members. This allows the whole family to participate in determining who should speak on behalf of the patient.

Different from the individual-oriented approach, the Confucian family-oriented model requires the involvement of the whole family in developing ADs. The living will and the appointment of a durable power of attorney should be signed by the elderly patient and other important family members including his/her children. As such, there are two advantages to the Confucian family-oriented model of ADs. First, it encourages elderly patients to participate in their own end-of-life decision making so that their family can accurately understand their wishes. Second, it also assists the parent and his/her children "engage in harmonious, interdependent cooperative interaction in order to arrive at an important decision" (Fan 2002, p. 359). Therefore, this approach might resolve conflicts between the surrogate and other family members, which is not uncommon in the Western individual-oriented model.[11] Consequently, since this approach focuses on the harmonious interaction between family members, it may not be best to legally establish an AD policy in China instantly, but rather to promote such a policy gradually and culturally, advocating for better ethical decision making at the end of life in a family-oriented society.

[10] Actually, in China, it is unusual for a patient to appoint a non-family member as a surrogate decision maker. As Fan (2011) has pointed out, the family is at the center of surrogate decision making in mainland China and, therefore, appointing a durable power of attorney outside the family might disturb the unity and harmony of the entire family. However, a Confucian family-oriented model does not imply that the patient cannot appoint a non-family member as a durable power of attorney. Under the principle of *xiao*, adult children should respect their parent's will. If the parent insists on appointing a non-family member as surrogate decision maker, the adult child should not violate the will of his/her parent. On the other hand, if the adult child deems the surrogate decision maker distrustful or incapable of representing the interests of one's parent, they are required by duty to persuade the parent to act in accordance with the virtue of *li*.

[11] For an example see: Parks et al. 2010.

References

Beširevic, V. 2010. End-of-life care in the 21st century: Advance directives in universal rights discourse. *Bioethics* 24:105–112.

Blank, R. H. 2011. End-of-life decision making across cultures. *Journal of Law, Medicine and Ethics* 39:201–214.

Blank, R. H., and J. Merric, eds. 2005. *End-of-life decision making: A ccross-national study*. Cambridge: MIT Press.

Cai, Y. 2013. On family informed consent in the legislation of organ donation. In *Family-oriented informed consent: East Asian and American perspectives on a cardinal moral practice*, ed. R. Fan, [THIS VOLUME].

Capron, A. M. 1998. Advance directives. In *A companion to bioethics*, eds. H. Kuhse and P. Singer, 261–271. Malden: Blackwell.

Chai, C., and W. Chai, eds. 1965. *The humanist way in ancient china: Essential works of confucianism*. New York: Bantam Books.

Chan, A. K. L. 2010. The Chinese classic of family reverence: A philosophical translation of the *Xiaojing* (Book Review). *Journal of Chinese Studies* 50:335–341.

Chattonpadhyay, S., and A. Simon. 2008. East meets West: Cross-cultural perspectives in end-of-life decision making from Indian and German viewpoints. *Medicine, Health Care and Philosophy* 11:165–174.

Chen, X., and R. Fan. 2010. The family and harmonious medical decision making: Cherishing an appropriate Confucian moral balance. *Journal of Medicine and Philosophy* 35:573–586.

Cong, Y. 2004. Doctor-family-patient relationship: The Chinese paradigm of informed consent. *Journal of Medicine and Philosophy* 29:149–179.

Cox, C. L., E. Cole, T. Reynolds, M. Wandrag, S. Breckenridge, and M. Dingle. 2006. Implications of cultural diversity in Do Not Attempt Resuscitation (DNAR) decisions. *Journal of Multicultural Nursing and Health* 12:20–28.

Deng, R. 2013. The Chinese mainland informed consent of human medical research: A multi-decision model of family-oriented. In *Family-oriented informed consent: East Asian and American perspectives on a cardinal moral practice*, ed. R. Fan, [THIS VOLUME].

Fan, R. 1997. Self-determination vs. family-determination: Two incommensurable principles of autonomy. *Bioethics* 11:309–322.

Fan, R. 2002. Reconsidering surrogate decision making: Aristotelianism and Confucianism on ideal human relations. *Philosophy East and West* 52:346–372.

Fan, R. 2006. Confucius filial piety and long term care for aged parents. *HEC Forum* 18:1–17.

Fan, R. 2007. Which care? Whose responsibility? And why family? A Confucian account of long-term care for the elderly. *Journal of Medicine and Philosophy* 32:495–517.

Fan, R. 2011. The Confucian bioethics of surrogate decision making: Its communitarian roots. *Theoretical Medicine and Bioethics* 32:301–313.

Fan R., and B. Li. 2004. Truth telling in medicine: The Confucian view. *Journal of Medicine and Philosophy* 29:179–193.

Fan, R., and J. Tao. 2004. Consent to medical treatment: The complex interplay of patients, families, and physicians. *Journal of Medicine and Philosophy* 29:139–148.

Fung, Y. 1952. *A History of Chinese Philosophy*. Vols. 1, trans. D. Bodde. Princeton: Princeton Univ. Press.

Gochman, D. S., and G. S. Bonham. 1990. The social structure of the hospice decision. *Journal of Hospice* 6:15–36.

Haley, W. E., R. S. Allen, S. Raynold, H. Chen, A. Burton, and D. Gallagher-Thomson. 2002. Family issues in end-of-life decision making and end-of-life care. *American Behavioral Scientist* 46 (2): 284–298.

Ho, Z. J. M., L. K. R. Krishna, and C. P. A. Yee. 2010. Chinese familial tradition and western influence: A case study in Singapore on decision making at the end of life. *Journal of Pain and Symptom Management* 40:932–937.

Hui, E. 1999. A Confucian ethic of medical futility. In *Confucian Bioethics,* ed. R. Fan, 127–163. Dordrecht: Kluwer Academic.

Kelley, A. S., N. S. Wenger, and C. A. Sarkisian. 2010. Opinions: End-of-life care preferences and planning of older Latinos. *Journal of the American Geriatrics Society* 58:1109–1116.

Kim, S., K. Hahm, H. W. Park, H. H. Kang, and M. Sohn. 2010. A Korean perspective on developing a global policy for advance directives. *Bioethics* 24:113–117.

Kwon, Y. C., D. W. Shin, J. H. Lee, D. S. Heo, Y. S. Hong, S.-Y. Kim, and Y. H. Yun. 2009. Impact of perception of socioeconomic burden on advocacy for patient autonomy in end-of-life decision making: A study of societal attitudes. *Palliative Medicine* 23:87–94.

Lau, D. C. (trans.) 1970. *Mencius*. Harmondsworth: Penguin.

Lau, D. C. (trans.) 1983. *Confucius: The Analects*. Hong Kong: The Chinese Univ. Press.

Lee, S. C. 2013. On the practice of family consent in medical decision-making in Taiwan: The implications of the new revision of the law of Hospice Palliative Treatment and related issues. In *Family-oriented informed consent: East Asian and American perspectives on a cardinal moral practice*, ed. R. Fan, [THIS VOLUME].

National Bureau of Statistics of China. 2011. Announcement of the main data from the sixth national census. http://www.stats.gov.cn/tjgb/rkpcgb/qgrkpcgb/t20110428_402722232.htm. Accessed 13 Nov 2013.

Parks, S. M., L. Winter, A. J. Santana, B. Parker, M. Rose, and R. E. Myers. 2010. Family factors in end-of-life decision-making: Family conflict and proxy relationships. *Journal of Palliative Medicine* 14:179–184.

Patient Self Determination Act. 1990. 101st Congress, H. R. 4449. IH. http://thomas.loc.gov/cgi-bin/query/z?c101:H.R.4449.IH. Accessed 13 Nov 2013.

Richter, J., and M.R. Eisemann. 2001. Attitudinal patterns determining decision-making in severely ill elderly patients: A cross-cultural comparison between nurses from Sweden and Germany. *International Journal of Nursing Study* 38:381–388.

Slomka, J. 1992. The negotiation of death: Clinical decision making at the end of life. *Social Science and Medicine* 35:251–259.

Tao, J. P. L. 1998. Confucianism. In *Encyclopedia of Applied Ethics*. 2nd ed., ed. R. Chadwick, 597–608. Salt Lake City: Academic Press.

Tse, C., and J. Tao. 2004. Strategic ambiguities in the process of consent: Role of the family in decisions to forgo life-sustaining treatment for incompetent elderly patients. *Journal of Medicine and Philosophy* 29:207–223.

Wang, L., Y. Hu, J. Lu, and W. Gu. 2012. A survey of the attitude of advanced cancer patients and family members towards advance directives. *Chinese Journal of Nursing* 47:197–200.

Werth, Jr, J. L. D. Blevins, K. L. Toussaint, and M. R. Durham. 2002. The influence of cultural diversity on end-of-life care and decisions. *The American Behavioral Scientist* 46:204–219.

van Wijmen, M. P. S., M. L. Rurup, H. R. W. Pasman, P. J. Kaspers, and B. D. Onwuteaka-Philipsen. 2010. Design of the advance directives cohort: A study of end-of-life decision-making focusing on advance directives. *BMC Public Health* 10:166–174.

Winzelberg, G. S., L. C. Hanson, and J. A. Tulsky. 2005. Beyond autonomy: Diversifying end-of-life decision-making approaches to serve patients and families. *Journal of American Geriatrics Society* 53:1046–1050.

Wong, H. C. (2014) Towards a good practice of family-oriented consent: Reflections on medical practice in Taiwan. In *Family-oriented informed consent: East Asian and American perspectives on a cardinal moral practice*, ed. R. Fan, [THIS VOLUME].

Yang, Q. 2009. Who determines my life. *Health Newspaper,* 12 June: 003.

Zhai, X., and R. Qiu. 2007. Perceptions of long-term care, autonomy, and dignity, by residents, family and caregivers: The Beijing experience. *Journal of Medicine and Philosophy* 32:425–445.

Part VII
Epilogue

Chapter 18
Families and Individuals in Medical Decision Making

Lisa M. Rasmussen

> *In the process of practicing a traditional way of life, our understanding of the world is translated into a deeper conviction and belief about how the world really is.*
> (Zhao, this volume, p. 9)

18.1 Introduction

This volume poses fundamental challenges to the Western bioethical decision-making framework. It can be viewed as a sally in the culture wars. Authors in the volume argue, *inter alia*, that the fundamental ethical unit to which questions and decisions in medicine ought to be directed is the family, not the atomic individual. This is a message requiring our attention, because it highlights a truth about the beliefs of many people—even those, like me, who ultimately still side in favor of individualism—that the family *is* a—fundamental unit of concern.

Sometimes to be heard against countervailing messages, one must articulate themes in broad strokes rather than nuance. Think here of feminism or the civil rights movement; the first step in having one's voice heard is to be disruptive, to refuse to be ignored. This volume is meant to be a disruptive text, not a comparison of two possible models of bioethical decision-making. It starts by assuming that most of its audience (those of us in the global West[1]) subscribes to the view that autonomy is a fundamental value, best instantiated and respected via individual decision-making.

[1] The terms "West" and "East" are problematic for a number of reasons, not least because they imply homogeneity of belief where it does not exist. However, the authors in this volume adopt these terms as heuristics for describing general cultural norms, so I likewise make use of them.

L. M. Rasmussen (✉)
Department of Philosophy, University of North Carolina, Charlotte, NC, USA
e-mail: lrasmuss@uncc.edu

R. Fan (ed.), *Family-Oriented Informed Consent*, Philosophy and Medicine,
DOI 10.1007/978-3-319-12120-8_18

One of the volume's aims is to articulate other approaches to decision-making in different contexts; these authors for the most part characterize a Confucian and/or traditional cultural context and outline some of its implications for decision-making in bioethics. They do not strongly urge those of us in the West to adopt this practice, though as I argue, there are changes we could profitably make in light of these arguments. But they do collectively find fault with the way in which Western bioethical decision-making policy is implemented or assumed in foreign contexts. As some of the contributors point out, a focus on individualism in bioethics can make it hard for families to participate even when patients *want* them to. If overall these authors do not give equal time to the value or benefits of individualism, it may be because they struggle to be heard against what they see as a dominant ideology. These essays forcefully articulate and defend the value of family decision-making in the face of an exported model of bioethical decision-making whose laws, media, and regnant ideology is clearly supportive of the individual rather than the family.

There are two main aspects of the familism vs. individualism debate. First, there is the empirical difference that the "family" approach is the norm in East Asian countries, whereas the individual approach is the norm in the West. Second, there is the normative difference that while Western bioethicists argue for individualism, and assume this is the superior approach, authors in this volume argue for familism. As I argue, however, there may be some agreement between Eastern and Western approaches about optimal decision-making with a patient and her family. Where significant differences between the approaches become clear is when we decide how one could craft policy for cases in which the optimum is not achievable.

18.2 Medical Decision-Making East and West

The authors in this volume all argue for the importance of the family in medical decision-making (in both the clinical and research contexts), but the ways in which they interpret the importance and implications of this vary. One common theme among the authors is the empirical observation that the Western cultural model of medical decision-making doesn't necessarily work in the Eastern context. When one culture's practices are sufficiently different from another's, one should not expect that the practices and values of that culture can be easily accommodated in another (and of course, neither should one expect that they cannot be accommodated at all). The force of these points is that the cultural lifeworld in largely Confucian countries and regions like mainland China, Hong Kong, Taiwan, and South Korea often don't find the Western model of individual decision-making easy to incorporate (see, e.g., Chap. 13, this volume, which articulates some of the benefits of family involvement in consent to participation in research in a Confucian context).

What should we make of this phenomenon? With a little "educating" of the populace, could it be made clear that in fact the individual model *does* fit the Eastern context? Many of the authors in this volume argue that this gap between East and West does not reflect a misunderstanding, but instead a different moral ontology and cultural history that cannot (and perhaps should not) be "educated" away. They articulate various strengths of and reasons for this difference, and are not in agreement on these points—some argue that Confucianism specifically *does* emphasize the importance of the individual's voice, for example in the obligation to respect one's parents and their wishes. They also disagree about the proper interpretation of Confucianism with respect to family decision-making and filial piety. That is to say, there is no univocal view expressed here; there is nuance and both textual and normative disagreement.

Some potential shortcomings of family decision-making are acknowledged, but some of these authors also argue that these shortcomings may be less problematic than the shortcomings in individual decision-making, a point we in the West should consider seriously. This is particularly true when the basis of the argument is that family decision-making actually *enhances* autonomy. Rui Deng, for example, proposes that "family or friends should participate in the informed consent process, to avoid the medical paternalism in human research and to eliminate the risks raised by value guidance from investigator" (p. 7), and stresses that "the purpose of the family's participation in the decision-making is to consider the risks of the trial from more angles" (p. 5). Similarly, Wenqing Zhao (Chap. 15) argues that since evidence is mounting that "having too many choices does not necessarily lead to satisfaction or happiness," and that "the obligation to choose in a vital situation may very likely increase the level of stress and depression for the subject," (p. 2) the family decision-making approach better supports the patient than does treating him as a lone individual. This is compatible, as Deng points out, with maintaining a focus on individual rights, but may in fact better succeed in fostering autonomous decision-making than the individual approach.

There are two main ways in which the authors in this volume articulate the case for the inapplicability of Western norms of biomedical decision-making in the East. The first is simply the descriptive point that because different historical and ideological forces (in this case, Confucian) have contributed to the current cultural context in these countries, residents of Confucian countries do not find Western ideas of medical decision-making suitable. Instead, they can find it unnatural to make important decisions within the individual framework rather than as family units, for a wide variety of reasons described in this volume, especially a family's moral and financial obligation to protect its members' health. This is a merely descriptive point, but it still has important implications for procedures of consent, because the goals of the activity will not meet with success if the methods are insufficient to the purpose. In addition, this cultural context contributes to other societal laws and policies that may not seem to affect medical decision-making, but actually do. For example, as Ilhak Lee points out, one study demonstrated that typically 80% of out-of-pocket end-of-life expenditures in South Korea are covered by children of the patient, with the result that "[a]s children bear the burden of care of their parents, they also play the role of surrogate decision-maker for determining end-of-life care" (p. 8). In this way, existing financial or other practices in a given society might have a significant effect on how we think it best to implement laws and practices in areas like medical decision-making. It would be easy to imagine pressure in the US for greater family inclusion in decision-making if we knew that the family rather than the individual paid more out-of-pocket expenses, even despite our historical focus on the individual.

But an even more profound claim found in this volume is that the lived experience and moral convictions of Confucian societies are not the same as Western societies. Unless one believes that the basic units through which we conceive of our lives and moral obligations are self-evident and universal, it is clear that the conceptualization of moral ontological units is a normative act. In the postmodern period, we should be familiar with the notion that universalist normative claims are contested, so it should not be surprising that other cultures vary in how they conceive of the basic unit to whom, for example, physicians owe the duty of respect for decision-making. Still, for those of us in the global West, the individual is so fundamentally a unit of moral consideration that it is hard to conceive of a different starting point. But as some of the authors in this volume demonstrate, there *are* alternatives. For example, as Zhao puts it, "Confucianism views families, instead of individuals, as the fundamental units of the society. From a Confucian perspective, an individual and her close family members are fundamentally 'one' bonded by the intimate emotional ties of caring and love" (p. 11). Yu Cai puts it most strongly when she argues that in a traditional Confucian family, "no individual personality exists because the family unity is the ethical entity" (p. 5), and "...personal belongings do not exist ..., only common assets of the family exist. Hidden individual assets (casually called personal funds) are forbidden and, of course, an individual does not have the right to dispose of family assets" (p. 9). The implication of this view for organ donation, she argues, is that "as material property due to separation with the body, organs are family property, a common asset, and the individual, therefore, does not have the right to dispose of one's organ and the whole family must decide together" (p. 9).

Though there is disagreement on this point within Confucian societies (and increasingly so in the modern world), many of these authors argue that in the Confucian context, the fundamental decision-making unit—the unit to which policies ought to be addressed—is not the individual, but the family. Although the authors in this volume by no means agree on interpretations of Confucianism or on the implications of Confucianism for bioethics, they do collectively describe a cultural context sufficiently different from the Western context that it forces us to reevaluate our own moral premises.

In the remainder of this essay, I critically consider some of the features of this collection of essays that I think will be striking to many Westerners.

18.3 Problems with the Family Assumptions

Having acknowledged the importance of considering the arguments in favor of family decision-making, I would like to comment on some of the challenges inherent in a family-centered policy. This is not to say that the individual focus is without problems (and perhaps they are even more serious), or that the authors in this volume do not readily acknowledge some of these the challenges. But there are general concerns about family decision-making (both the possibility of adopting it in the West and inherently in the family model itself) that are worth articulating in the context of this volume.

One of the points made by some of the authors in this volume (for example, Mark Cherry and Jeffrey Bishop) is that practices such as individual decision-making which do not instantiate a respect for the family as the fundamental decision-making unit contribute to the modern fracturing of the family, with increased rates of divorce and single parenting leading to worse outcomes for children and society. Let us assume for present purposes that this fracturing of the family in fact does lead to worse outcomes. Even so, the family ecosystem is so complex and affected by so many external factors that it is not clear what effect changing the medical decision-making model in the West to a more familist approach would have. Absent a radical change in society, it would at best be at odds with the rest of our societal laws and policies, which still center on an individual decision-making model. (Imagine, for instance, adopting a financial policy that prevented individuals from opening bank accounts, retirement accounts, etc.) It is therefore important to distinguish between the plausibility of the more general point that we would do well to foster family integrity and communication, and the specific point that the family should be included by policy in medical decision-making with competent patients. One could acknowledge and support the former while recognizing that the latter is not feasible in existing contexts.[2]

It is also worth considering how the notion of filial piety, which undergirds the family decision-making model, can go wrong. Many of the authors explain the structural position of the family in Confucianism and its role as the basis for the order of society, which both makes possible and strengthens the existence of the state.[3] Its focus on filial piety and loyalty can serve to discourage challenging extant relations of power, with potentially serious negative consequences. For example, Ilhak Lee cites the (ab)uses to which a government put the concept of filial duty in Korea during and after Japanese occupation. As a result, he says, "we can imagine that the practice of filial piety is also often misled by factors outside of its original meaning" (p. 6). Even if these are *misuses* of Confucianism, it is worth considering what misuses are possible in a given culture. If the way of life at the heart of this cultural context is respect for what *is*, how can citizens also be encouraged or relied upon to *challenge* what is, or what is proposed, or what is forced, without also violating filial piety? Consider Zhao's characterization:

> Liberal individualists see autonomous agents as the fundamental units of the world. The Western liberal conceptual schema puts much weight on the tension between individuals and the rest of the world. It is often suspected that "others" (society, government, institutional authority, family and etc.) are always seeking to control and restrain individual freedom and to prevent individuals from exercising their wills, whereas the individual is always struggling to break free. This is not to say that Chinese people do not face tensions between themselves and their government, however, *the government is never assumed to be evil in the first place*. In a Confucian conceptual schema the idea of individual cannot be separated from one's family and community (Zhao, p. 10, emphasis added).

[2] I do not mean to suggest that authors in this volume are arguing that the West should change its model—rather, most of them are arguing against the exported Western individualist model being applied to—or imposed on—East Asian contexts.

[3] For example, Ilhak Lee quotes the seminal Confucian text and explains: "In *The Analects*..., filial duty is mentioned as the foundation of all human relationships: 'filial piety and fraternal submission! Are they not the root of all benevolent actions?'" (p. 4).

I do not wish to overstate this point; clearly there have been revolutionary and resistance movements across the Confucian East. But if we are to consider the shortcomings of individualism, we must likewise bear in mind potential shortcomings of familism.

It can also be difficult to know what filial piety requires in end-of-life-decision-making. As Hon Chung Wong and Kam-Por Yu explain, if filial piety is understood to be properly demonstrated by fighting at all costs for a parent's survival, then it is hard to imagine conditions under which it would be permissible for a family to withdraw end-of-life care. Wong illustrates this with a news story in which a family who insisted that a medical team respect their elder's DNR wishes was later criticized "as lacking filial piety toward [him]" (Wong, p. 4). In this case, it was very clear that this was what the patient wanted, but even so the family was criticized for their decision. And as Yu points out,

> Surveys have been done to find out whether people would support some form of euthanasia under some specific conditions for themselves as well as for their family members ... It was found that most people would prefer ending their life under some specific conditions (such as in vegetative state, or in a painful dying process) but they would not prefer the same for their parents. There seems to be some apparent contradiction here. If they prefer what they regard as the best for themselves, why would they want to have something else for their parents? The major reason given was that people were influenced a lot by the idea of filial piety. They would like to be regarded as filial, and withdrawal of treatment may be interpreted as saving resources and not caring for their parent enough (Yu, 11–12)

In contrast, Yaning Yang points out situations in which true filial piety is demonstrated by respecting a patient's wishes:

> [T]he implication for end-of-life decision-making is that prolonging the life of an incompetent terminally ill parent without caring for their mental needs and dignity cannot be considered *xiao* [i.e., filial piety]. ... Moreover, if a healthy parent establishes an [advance directive], children with the virtue of *xiao* should respect their parent's will (p. 9).

> A Confucian family-oriented model of [advance directives] invites the family as a whole to participate in the advance planning of the patient's end of life decisions. However, it does not mean that the patient's own will should be subject to the interests of the family. On the contrary, this model encourages the elderly patient to express his/her wishes and to participate in his/her end of life decision-making (p. 11).

Obviously, family and individual models of decision-making can both be difficult to navigate. Both can also be open to misinterpretation, and it may be argued that those who critiqued the family in the example Yu gives above simply misunderstood what family decision-making ought to be. That may be true, but if we are to grant that family decision-making can be misunderstood and misinterpreted, we should acknowledge that the same is true of individual decision-making. Instances of poor execution of a model do not necessarily invalidate that model, and neither these stories nor stories of the paucity of some instances of individual decision-making defeat the approach being used.

Among other potential problems with family decision-making, as Ilhak Lee observes, is the fact that just as individuals' decision-making may be hampered by illness and proximity to death, so too may the decision of family members. They are often overwhelmed by the possible loss of their loved one or exhausted from the long, burdensome care that is required. Confusion, a lack of pertinent medical knowledge, and the unavailability of legal and ethical counseling can lead to tragic decisions. Family members of dying patients are often in need of help and the moral justification for the authority of one's children needs to be explored.

What I next want to argue is that one point of disagreement between the family and individual decision-making models may be what should be adopted as an appropriate default strategy when there is disagreement or when it is unclear who the appropriate decision-maker is. I do not mean this as a mandate for all contexts, East and West. As I argue below, policy decisions are *political* decisions, and must take account of the context in which they will be implemented. As the authors in this volume argue, the East Asian context is different from Western contexts, and the *imposition* of individualism in that context would be wrong.

18.4 Default Policy

Two of the authors in this volume point to empirical surveys demonstrating that in two different Confucian societies, the overwhelming majority of individuals surveyed say that they want to and/or have a right to make decisions for themselves.[4] Many of the authors in this volume specifically claim, contrary to straw versions of the family approach, that most of those who approach medical decision-making in the family model would not think of excluding and would honor (even if not necessarily exclusively) the voice of the patient. For their part, Western citizens generally want their families to have a say in their care.[5] If all of this is right, then there is substantial convergence on the optimum: taking advantage of one's family's good will, knowledge, and diversity of perspectives to make an informed decision. The real disagreement comes at the margins and in uncertain cases (for example, where family is non-existent or non-functional; not in communication with the patient; in substantial disagreement with the patient; or when the patient does not wish to inform and/or include the family), and in what follows, I will argue that the individual approach to medical decision-making better accommodates these uncertainties and what I think are the on-balance greater dangers of not providing space for individual decision-making.

---

[4] Wong points out that in a survey done in Taiwan, "the results showed that patients strongly claimed their own right to be informed about their disease to be superior over their family's wishes to keep them uninformed" (p. 4). Yang cites a survey in China showing that "Ninety percent of patients indicated that they wanted family members and physicians to understand and respect their end of life wishes" (Yang, p. 11). In the Chinese context, this may mean that they want their families to consider their wishes, not that their wishes must be followed.

[5] For example, as Yang points out, "In the United States, studies have indicated that patients' family members are involved with end-of-life decisions 60–80% of the time and patients often prefer to make end-of-life decisions within the family context (Haley et al. 2002)" (Yang, p. 45).

The benefit of defaulting to the family, according to the arguments in this volume, is that this is both the way the majority of patients prefer to make medical decisions, and also the way that best supports individual flourishing. In his contribution, Jeffery Bishop asserts that in the West, and particularly in the US, "the structures and practices of medicine, including informed consent and decisions about removing life support in the critically ill, organize decisions such that the family is structurally marginalized" (Bishop, p. 2). We can debate whether or not this is true, but suppose that it is. Just how many families are marginalized, and how corrosive of the family is individual decision-making, if most individuals choose to include families in such deliberations?

Imagine, in contrast, that we were to design the structures and practices of medicine "such that [the individual] is structurally marginalized." Which is worse? As might be predicted for someone hailing from an individualist context, I want to argue for the importance of the individual and call to mind why the focus on the individual became important. I am acutely cognizant, for example, of the long human history of preventing women from substantially determining their own lives. It cannot be maintained that men who had women's good in mind have on the whole made choices for women that historically led to their flourishing—unless the supernatural and its plan for appropriate hierarchy in this world is invoked, promising rewards only in the afterlife as compensation for subordination in this one. Consider, through a woman's experience, how this quotation sounds given the overwhelming likelihood that families are patriarchal: "…all decision-making, even within the structures of informed consent, ought to emerge from the context of the family, where the frailties and dependencies of human life have always been the site of care…. Informed consent ought therefore to emerge from the family of care and not from the corrosive myth of the individual" (Bishop, p. 2). Should this choice be imposed on those unwilling to subscribe to it? To put it more starkly, should we support a default model that allows an individual to exit the family space if s/he chooses, or rather one that forces an individual to accede to the decisions of others about the life that she must live?

Consider too the case of parental decision-making for children. As Bishop points out, "…the state typically expects the family to make decisions in the best interests of any individual, especially a child, even if the decision might undermine the familial structures" (Bishop, p. 10). Think here of donor siblings, and how hard it is to separate the best interests of the child from that of the family; Bishop is right to point out the tension in this context. How will a girl whose brother dies feel if she realizes years later that she could have saved her sibling's life with a kidney or bone marrow, but was prevented from doing so due to a stranger's interpretation of what is permissible? Isn't it more likely that as an adult, she would elect to save his life even at a cost to herself, and that her family would best know this about her? But even this story illustrates exactly what a focus on individualism is meant to combat in such a situation: the instrumentalization of one individual for the sake of another, or of a collective. Should we, in such a case, stand by when it is fairly clear that a family is using one child as a mere means to another child's end? I have no solution to this problem; again, I mean only to match assertions of the benefit of family control in medical decision-making with examples of its dangers.

One of the worries expressed in this volume that must be taken seriously is that a focus on the individual can lead to a cultural disvaluing of family units. It is important to keep in mind the trajectory of family in the West, as described by Bishop and Cherry: marriage and birth rates are down, single parenting rates are up, and a variety of social ills exist. The claim is made that it is the cultural disvaluing of family units that contributes to these social ills, and its expression in medical decision-making is part of the problem.

Even if this is true, though, the proferring of these statistics doesn't give us reason to think that simply changing the decision-making model in medicine would fix the problem. It may be that both family unity and social ills stem from other factors which, if addressed, might considerably affect both. For example, it is not clear that it is the cultural disvaluing of the traditional family that leads to the dissipation of family decision-making rather than, say, global economic forces and longer lifespans affecting both of those factors. This may be true even in a Confucian context: Ilhak Lee observes that "...most Koreans now live separately from their parents, resulting in the loosening of family ties" (Lee, Chap. 9, p. 9). And, importantly, even family-centered decision-making may not resolve conflict. As Shui Chuen Lee notes,

> [h]ealth professionals and hospice services in Taiwan will not accept a decision made by one or some family members without the consensus of all known family members because they are afraid of endless protests against them if a family member was not told of the decision and would have opposed the DNR if informed (p. 5)

If this is the case in a country with even better family coherence than in the United States, how much worse might family decision-making go if implemented in the litigious American context?

One striking contrast in the volume occurs in a consideration of two similar anecdotes discussed by Bishop and Zhao. Bishop outlines a case between "Bill" and "Mildred" in which Bill refuses to let Mildred into discussions about his impending death from cancer. He wants to stop aggressive care, but thinks his wife could not handle his death, so he continues to request treatment. Meanwhile, Mildred understands that Bill is dying and doesn't want him to suffer, but refuses to challenge his decision. Bishop concludes by observing that "[t]he myth of the individual had become so much a part of their individual identities, that Bill would not authorize a shift to hospice care because his wife could not cope with this decision, and Mildred would not challenge Bill, as it was his decision" (Bishop, p. 5). In Zhao's example, reported in the news as "a touching story of true love," a husband with cancer does not improve with treatment, but his wife fabricates his test results so that he seems to be getting better, in order to keep his spirits up. As Zhao observes, "[t]he commentary surrounding this story demonstrates that Confucian-influenced Chinese accept the practice of concealing diagnoses from patients with understanding, approval, and even praise" (p. 4).

Why does the former illustrate the deficits of the myth of individual decision-making while the latter illustrates the benefits of family decision-making? There may be missing details from the abbreviated version of these cases, but it seems quite plausible that Bill and Mildred are engaged in a loving deception just as the Chinese couple were understood to be. And, it may have been that the husband in the story Zhao relates was struggling on, despite knowing his true condition, for the sake of his wife. It also seems plausible that in this latter case, the wife's deception could have been motivated for the wrong reasons. If we think that in both cases, due to the decision-making model in operation, conversations were not happening that could have led to a better outcome, this simply demonstrates potential problems with both models. It does not illustrate the superiority of one to the other.

I am not arguing that the individualist approach is morally neutral. Neither should advocates of family decision-making argue that that approach is morally neutral. The irony is that a focus on individualism is in part meant to address the dilemma of pluralism—and yet it itself is a form of cultural assumption. Both are positions informed (or determined) by values. The family may generally care more for the patient and have his interests more in mind than the state or other institutions. (However, we should at least bear in mind that various studies have shown that families and other surrogates may not be very good at articulating what the patient would have wanted.) But whether or not the family means well, and even conceding that on average the family might be better able to advance the individual's interests, that is not the central issue in the consideration of family vs. individual decision-making in medicine. The question is what approach best accommodates the existence of disparate views (such as the fact of non-intact or poorly-functioning families). What we should want in formulating a policy on medical decision-making in the West, I argue, is a way in which to acknowledge the individual's right to *include or exclude* family decision-making. Since it is precisely this right that is rejected by the familist model, this default is not likely to be acceptable to those who support that model.

18.5 Lessons for the West, Challenges for the East

Even a policy focused on individual decision-making can be better designed to include families. From a Western perspective, possibly the strongest argument in favor of supporting family input in medical decision-making is that it fosters autonomy. One argument given in this volume is that family decision-making may help to protect patients in contexts in which others may not have their best interests at heart, for example in consent to research participation. Changing consent procedures to facilitate family involvement, if desired, could help to protect human subjects. For example, if we took seriously the likelihood that most potential subjects want their families to be informed and included in consent discussions, we would make it the structural default that, e.g., "consent appointments" are set up with appropriate family from the beginning. Obviously this is less streamlined than our current practice, but perhaps it will do a better job of ensuring robust consent discussions and supporting autonomous choices to participate in research.

Another way in which family participation in decision-making may foster autonomy is by combating the fact that one mind may not be enough: as scholarship in behavioral economics has demonstrated, our perception of risk is frequently colored by framing heuristics and biases. This is particularly true in stressful situations, so encouraging patients to make decisions individually may result in worse outcomes than if we worked to include families as well. There has been ongoing criticism that we in the West overemphasize respect for autonomy, with the resulting goal becoming simply "getting the consent" on a form. Taking seriously the psychological limits of individuals, while also respecting their right to make decisions for themselves, could result in retaining the policy of individual decision-making but also empirically studying the effect of family or other surrogate decisionmakers on patient well-being, outcome, satisfaction with consent processes, etc. and implementing successful approaches.

It is also important to keep in mind, as Bishop outlines, the demands of individual decision-making with respect to advanced-care planning, which optimally require individuals to have completed forms guiding care if they become incapacitated:

> The social apparatus of the ICU, along with all the documents—living will, durable power of attorney, and consent forms—that an individual ought to have completed before going into the ICU, have at their heart a sovereign individual who sets in motion the aggressive technological features of medicine, that once deployed cannot be easily brought back under his own control (Bishop, p. 2).

Designing a system that requires all of this in order to robustly respect individual wishes in the case of incapacity seems unsuccessful, in that the majority of patients have completed no end-of-life or medical proxy forms. More importantly, the "social apparatus" that Bishop cites seems to require an individual—a strong, forceful individual—to say that enough is enough. Designing a system in which family voice is better heeded may also diminish such demands on individuals.

What we make of these arguments depends on whether we think that they make empirical or normative points about the importance of family decision-making. Any assertion that family decision-making is better than individual decision-making should be accompanied by an account of what constitutes better or worse situations. If it is asserted that family participation better supports autonomous decision-making, this is an empirical claim that can be studied. If we are committed to a respect for autonomy, we should care whether these claims are correct, and better implement them where we need to. Even if the empirical claim is only that less disagreement about cases occurs when families are included, that may give us a reason to work harder at including families. For example, as Cai suggests, having individuals *and* families sign organ donor consent forms could lead to less disagreement about deceased organ donation.[6]

[6] Of course, in a Western context we might not decide to require this, but we could design a form that strongly encourages and enables it. In both Eastern and Western contexts, though, the legal status of such a consent form would need further delineation. For example, could only donors with family signatures successfully donate? What happens in the case of disagreement among family members? Which signatures need to be on the form?

Another strand of arguments in this volume is that family *ought* to be taken as the fundamental decision-making unit, regardless of cultural tradition or the preference of individuals. Some of this is based on assertions of biological kinds (i.e., families), which can surely be questioned historically. What we take to be the "traditional" family is probably not representative of how humans have always existed: polygamy, polyandry, extended families, kibbutzes and other family arrangements have all been the norm at one time or another. We may not yet have implemented the arrangement that is "best" for humans. Outside of theological or other supra-rational assertions about the natural order of things, it is no more neutral or natural to default to family decision-making than it is to default to individual decision-making (though I have argued for the latter on the basis that it can account for the former better than vice-versa).

Clearly there are improvements to be made in the Western medical decision-making model. It is worth also asking whether concessions can be made in the Eastern decision-making model. I said at the beginning that the message in this volume "highlights a truth about the beliefs of many … that the family *is* a—or *the*—fundamental unit of concern". I want to shift the emphasis of this point now, to ask whether it should be *the* fundamental unit of concern, or merely *a* fundamental unit of concern. To say that it should be *the* fundamental unit of concern is to make a moral assertion that can be contested. In addition to counterarguments offered above, there is the question of whether we wish to respect the freedom of individuals to exit the context into which they are born. Medical decision-making, encompassing grave choices, is one moment in which individuals can voluntarily try to exit the situation in which they live. For example, consider the case of a doubting Jehovah's Witness who agrees to receive a blood transfusion to save his life. A family insistence that his wishes be overridden because he has false consciousness about what is at stake is just as problematic as a case in which a woman who has chosen a traditional lifestyle is judged to have false consciousness by capitulating to patriarchy. Individualism has been defended as a value in part because it resolves accusations of false consciousness in favor of the person who must live the ensuing life. However, a focus on autonomy, hard as it can be for those of us in this context to see it, may be a form of cultural imperialism. We value individual decision-making, and have robust and long-lived justifications for doing so, but other cultures may not. The ensuing challenge is how to navigate issues like medical decision-making in cultures whose values are different. The horns of the dilemma are on the one hand, cultural imperialism and imposing a set of values, or, on the other hand, relativism, and allowing cultural context to determine the values adopted. In the end, at least in countries responsive to the will of the people, which policy will govern will be a result of a political decision. In the current context in East Asian countries, this might mean the family model is preferred.

And yet, a deep existential challenge this volume poses is to ask whether freedom of exit is more important, or better, than freedom to belong. One of the root defenses for individualism is that individuals should be allowed or fostered, via exposure to many ways of life, to make an informed decision about which they want to make their own. However, one of the arguments offered in this volume is that radical freedom is not possible. It is not the case that we can simply choose, from "the view from nowhere," among a selection of lifestyles to pursue. It may be impossible to know the richness of some ways of life—in the accounts given here, traditional family decision-making—from outside. Thus, to impose individualism is potentially to exclude the possibility of living some ways of life. Similarly, it may be impossible to know the richness of an individualist approach without having been granted it and raised with the expectation of it. We may have good reason to praise the benefits of individualism, but the authors in this volume have given good reason to praise the benefits of familism, and the arguments in favor of individualism do not trump. This volume clearly articulates the arguments in favor of retaining the family decision-making approach in East Asian contexts, and imposition of individualism in this context would be unfounded.

Index

R. Fan (ed.), *Family-Oriented Informed Consent*, Philosophy and Medicine,
DOI 10.1007/978-3-319-12120-8

CPSIA information can be obtained at www.ICGtesting.com
Printed in the USA
BVOW02*2238260315

393576BV00006B/13/P

9 783319 121192